AF679017

Dynamical Psychology:
Complexity, Self-Organization and Mind

Dynamical Psychology:
Complexity, Self-Organization and Mind

Written by
Jay Friedenberg

Cover images were provided by Genaro Martinez, and reproduced with permission.

Dynamical Psychology: Complexity, Self-Organization and Mind
Written by: Jay Friedenberg

Library of Congress Control Number: 2009924499

ISBN13: 978-0-9817032-9-9

© 2009 ISCE Publishing.

All rights reserved. No part of this publication may be reproduced, stored on a retrieval system, or transmitted, in any form or by any means, electronic, mechanical, photocopying, microfilming, recording or otherwise, without written permission from the publisher.

Printed in the United States of America

Contents

CHAPTER 1
Introduction—What Is This Book About?

CHAPTER 2
Systems and Complexity

CHAPTER 3
Self-Organization

CHAPTER 4
Dynamical Systems

CHAPTER 5
Networks

CHAPTER 6
Neurodynamics

CHAPTER 7
The Fractal Mind

CHAPTER 8
Statistical Mechanics and Noise

CHAPTER 9
Dynamics and Psychology

CHAPTER 10
Personality, Pathology, and Therapy

CHAPTER 11
Cognition

CHAPTER 12
Cognitive Processes

CHAPTER 13
Problem Solving and Evolution

CHAPTER 14
Mental Ecology

CHAPTER 15
Models

CHAPTER 16
Conclusions

CHAPTER 1
Introduction—What Is This Book About?

Dynamics, Complexity, Self-Organization and Mind

The world is a complicated place. Obviously we haven't figured it all out yet. However, we certainly have come a long way. Modern science and technology have yielded a remarkable understanding and control of the physical universe. A person transported in time from the beginning of even the 19th century to now would be in awe of humankind's current achievements.

But let's not get carried away. There are still many fundamental mysteries that have yet to be solved. One of these is very close and personal. In fact, you are using it right now. It's called the human brain and much about the way it operates is still a mystery. Part of what makes the brain so hard to understand is its dynamic nature. Brains are never the same from one instance to the next. Messages are always being sent back and forth between billions of cells in an intricate dance. This activity is what gives rise to our mental selves. It underlies what we see, think, and feel.

It is difficult to get a handle on this activity in part because of its complexity. There are an estimated 100 billion neurons in the human brain. Each of these can have as many as ten thousand connections to other cells. At any given time, there are multiple patterns of activation that course through this network. These patterns are constantly changing in response to the environment and to each other.

This complexity is compounded by the fact that brain processes are simultaneously happening at many different spatial and temporal scales. At the level of the neuron, neurotransmitter molecules are being packaged, released and recycled, in some cases, many times a second. At the level of a society, brains make decisions that affect how we interact with each other. Is there an appropriate level of organization that we should study? One theme of this book is that we must study complex phenomenon at multiple levels. No level is privileged. What's more, we need to fathom the connectivity between these levels. No complex system can be fully understood by focusing only on one level, it is often the coordination and interaction between levels that yields new and interesting insights.

Another problem with studying complex systems has to do with organization. Many natural processes seem to develop and organize themselves. There is no single external "boss" to tell a brain how to grow or what to do. The brain figures it out on its own. This runs counter to traditional conceptions in psychology and cognitive science, where there is a central executive or processing unit that monitors ongoing functions, issues commands and generally keeps things operating smoothly.

When we look at nature, we can never seem to find this boss. Nature instead builds and operates using the principles of self-organization by which systems come into existence and function under their own command. The interaction of parts with each other and with the environment is enough to give rise to ordered behavior. These processes are an integral part of who we are. They are taking place in the chemical reactions inside our cells, governing how our cells grow and differentiate during development, and probably underlie mental and conscious phenomena. To explain the mind we therefore need to understand self-organization.

So far we have touched on several themes. These are change over time, intricacy, and how ordered behavior emerges and operates. Each of these themes corresponds to a particular scientific-theoretic approach; namely, dynamics, complexity, and self-organization. In this book, we discuss these approaches in great detail and then apply them to the study of mind. Separate chapters are devoted to each, where we first introduce the topic from a theoretical standpoint and then examine how it helps us to understand mind.

Networks and Neurodynamics

The brain is a network made up of units and connections between them. There are many other examples of complex networks, both natural and man-made. The society we live in is a network, made out of individuals that move around and exchange information. The Internet is also a network. In recent years we have seen the development of a new approach to the study of networks. Called network science, researchers in this discipline attempt to understand the abstract principles by which networks operate.

After introducing the ideas of dynamics, complexity, and self-organization, we therefore next turn our attention to the study of networks. We start by examining Boolean networks, whose nodes operate in a very simple way, but whose overall behavior when wired up together is quite complex. We then examine random graphs, whose nodes are connected randomly but which still display ordered and interesting behavior. Following this, we discuss how self-organizational processes can lead to ordered connectivity, even in networks that are mostly random to start.

One problem in networks is how activity gets coordinated, especially if there is no "leader". How for example, does a message spread from one part of a net to another? We look at different ways that this can occur, along the way discussing networks like corporations and the human brain that show hierarchical organization.

A fascinating discovery, first noted by social psychologists some years ago, but only recently studied with more rigor, are small-world networks. In these nets, any two nodes can communicate with one another through a small number of connections, even though the number of nodes can be very large. In popular culture, this phenomenon is referred to as "degrees of separation". We look at the scientific study of small-world networks and what the implications of these findings are for mental processing.

It turns out, perhaps not unsurprisingly, that brains are small-world networks. This makes sense, because animals need to react quickly in order to survive. If information has to flow through too many connections, reaction times are slow and the proverbial tiger will eat you before you can react quickly enough. A number of other interesting properties can be instantiated in small-world networks, including signal synchronization and collective computation.

With this foundational knowledge in mind, we turn next to the specifics of how brain networks operate. The focus is on dynamical activity inside neural networks, what we call neurodynamics. Neurons are oscillators because they exhibit repetitive wave activity. We look at neural oscillation and how this can code for information. Then we summarize some of the research on synchrony and coupling, the means by which brains coordinate their activity between different regions. Another interesting feature of brain networks is metastability, a state that allows both for pattern formation and dissolution. Metastability may explain many cognitive processes like memory and free association.

Levels of Organization, Fractals and Noise

Classical methods in science have been useful at getting nature to reveal her secrets. Traditional science adopts an analytical and reductionist approach. It takes a whole such as the brain and breaks it down into its component parts. Once we know what the parts are and how they work, we can then figure out how they combine and interact to produce the behavior of the whole they make up.

Unfortunately, many wholes in nature defy reductionist explanation. The behavior of these wholes, to borrow a phrase from the Gestalt psychologists, is "more than the sum of their parts". There are several reasons for this. The sheer number of parts and the possible ways they can combine is vast in even moderately complex systems. Another obstacle has to do with our conception of what a whole is. Many investigators draw arbitrary lines around what they investigate. They agree to study a certain aspect of a whole, such as a neuron, person, or society, while neglecting that the "piece" they look at is both part and parcel of larger and smaller systems respectively.

The holistic emergent paradigm arose in reaction to these problems. In this approach, one takes care to examine interactions between levels and to consider any phenomenon in light of the other systems of which it is a part. In this book, we favor neither the reductionist or emergent perspectives. Rather, we advocate that both approaches are necessary to understanding complex systems. They must be used in conjunction with one another as part of a proper cognitive science. However, we spend more time examining the emergence perspective because it is newer and has yet to be applied in the same general and systematic way as its reductionist cousin.

A fractal is a pattern that is self-similar at different scales. Fractals exist everywhere in nature. We describe what fractals are and summarize some of the literature on psychology and fractals. This literature includes magnitude estimation, perception of fractal stimuli and aesthetic judgments. The brain has a very definite fractal organization, so we describe this activity and how it relates to mind and human nature.

A persistent problem in statistics and research methods is what to do with noise. Traditionally noise has been treated as error, the proportion of variance in the dependent variable that cannot be accounted for by the manipulation. A dynamical account views noise very differently. In this view, noise should be studied as it reflects some mechanism or aspect of the system under investigation. So we next describe sources of nervous system noise, different types of noise, including white, pink, and brown noise, and how they can account for neural and cognitive processes.

Dynamical Psychology and Cognition

The scientific study of how complex systems like the brain change is called dynamical systems theory, although it is has also been referred to as chaos theory. This relatively new paradigm provides us with a novel theoretical orientation and a set of tools and techniques that we can apply to things like the mind and understand them better. Unfortunately, psychologists do not seem ready for it just yet. In the section on dynamics and psychology we mention a number of reasons for this. We then summarize dynamical theory and practice for several major areas of the discipline, including development, social behavior, and industrial and organizational psychology. The amount of research that has gone into clinically relevant fields requires a separate chapter on personality, pathology and therapy.

Perhaps the greatest application of dynamical ideas in psychology has occurred in the cognitive area. There is much to talk about here. The dynamical approach forces us to rethink our traditional notions of representation, concepts and learning. We contrast classical views in cognitive science with the dynamical approach. Classical cognitive science advocates a mind that is modular, component-dominant, closed and feed-forward. The dynamical account sees mind instead as distributed, interaction-dominant, open, and recurrent. The recurrence theme is a particularly important one, so we describe several theories of cyclical processes in cognition.

This is followed by an in-depth analysis of studies in different areas of cognition. Each area corresponds to a distinct form of cognitive processing. We cover visual perception, including perception of ambiguous figures, Glass patterns and apparent motion. Then we look at attention in a temporal context. Following this we examine dynamical processes in memory, looking at the stability-plasticity issue. Language has been analyzed from the dynamical perspective, so we look at dynamics and the lexicon, syntax, and conversation. This chapter ends with a discussion of human decision-making and problem solving.

An Ecological Approach

People are not the only systems capable of problem solving. The process of biological evolution can viewed as a continuous solution to the problem of adapting organisms to an ever-changing environment. We discuss evolution in terms of optimization on a fitness landscape. This model is not complete however, without the introduction of self-organization. The key ingredients of selection, variation, and reproduction explain how evolutionary change happens. But the tendency for the world to form into ordered and stable forms is necessary to produce the structures upon which such change acts.

Evolutionary optimization on a fitness landscape can also serve as a model of human problem solving. We outline some of the work on human cognitive problem solving such as analogical reasoning, functional fixedness, insight learning, means-end analysis, heuristics and their relation to optimization processes.

The analogy between dynamic processes in ecological and human systems does not stop here. Ecological systems can be viewed as dynamical systems. They have web-like structures, consist of multiple differentiated competing/cooperating units, and play an active role in regulating themselves and their environments. Both ecologies and the mind defend against intruders and reorganize themselves after damage. It seems that similar underlying forces are at work in both, and perhaps in all complex systems with a given nature.

Models and Recommendations

A hallmark of the dynamic approach is the use of models. These are created and run in order understand some aspect of a complex real world system. Models of course cannot capture every aspect of a system, but if done correctly, they can reveal many interesting features and processes. Models are best used in concert with empirical methods. Experimental validation of a model is a crucial technique that all researchers should practice.

There are many sorts of models. Traditional cognitive science models are either statical or fail to capture important dynamical aspects of the process under consideration. We briefly summarize some of the basic types of models, including deterministic and stochastic, oscillatory, differential equation, cellular automata, and agent-based models. Examples are given for how some of these can be used to model brain activity and cognitive processes. We then present the advantages and disadvantages of various models and of modeling in general.

To conclude, we summarize the salient points made in this book and analyze the concept of emergence levels. We then evaluate the reduction-emergence debate and make suggestions for how to deal with some of the issues raised. The book concludes with a discussion of how the dynamical approach can serve to unify psychology and cognitive science.

Classical Science vs. the Dynamical Approach

Before describing the dynamical approach in any depth we need to first compare it to classical notions in science. The dynamical systems approach represents a sharp break with these views (Goerner, 1995). In this section we will summarize the classical scientific perspective and show how it contrasts with the ideas of dynamical systems theory. Table 1.1 shows a summary of these differences. Many of these distinctions are discussed in greater detail in other sections throughout the book. We introduce them here now to provide the reader with an overview.

Assumption	Classical Science View	Dynamical Systems View
System's Output	Linear	Nonlinear
System's Parts	Independent	Interdependent
Methodological Approach	Analytic	Holistic
System	Closed	Open
Causality	Linear	Circular
Prediction	Complete	Limited
Change	Slow and Gradual	Can be Fast and Rapid
How Order Arises	Unspecified or Agents	Spontaneously
Philosophical Perspective	Reduction	Emergence

Table 1.1 *A comparison of the differences between the classical and dynamical views of science.*

Linearity and Independence

Historically, scientists have focused on *linear* explanations, where the output of a system is proportional to its inputs. This type of relationship assumes a simple sort of system whose processes can be measured and understood. Of course, most if not all systems in nature are *nonlinear*. Outputs in this case are not proportional to inputs. A small input to a complex system can produce a small, moderate or even large effect. The processing taking place in these systems is intricate, convoluted and difficult to explain.

The parts of a system are in many cases assumed to be independent in the classical view. *Independence* assumes that a part functions with little regard for what is occurring in other parts of the system. It is compartmentalized, insulated or "boxed off" from the rest of the system to which it belongs. The cognitive science postulation of separate modules in the brain is a good example of independence. These modules are dedicated to processing specific information. When they receive an input, they crank through a computational routine uninfluenced by other modules. The result is that we always know what the result will be regardless of the overall state of the system. We discuss modules in greater detail later.

Modularity makes the brain easier to understand and model, but it throws the baby out with the bath water, since it fails to account for complex interactions between modules. The dynamical systems view can account for this. It sees the parts of a system as *interdependent*. In interdependent systems a processor in one location may be monitored, modulated or regulated by other processors. In this case, a processor's performance is dependent on the state of the entire system. The way it operates in one state is different from the way it might operate when the system is in another state.

If we assume independence, then a system can be understood *analytically*. One can examine the behavior of the different pieces of a system and then put these separate descriptions together to explain the behavior of the whole. This is the reductionist approach. However, this approach often fails. Many systems in nature, including the human brain, resist reductionist explanations. This suggests that they function in a holistic manner. A system is *holistic* if it cannot be broken down piecemeal. Each piece relies on the parts around it to function properly. In the words of the gestalt psychologists, "the whole is greater than the sum of its parts". Emergence is a property of these kinds of systems.

Systems in the traditional view are considered closed. That is, the system as a whole is assumed to be shut-off from the larger world around it. So when considering the brain as a *closed system* we treat it as distinct from the body in which it is embedded. The dynamical systems view instead sees the brain as an *open system* that cannot be considered independent of its surroundings. The brain is part of a body that is in turn part of the world. Treating the brain as an open system means taking into account "external" factors such as hormones from the body or visual stimuli from the outside world. It also postulates a two-way street between brain and its context, where the brain influences the environment and the environment also influences the brain. We discuss open and closed systems later in the section on the continuity of mind.

Causality and Rules

The independence approach has a linear view of causality, where variables affect others in a sequential manner. The best way to conceive of this is as a chain of arrows where variable A affects B, which then affects variable C, etc. (Figure 1.1). The dynamical systems view is that variables can have all sorts of causal relations that act concurrently. Much of this occurs as the result of feedback, where a variable's output at one point in time can alter the future activity of both itself and many other variables. We can diagram this type of *circular causality* by showing loops and networks as shown in Figure 1.1.

The difference between linear and circular causality is to some extent reflected in the serial vs. parallel processing approaches in cognitive science. In the serial approach, information processing happens one step at a time. Information processing must finish in one module before it can be passed on, as is the case in most contemporary computer architectures. In the connectionist or parallel-distributed processing perspective, streams of information are processed simultaneously. There is no need to wait for one unit to finish before proceeding.

Dynamical systems takes a long-term view of system action, since feedback can alter not just what happens in the immediate moment following a cause, but over the entire future course of the system's behavior. Feedback mechanisms can promote the maintenance of homeostasis, but they more often produce increases, decreases, oscillations, fluctuations, and other sorts of dynamic processes that change with time. So while the traditional view sees short-term immediate effects, the dynamical systems view sees long-term dynamical effects.

The classical view in science was that the world operated according to deterministic rules. These rules can be summarized in general as the laws of physics and more specifically as the laws that govern each of the different branches of

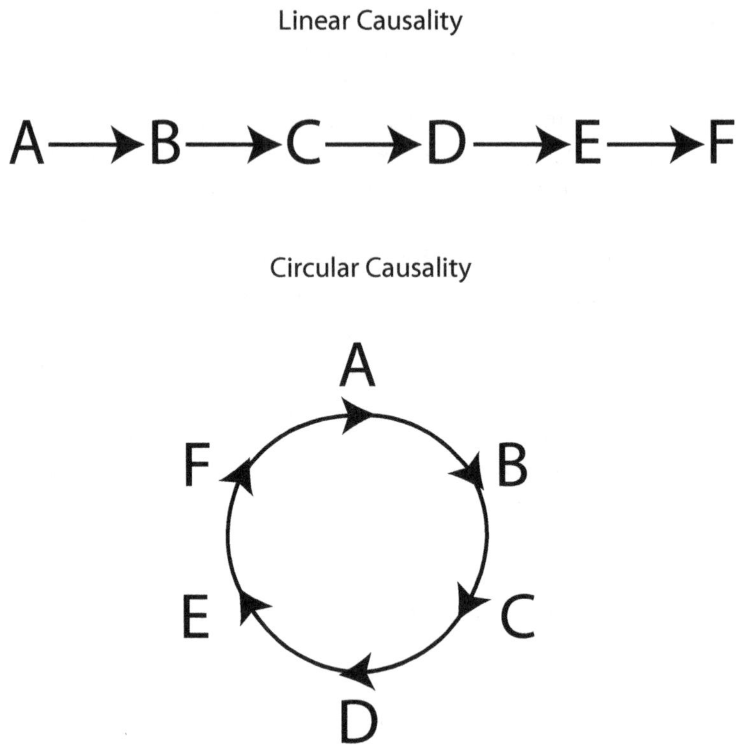

Figure 1.1 *The classical view in science is that causality is linear. The dynamical perspective allows for circular causality.*

science like chemistry and biology. If we can know these laws and have sufficient information as to the state of a system, then we can predict with accuracy what the system will do far into the future. We have *complete predictability*. Knowledge of physical law and conditions not only enables prediction, it implies also that complete understanding and control of a system are possible.

Dynamical systems theory also acknowledges that there are deterministic rules that govern a system's behavior. However, these rules and knowledge of starting conditions are not sufficient to allow complete prediction. The more accurate our knowledge of a system's starting state, the longer we can predict its future course. Eventually over time though, the behavior will diverge from our prediction no matter how accurate our starting observations. This is because the system is sensitive to initial conditions. Whereas the classical science view postulates long-term predictability, the dynamical systems view acknowledges only *limited predictability*. It allows for only partial understanding and control.

Change and Bifurcation

Change in the traditional view occurs gradually and slowly. In natural selection, for example, it was believed that species evolved continuously over long time periods and many generations. Evidence is now available though, that supports faster evolutionary change. This theory, called punctuated equilibrium, states that species evolution can happen fairly quickly, in a short time period with a smaller number of generations.

The dynamical systems perspective allows for such sudden and rapid change. Oftentimes, it is driven by the system's arrival at a *critical point*. These are conditions where a system undergoes a complete and dramatic alteration. For example, water molecules undergo a phase transition when they reach 100° C / 212° F. At this point they transform from water to steam and exhibit many new properties.

Chaotic systems can reach what is called a *bifurcation*, where they can switch to one of two alternate states. A state in this sense is large-scale or macroscopic. The bifurcation is a sizeable qualitative change in behavior. In perception of ambiguous figures such as the Necker cube, the perceptual organization can alternate back and forth between each possibility. The transition from one to the other can be considered as a bifurcation. We should also mention that there is an entire field devoted to the study of rapid change, called catastrophe theory. This discipline tells us that a particular combination of conditions, such as temperature and the degree to which steel is flexed, can give rise to dramatic alterations such as the collapse of a bridge.

Order and Energy

Perhaps one of the most important differences between the classical and chaotic perspectives is their view of order. The traditional science view has difficulty explaining how order in the universe comes to exist. It is accounted for by various factors such as accident, anomaly, or certain "agents of order" such as selfish genes, the human brain, or life itself (Goerner, 1995). According to the supernatural and unscientific view of vitalism, there is a mysterious force, vital spark or energy that creates phenomenon such as life and consciousness.

To a classical scientist, the world is passive and directionless. As we mention elsewhere, it acts much like a giant machine or clock that inexorably ticks away following known physical laws. How then, does order and organization emerge from this machine? According to the principle of self-organization, certain systems spontaneously organize themselves. Numerous examples of this are evident in nature. For instance, in chemistry, there are autocatalytic reactions that sustain themselves based on the amount of available precursor molecules. In biology, we see emergent behaviors when birds flock and fish school. Consciousness and cognition may also be examples of self-organization. This provides a very different conception of the universe, one that is active and directed toward the creation of ordered, complex structures and processes. The idea that natural ordered processes can arise from disorder is not entirely new and was proposed early in the 20th century (Schrödinger, 1944).

Another important distinction between the classical and dynamical systems views has to do with equilibrium. Systems that are in equilibrium have a homogenous distribution of energy and are incapable of driving processes. A process in this sense means any activity that requires energy. In biological organisms energy-requiring processes include locomotion and cognition that rely on metabolism. Self-organizing systems in contrast are *far-from-equilibrium*. They have a large energy concentration that has built up. This energy concentration creates a pressure to flow and drive operations.

According to thermodynamics, energy in any system always "seeks" to flow as fast as possible (Odum, 1988). It turns out that structured energy flow moves more quickly than that which is less structured. For example, energy flow is faster in a chimpanzee than it is in an amoeba because the chimpanzee is more structured and ordered. We can therefore see the creation of order and organization in the universe as driven by energy flow. This property of dissipative efficiency then could cause systems to become more ordered with time. If we plot energy flow over the time course of the evolution of life we see a dramatic increase. Energy flow, measured as metabolic rate, gets faster as organisms evolved over several billion years on earth (Swenson, 1989).

Energy flow can also explain why reorganization occurs within a given system (Goerner, 1995). If the pressure to flow builds up, energy cycles faster and faster within the system. The energy eventually reaches a limit on how fast it can flow. If this limit is reached and the resistance to flow is still less than the pressure driving the flow, a reorganization takes place. A bifurcation occurs and the pattern restructures itself into a more efficient configuration that can dissipate energy better.

A good example of this are Rayleigh-Bénard cells. If a thin layer of silicon oil is heated in a pan, the oil particles at first move randomly. As the temperature is increased, this pattern no longer dissipates heat well. There is a sudden reorganization into hexagonal-shaped cells where particles rise in the center of each cell and fall near its edges. This new arrangement is better at dissipating heat. If the temperature is increased further even more complex arrangements, such as rotating spiral patterns develop.

Reduction and Emergence

We have already mentioned reduction and emergence. Whereas reduction is analytic and seeks to break wholes into parts, emergence is holistic and seeks to build wholes out of parts and part interactions. Reduction is part of the classical approach while emergence is more recent. In what follows, we introduce these perspectives in greater detail and discuss the problems that are inherent in each.

Reductionism

In a general sense, *reductionism* is the belief that everything about a whole can be explained entirely by understanding its parts. If we can understand all there is to know about the parts that make up a system, then this enables us to completely understand the whole that they make up. In reduction we attempt to say that a given phenomenon X is "nothing more than" or "nothing above and beyond" phenomenon Y. X thus becomes reduced to Y.

Silberstein (2002) describes two types of reduction. *Ontological reduction* is concerned with the nature of reality and posits that everything in the universe consists of the fundamental constituents of reality, some basic particles or entities, or is determined by them. *Epistemological reduction* is concerned with descriptions of reality and posits that scientific theories or common sense conceptions can be reduced to other scientific theories or conceptions about the fundamental features of the world.

In ontological reduction, the things that need to be related are real-world items such as entities, events, and properties. For instance, we might want to relate the property "heat" from thermodynamics to a more basic global description derived from the properties of individual molecular elements, as is done in statistical mechanics. In epistemological reduction, the things that are being related or reduced are representational items like theories, concepts, models, frameworks, and schemas. A theory in biology for example, might be reduced to another theory in chemistry.

Inherent in the notion of reductionism is the idea of *levels of analysis*. In this view, natural phenomena can be analyzed at different levels of spatial and temporal scale. Usually these two are correlated, with things happening faster at smaller sizes. We can take a given phenomena, such as the mind, and examine it from the large scale of psychology, which examines abstract concepts such as thoughts and emotions, all the way down to the quantum scale of subatomic particles. Because each level contains different phenomena and is organized and operates differently, there are distinct disciplines that have developed to study each. From large to small, we could study the mind from the disciplines of psychology, biology, chemistry, and physics, with multiple sub-disciplines in-between each of these.

Problems with Reductionism

Reductionism was one of the major goals of the traditional scientific endeavor. At one point, it was thought that even the largest scale phenomena could all be reduced to physics. Each level of explanation would eventually reduce to the one below it like the floors of a collapsing building until we would be left with nothing but the "basement". Sociology could be explained by psychology, which could be explained by biology, which could be explained by chemistry, and so on. Findings from modern scientific investigation however suggest that this goal is untenable.

One of the central problems with reductionism is that there are very few actual cases of successful inter-theoretic reduction, where the explanatory constructs of one discipline are explained by those beneath it. Examples of failed or incomplete cases concern the reduction of thermodynamics to statistical mechanics, of thermodynamics to quantum mechanics, and of chemistry to quantum mechanics. In the last instance, the concept of a molecule cannot be accounted for using quantum formulations.

These are troubling because they come from the physical sciences. Reduction is even more problematic in the social sciences. It is not at all clear for example how to link the "hard" problems of consciousness, i.e., those that explain the subjective character of mind, with the "easy" problem of neural states. There are no clear linkages between mental or psychological terms like "perceptions", "thoughts" and "feelings" to particular physical brain states and processes.

It is very difficult to start with the characteristics and properties of parts and use them to explain the features of the whole, even in simple systems that

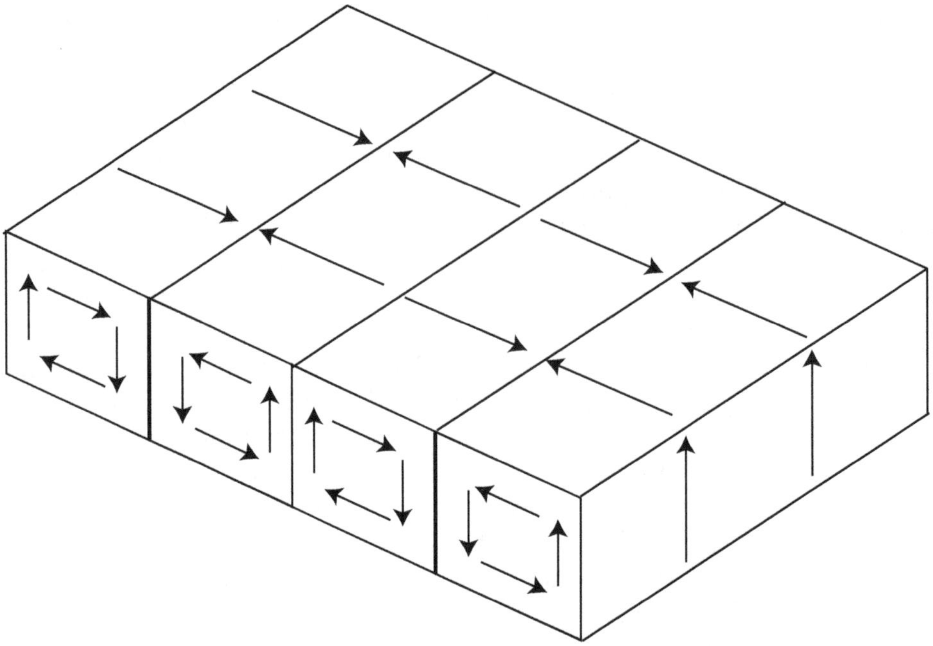

Figure 1.2 *Order from disorder: liquid molecules spontaneously transition into Bénard cells from random motion when heated.*

are just aggregates of a single kind of part. Take water for example. It is made up of the molecule H_2O. But knowledge of this molecule alone does not easily explain the interesting changes that take place when we heat large numbers of them. In this situation, the system undergoes dramatic qualitative changes, turning from ice to water to gas.

Many natural systems are like this. Their behavior doesn't change slowly, gradually, or predictably with a change in conditions, but instead undergoes dramatic alterations. These critical shifts produce a whole new organization, one that is best explained by resorting to different concepts and principles. It may seem disconcerting, but many phenomena in nature act this way. It is as if somebody flipped a switch that changed all the rules, forcing us to constantly come up with a new explanatory language. The failure of reductionism has cast doubt on whether the sciences can be united and has in part contributed to the current popularity of the emergence movement.

Bénard Cells

To show how difficult reductionism is, we will describe one phenomenon in detail. A Bénard cell is a convection cell, a region where gas or liquid matter rolls about upon itself. These cells appear spontaneously in a liquid layer that is progressively heated from below (Figure 1.2). Why should these ordered patterns appear, as it were, out of "nowhere"? It seems that a simpler outcome would be simply to have the liquid molecules move faster and faster, continuing to bump into each other in random fashion as more heat is applied.

The answer has to do with how temperature spreads in a medium. At lower temperatures, heated molecules can get rid of their heat through dissipation. They simply transfer the heat to neighboring molecules, rising from the bottom layers where it is hottest to the top layers where it is cooler. But at a critical value of temperature, this movement cannot get rid of enough heat. A new movement pattern develops consisting of the Bénard cells. These are more efficient at reducing heat because hot particles near the bottom move quickly to the top, lose their heat near the surface, and then plunge back down to heat up once more. The Bénard cells in this capacity can be thought of as a heat "conveyer belt".

From a global perspective, the liquid has spontaneously organized itself into a new and more ordered spatial configuration. Instead of random motion, we see cyclical motion. When viewed from above the cells form into a variety of shapes depending on the geometry of the container and can be in long cylinder-like forms or more spherical shapes. A three-dimensional perspective shows that the direction of flow within a cell alternates clockwise and counterclockwise directions every other cell. This preserves the same flow direction at cell borders, again maximizing the efficiency with which the heat passes out of the liquid.

The Bénard cells seem to emerge spontaneously from the liquid system. They are considered an example of self-organization, where order can arise naturally from disorder. Much of the order we see throughout the natural world is of this sort. It can be seen in weather patterns, chemical reactions, biological development and even human societal organizations.

For our purposes, Bénard cells provide us with a simple but useful guide for understanding how mind and consciousness might arise from the brain. It could be that mental properties are the result of self-organizing principles at work in the brain. Just as the dissipation of heat, the liquid structure and the geometry of the container create a context that gives rise to rolling convection, energy laws, neural structure, function and other properties naturally give rise to mind. The challenge for cognitive science is to describe how this happens.

Emergence

The concept of *emergence* is that the features of a whole or complex are not completely independent of the parts that make them up. They are said to go beyond the features of those parts or emerge from them. Many phenomenon in the natural world seem to be emergent. In fact, emergence has been used to explain almost everything from the big bang to mind. Table 1.2 outlines a fairly comprehensive list (Morowitz, 2002). Just as was the case with reduction, there is both ontological and epistemological emergence. The first is concerned with real world items such as parts/wholes, properties, events/processes, laws and entities. The latter deals with concepts, theories, models, frameworks, and states of a system (Silberstein, 2002).

Holland (1998) describes several features of emergence. Emergent systems are made up of interacting parts. The function of these systems is rule governed. These rules are invariant even though the parts or components of which they are made do change over time. In fact, the states or processes of these systems are in constant flux; we say they are dynamic. Their dynamic and constant-

Step	Process	Description
1.	The Primordium	Origin of the universe
2.	Large-Scale Structure	Structured matter like galaxies
3.	Stars	Formation of stars like our sun
4.	The Elements	Creation of chemical elements
5.	Solar Systems	Matter organized around suns
6.	Planets	Condensation of planetary bodies
7.	The Geospheres	Geological structures like rocks, water, and gas
8.	The Biosphere	Chemical combinations, replication, autocatalysis
9.	The Prokaryotes	Primitive biological cells
10.	The Eukaryotes	Cells with organelles
11.	Multicellularity	Multicellular organisms like plants, animals and fungi
12.	Neurons	Cells allowing the transmission of information
13.	Protostomia and Deuterostomia	Evolution of animals with single and dual gut openings
14.	Vertebrates	Animals with brains
15.	Fish	Various forms of fish in aquatic environment
16.	Amphibians	Transition to land-based life
17.	Reptiles	Animals with shelled eggs
18.	Mammals	Emergence of these major terrestrial vertebrates
19.	Arboreal Animals	Forest-Dwelling Animals fill this ecological niche
20.	Primates	Enlarged cerebral cortex, extended maternal care and training of young
21.	The Great Apes	Adaptation to savannah ecology
22.	Hominids	Full bipedal locomotion, cooperative food acquisition, and increased cognitive skill
23.	Toolmakers	Creation of effective instruments for altering the environment
24.	Language	Cultural transmission of information
25.	Agriculture	Domestication of animals and plant cultivation
26.	Technology and Urbanization	Applied mathematics and city-dwelling
27.	Philosophy	Efforts at self-understanding
28.	The Spiritual	Attempts to ascribe meaning to other emergences

Table 1.2 *Natural processes attributed to emergence in the temporal order in which they occurred (from Morowitz, 2002).*

ly changing nature gives rise to novelty; a non-stop parade of new patterns or behaviors that are difficult if not impossible to predict.

There is another sense in which things change but stay the same in emergent systems. Holland (1998) calls this persistence. This is where patterns of interaction persist despite a continual turnover of their constituent parts. He cites as an example the standing wave in front of a rock in a white-water river. The water molecules making up the wave are constantly changing, but the global wave pattern remains. So we can say that although the rules and particular patterns of emergence are invariant, their parts are always on the move.

Emergent systems, even though they are complex, demonstrate regularity. Under certain conditions, the system will show patterned behavior. Although the system may not repeat itself exactly, it can be shown to act similarly in a given situation. A random process would not be like this. Random processes cannot be predicted and cannot be guaranteed of acting in similar ways in a given context. Emergent systems are therefore not random or stochastic.

Problems with Emergence

Like reduction, emergence is not without its own set of difficulties. To start, a philosophical account of emergence doesn't count as an explanation. It merely says that if one thing cannot be reduced to another then it must be something more than what it is made of. In most cases there is no attempt to say exactly what happens. What is needed is some sort of explanation or theoretical account of how this something extra comes into being. Given this explanatory vacuum, some may come to equate emergence with magical, mystical, or theological workings. If relationships between parts are considered important in emergent phenomena, then a theory is needed that explains how relational processing gives rise to emergent properties.

CHAPTER 2
Systems and Complexity

Systems

A *system* is a set of entities that together make up a whole. Each entity or component interacts with or is related to at least one other component. All of the parts work together to allow the whole to perform some function or set of functions. An object that has no relation to any other element of a system is not a part of that system but rather a part of the system's environment. Systems can be more or less connected to their environments. Those that are insulated from their environment are closed. Those that are richly interconnected with their environment are open. Most systems (in fact perhaps all) with the exception of the universe as a whole are embedded in an environment. It is also doubtful whether any system can ever be completely closed off from its environment. So the use of the term closed system really refers to those that have minimal rather than no interaction with their environments.

Types of Systems

Before we delve into a description of complex systems we must first lay some groundwork. We need to know the conditions under which complex systems arise. This requires a description of the different types of systems that exist in nature. A sketch of this typology allows us to see how complexity relates to these other systems.

Ordered Systems

An ordered system is made up of relatively simple structures that interact in simple ways. Because of this, the behavior of the system is also fairly simple. Ordered systems are characterized by either constancy, where a single behavior continues without change, or by periodicity, where behavior alternates back and forth between a small number of states. An old-fashioned clock made up of gears and springs is a good example of such a system. Winding these clocks builds tension in a spring, which drives a set of gears to turn the hour, minute, and second hands at different rates of rotation. The clock's behavior is thus simple and understandable. The various hands rotate at different speeds because of the size of the gears that drive them.

A clock of this sort displays *linearity*. This means that it produces an output or action that is proportional to its inputs. If we wind the clock a certain number of times, we can anticipate how long it will work. A few winds means it will operate for only four or five hours. More winding means that it will operate for a day or so. In any event, the number of winds can be mapped onto the duration of operation in a straightforward way. We call the relationship between these two variables linear. The same number of winds always produces the same length of operation. In the section on chaos we talk more about the concepts of linearity and nonlinearity and describe what they mean mathematically.

Ordered systems are predictable. If something is predictable, we can anticipate what its future actions will be. One way to do this is to base future performance on past performance. If a single wind causes the clock to operate for one hour and two winds cause it to operate for two hours, we can predict that three winds will cause it to operate for three hours. Even though we may have never wound our clock three times, we can anticipate how long it will work based on the observations we have made about its behavior in the past.

For many centuries, scientists viewed the universe as a giant clock. All we needed to do, they thought, was elucidate the mechanisms of a system and then we could completely understand and predict it. This raises the philosophical issue of free will versus determinism. If all systems are ordered, then with sufficient knowledge we can predict exactly what they will do out to any date in the future. Our failure to do this so far does not mean that the universe is not ordered, but that we may simply not understand it well enough.

Chaotic Systems

Chaotic systems contain many parts that interact in a complicated way. Their behavior is therefore complicated and difficult to predict. The behavior of these systems is *dynamic*, meaning that it changes over time. This is in contrast to a *static* state of changelessness that an ordered system might be in. A chaotic system can exist in many different states and it can be hard to tell what state it will move to next, given the state it is currently in. A good example of a chaotic system is the weather. There is some structure and organization to the weather, however, as we see in weather patterns such as tornados, blizzards, cold fronts, etc.

The weather is *nonlinear*. This means that outputs from a weather system are not proportional to inputs. If we consider a particular weather system such as a hurricane, we could count as inputs things like wind currents, temperature differences, and whether it is passing over the land or ocean. These would then affect the system's outputs, such as the hurricane's wind-speed and direction. However, even if we could control such inputs, we would see that if they were duplicated, the weather would not act the same way each time. Computational models that take these variables into account and attempt to predict hurricane behavior are often wrong.

Chaotic systems have limited predictability. They can be predicted in the short term but not the long term. We all know that the forecast for tomorrow's weather will be reasonably accurate but that a forecast for a week from today may be completely off. Surprisingly, this lack of long term predictability is due not to a failure to understand the mechanisms that drive it, but because these systems are very sensitive to their starting conditions. Two identical chaotic systems deviating just slightly when they start will down the road diverge and produce very different behavior. We discuss why this is in greater depth in the section on chaos later on.

Most systems in nature are chaotic. In fact, some of the systems that seem simple may just be chaotic systems in disguise, i.e., chaotic systems that are in a particular state space or set of conditions where they fail to demonstrate chaos.

The chaos paradigm came on the scene just a few decades ago, but has had a profound effect on the scientific community. Many researchers in fields such as biology, ecology, and climatology now realize that the phenomena they study are not as simple as they may have wanted to believe. The chaos view requires not just a new theoretical orientation, but a new set of investigatory tools. Later in this book, we discuss how this paradigm shift has affect the thinking and methods of one field in particular, that of psychology.

Random Systems

There is a third type of physical system, that of randomness. Random systems can be made up of simple parts, such as atoms or molecules. But this doesn't mean that the behavior of these constituent parts can be predicted with any accuracy. Random systems can only be described statistically. We can say there is a certain probability that a behavior will occur, but we cannot know for certain that it will happen.

The movement of gas molecules in a container is an example of a random system. We can determine certain global properties of such systems. For example, we know that reducing the volume of a gas container by one half results in a doubling of its pressure. But we cannot predict the exact position and movement of each of the particles in the container when this happens. Random systems therefore lack any predictability as they can only be described stochastically.

Randomness seems to be a hallmark of microscopic phenomena. In quantum mechanics, when all causally relevant parameters in an experiment are controlled, there are still some aspects of the outcome that vary randomly. This random element may be objective and not a by-product of the way the effect is measured or observed.

It is difficult or impossible in some cases to distinguish a random system from a chaotic one. Chaos may also sometimes disguise itself as random. A system that seems random may be a chaotic system passing through a long series of states that, if measured long enough, may repeat or exhibit some other type of order. We have more to say on randomness later in the section on chaos.

The Edge of Chaos

So how does complexity fit into all of this? Are complex systems ordered, chaotic, random, or something else altogether? It turns out that complex systems exist in a region between order and chaos, what has been called *the edge of chaos*. Complexity, because it is partway between these two states, contains some order but also some disorder. Complex systems, like chaotic systems, are everywhere in nature. Examples include certain chemical reactions, ant-hills, brains, and human economies.

It is helpful to use an analog to understand the edge of chaos. In chemistry, matter undergoes a transformation as it is subjected to a change in temperature. Water, when less than 0° Celsius, is in a frozen state and exists as a solid. Its molecules are aligned in perfect order, like a crystal. When heated past this point, the ice melts and turns to liquid water. Now the molecules are less ordered and free to move around. Further heating past 100° Celsius turns the water into gaseous steam. The molecules in this state are even less ordered. Each point where the water transforms from one state to another is known as a *phase transition*. Phase transitions are marked by critical points, values of some parameter like temperature at which a major reorganization takes place.

Ordered systems are very much like solids. They have a high degree of order and therefore can be understood and predicted. It helps to think of chaotic systems in this analogy as a gas. They have less order and are therefore more difficult to understand and predict. In between these two states, near the phase transition from order to chaos, is where complex systems exist. They contain structure, but also diversity. Complex systems need this balance. If they were too structured or "frozen" as it were, they would be too rigid and lack sufficient flexibility to carry out their processes. On the other hand, if they had too many degrees of freedom, there would be insufficient organization to allow their structures and functions to emerge. For our purposes, then, a *complex system* is one that is neither too ordered nor too random.

Complex Adaptive Systems

A *complex adaptive system* (CAS) is one that adapts or changes to the dynamic world around it. It does this by acquiring information about the environment, both passively through perception, but also actively through actions and interactions. The CAS then identifies regularities in this information and compresses these regularities into a schema. Actions in the world are then carried out on the basis of the schema. In each situation, there can be different competing schemas. The consequences of actions then feed back to influence the competition between them (Gell-Mann, 1994). Figure 2.1 shows each stage in this process.

A *schema* in this context is a model of the world that tells an agent such as a person what to expect and how to act in a certain situation. For example, the schema for restaurant can serve as a script, so that we know what to do when we go out to eat. The schema would specify the order of actions in this scenario, indicating that one is first seated, then given menus, then orders, eats, and pays the bill. Selection of the appropriate schema depends on the kind of restaurant. If it is a fast food restaurant, we may pick up the food and eat it back home. If it is a drive-thru, we know to pull up alongside the building and order from within our car.

It is important to note that a CAS is a collection of numerous agents that are each capable of making independent decisions and operating autonomously. An *agent* in the most general sense is something capable of taking in information about the world, processing it, and using it to act. An agent can thus be a cell, an animal or a person. The environment these agents inhabit consists not just

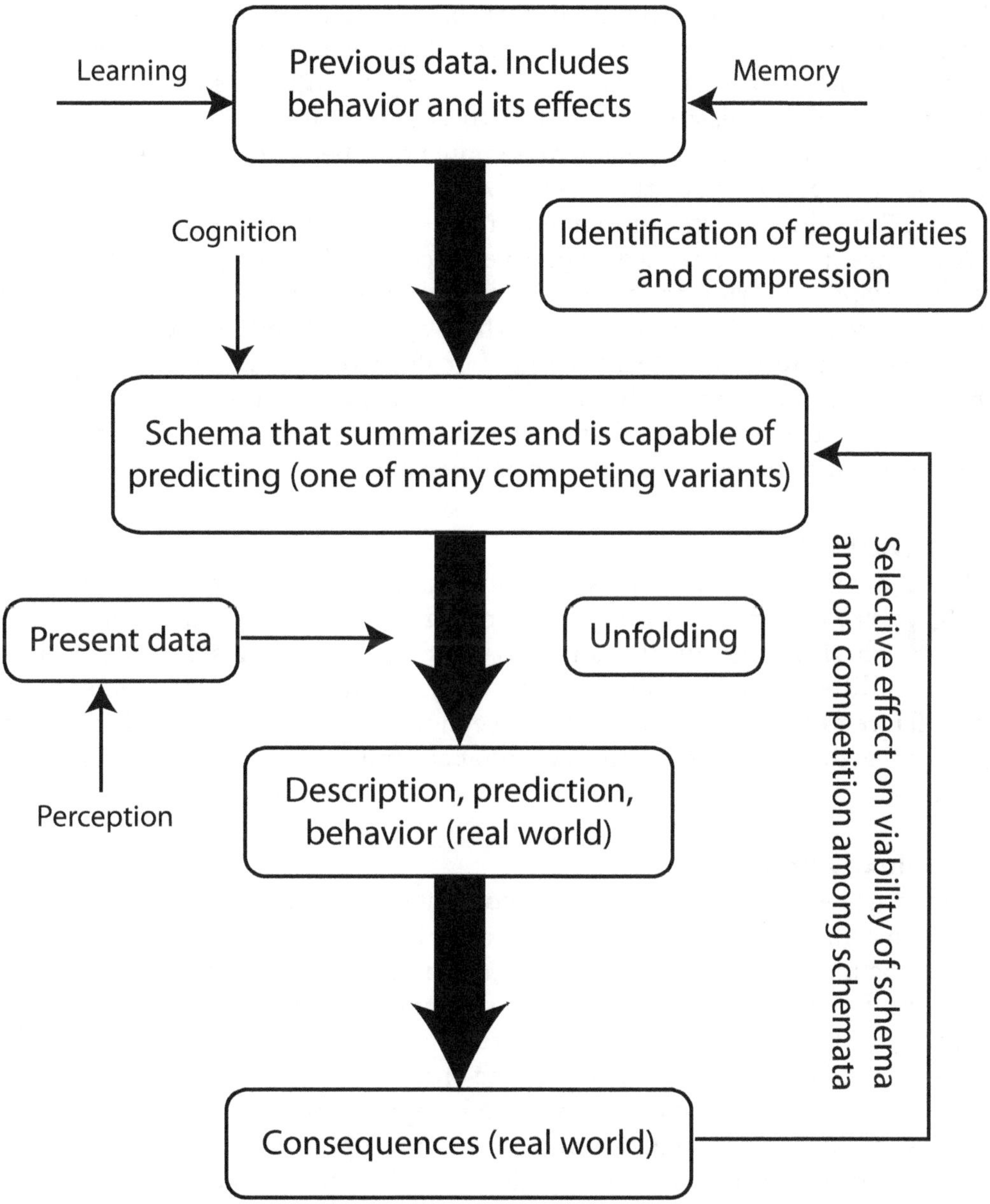

Figure 2.1 *A complex adaptive system and the cognitive operations that would support it (after Gell-Mann, 1994).*

of objects but of other agents as well. Agents can interact both cooperatively or competitively with one another. Examples of CAS can be found everywhere. If organisms are the agents then the ecosystem becomes the CAS. If people are the agents, then the CAS becomes society or culture.

Holland (1995) lists the significant characteristics common to all CAS: (1) All CAS consist of many component agents that continually interact with one another. (2) It is the aggregate behavior of the agents that we wish to describe. (3) The interactions between agents are nonlinear, meaning the aggregate behavior is not just the sum of the behaviors of the individual agents. (4)

The agents in a CAS are diverse, consisting of many varied types that can act differently. (5) If one of the agents or class of agents is removed, the system reorganizes itself with a series of changes that serve to "fill in the hole" or perform the function of the missing element(s). (6) The diversity of a CAS evolves. New agents come into existence that demonstrate novel behavior. (7) Agents employ models or schema to guide their behavior.

The mind and its component processes can of course be considered as CAS. Singer (1995) states that cognitive processes like perception and memory can be composed of rule-following agents such as feature-detectors and working memory structures. The environment these agents inhabit he says consists not just of stimuli from the external world, but also from our internal bodily state. Internal states include stimuli such as feedback from the muscles, heart rate and blood pressure. He also says that we should consider emotions as agents, since these also affect our decision-making and behavior.

Complexity

Now that we have discussed systems and complex systems, we need to get a better handle on what we mean by complex. What is complexity? This is a good question and it has no simple answer. As we will see, there are many different conceptions of complexity. No single definition seems to capture all facets of this elusive idea. In this section we will discuss complexity from a variety of different perspectives without attempting to provide a single overarching definition.

Complexity and Order

Weinberg (1975) provides us with a plot that clarifies much of the relationship between complexity and order. The x-axis in Figure 2.2 is labeled as complexity as may be measured in a standard way, such as number of parts or number of relations between parts. The left side of this axis corresponds to simple systems. Moving to the right we encounter more complex ones. The y-axis in the plot represents randomness and stands for the order or organization present in a system. Low down on this axis we have a high degree of organization. As we move up this axis we lose organization and correspondingly gain randomness.

Notice that there are three regions in this space. In the lower left corner are simple ordered systems. This is the realm of the mechanical. Constructed technological machines and devices would fit into this category. These systems are basic and regular enough that they can be described and understood analytically. The traditional methods of science work here. Wholes can be broken down into their parts and described in terms of the functioning of the parts. Reductionism is a valid means of understanding in this regime.

In the middle of the plot we have moderate levels of both complexity and order. Systems here are somewhat complicated but also function with some degree of order; they are in-between complete order and complete randomness. This is the realm of the biological. Cells, organisms, and ecosystems tend to fall into this category. Reductionist methods have difficulty here. We see phenom-

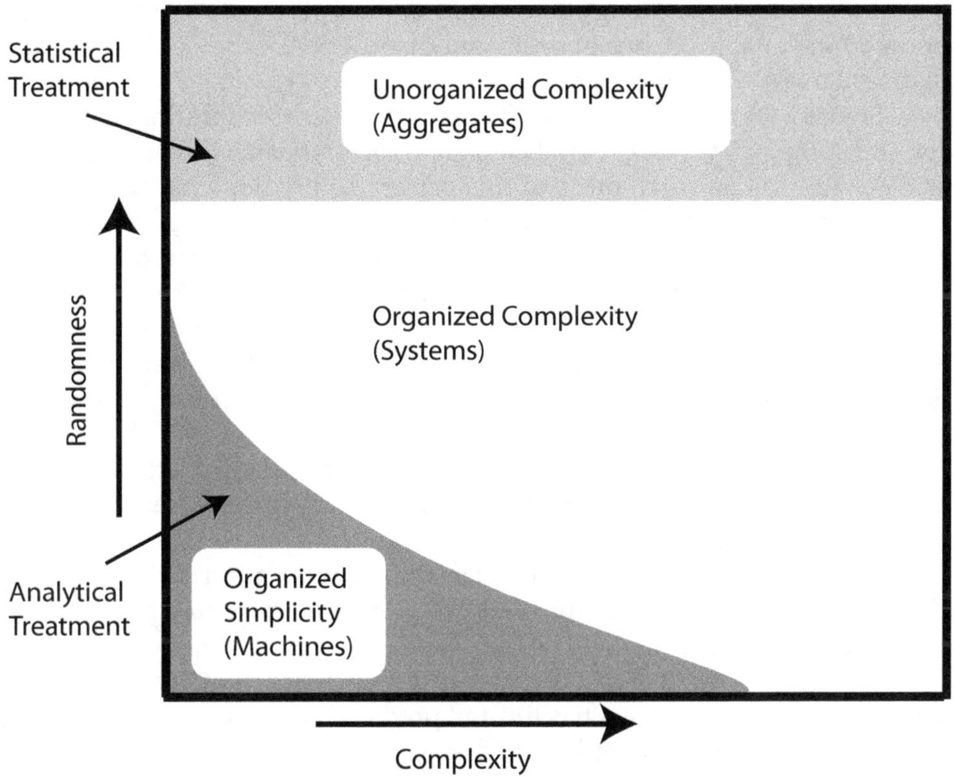

Figure 2.2 *The relationship between complexity and order (after Weinberg, 1975).*

ena that cannot be accounted for by part decomposition and description. These properties and behaviors seem to emerge from the interaction of the system's parts. This is the arena within which the relatively new tools of complexity, nonlinear dynamical methods and network analysis seem to offer us the greatest hope of understanding.

Across the top of the plot lies the third region. These are systems demonstrating a high degree of randomness. There is very little in the way of order or organization here. This is the realm of aggregates. An *aggregate* is a group or mass of distinct things gathered into or considered as a whole. Examples of aggregates include particles of sand or collections of molecules. The actions of the individual parts here are unpredictable. A gas molecule may bump into its neighbors, but we cannot predict where it will go next. These systems can best be fathomed using statistical methods that describe population rather than individual behavior. For instance, we can describe the average movement of a population of atoms, but not the movement of any of its constituent individuals.

Entropy and Complexity

At first blush, one might think that entropy is a good way to think about complexity. *Entropy* is a measure of how disordered a physical system is. The greater the disorder, the higher the quantitative measure of entropy. But complexity involves both order and disorder. Entropy only measures this unidimensionally and it is not clear where along the continuum complexity lies. Are all complex phenomenon equal mixtures of order and disorder? Are they 60% ordered and 40% disordered? We cannot yet say.

Figure 2.3 shows examples of three patterns corresponding to order, disorder, and complexity. The ordered pattern in the first panel is regular and repetitive. We can use one small piece of this pattern to predict what the remaining whole is like. In highly ordered systems such as this one, precise predictability is possible. The last panel shows a disordered pattern, one created by a random process. In effect, each pixel of this pattern has a 50-50 chance of being filled (made black) or unfilled (left white). Regions of this pattern can be described statistically, for example, we can say that roughly half of the pixels in any given region, large or small, are filled. But we cannot with any certainty make predictions about what the pattern in an unobserved region is like.

The middle panel of Figure 2.3 shows a complex pattern. It is neither completely ordered or completely disordered. However there is structure present. We can see squares of various sizes and other shapes too. This pattern displays fractal characteristics in that it is self-similar at different scales. Notice that with some knowledge of the pattern we can predict with limited precision what the remainder of the pattern will be like. We can, for instance, predict that there will be squares in other regions, but we cannot say what their exact size or location will be. This pattern thus has limited predictability, as we would expect of a system that is a mixture of order and disorder.

A Complexity Metric

Sole and Goodwin (2000) provide us with a simple quantitative measure of complexity. They use as their foundation the interactions between units in a system. In a simple system, the units that make it up will not interact. They will not exchange information, matter or energy. A container filled with gas molecules is a good example, although the molecules will bump into and jostle one another, their behavior is basically independent. In these kinds of systems, global behavior is simply the sum of the individual part's behaviors.

Let's take the case of two coins, A and B. If these coins are independent of each other, flipping one should have no effect whatsoever on the other. The global behavior of the two coins can be expressed as: $p(A, B) = p(A)p(B)$. In other words, the probability of obtaining two heads is equal to the probability of A coming up heads multiplied by the probability of B coming up heads. Since this is 0.5 in either case, assuming fair, unbiased coins, the probability of obtaining two heads in a row is $0.5 \times 0.5 = 0.25$.

But in complex systems, units interact. Two neurons in a brain can mutually influence the other's behavior. One neuron might inhibit or slow down a second neuron, the more activation it receives. The behavior of one unit in this case modifies the behavior of the other. In these situations, the probability of a given state of the system is more than the product of the separate probabilities. To use the coin example, if there is a dependency between the coins, they are more likely to both come up heads or tails. So we now have: $p(A, B) > p(A)p(B)$.

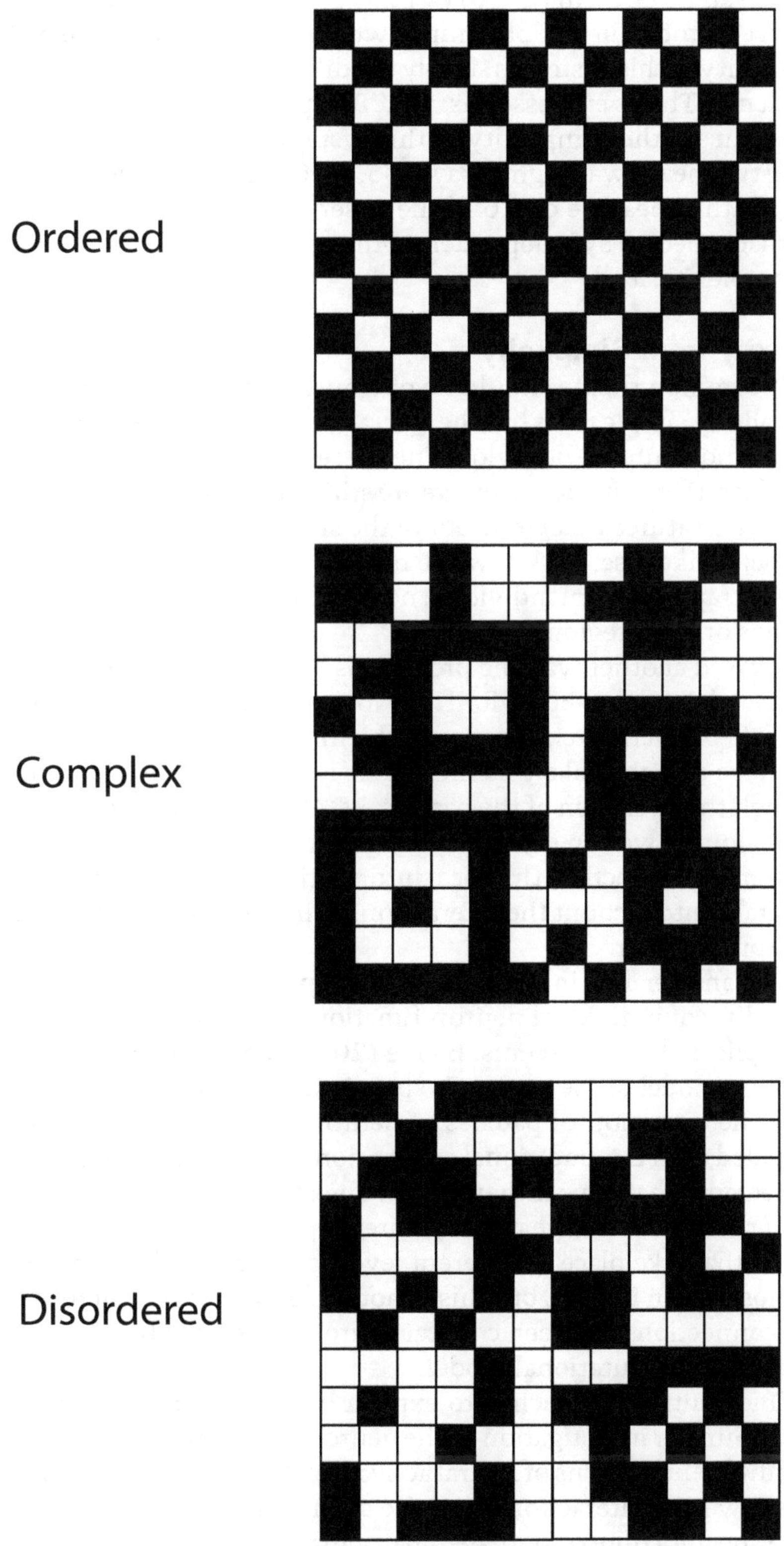

Figure 2.3 *Visual examples of ordered, complex and disordered patterns.*

Sole and Goodwin (2000) call this measure *distance to independence*. If the observed probability of obtaining two heads for biased coins is 0.75, then the complexity of this system is simply the difference between this outcome and independence. This is expressed as: $p(A, B) - p(A)p(B)$. So we have $0.75 - 0.25 = 0.5$. We can say the complexity of this system is thus 0.5. If the likelihood of obtaining two heads was higher, say 0.85, then the system would be even more complex, with a measure of 0.6. If the observed outcome for two heads is the same as that expected by independence, then the distance is zero and the system has no complexity at all.

Complexity from Simplicity

By now a common theme should be obvious. Many complex systems are governed locally by simple rules but the interaction of these rules gives rise to emergent order and complex behavior. The three-rule system in the Game of Life (described later) is quite basic, yet we are still not able to understand how it produces all the creatures that come out of the simulation and how they interact to create others. Likewise, we know the basics of neuron function, but we cannot perfectly extrapolate from individual neuron function to the operation of neural populations or brain regions.

Phrased another way the problem is "you can't get there from here", or you can't explain global behavior from local behavior. This is a concrete illustration of the reduction-emergence problem. Reductionists argue that we will eventually be able to make this linkage. It is just a question of understanding well enough the operation of the basic units. Once we understand neuron function well enough, we can say how they give rise to global action. According to the emergence perspective, this local information by itself will never be enough. We must take into account the interactions at multiple levels and how they each influence one another.

The answer lies in adopting both approaches. We still need a greater understanding of individual neuron function and are now realizing that this is more complicated than it seems. Bickle (2001) states that the traditional integrate and fire model of the neuron is simplistic. He has formulated a model that simulates the operation of patches of neuron membrane. This model is much more detailed than the traditional conception and can help to explain aspects of neuron behavior that have hitherto been unexplored.

At the same time, we need a greater appreciation of the complex forms of interaction that take place at different levels of spatial scale in the brain. In the cortex, most action is local, but this is not exclusively true. There are also long-distance connections between cortical neurons that must be taken into account in formulating computational models.

One fruitful approach is to explain how local rules give rise to higher-level rules. Further investigation in the neurosciences may result in the categorization of different patterns of neural activation. Once this is done, we could next examine how local interactions give rise to these patterns. It may turn out they occur in a specific context, perhaps a particular type of neural architecture with a given type and number of connections, etc. These new patterns could then be grouped and treated as basic units themselves that follow a new set of rules. These rules would then be the basis for explaining even more global, higher-order interactions, where these patterns interact to form larger and more complex phenomena.

This method of forming bridge rules from lower to higher levels has not met with much success. It has failed for instance, to link quantum mechanics to chemistry or chemistry to biology, i.e., to bridge the atomic to the molecular. But a new methodology taking both reduction and emergence into account would fare better. For any given system under investigation, this approach would be both analytical and holistic. It would seek to link rules while factoring in influence from other levels. We talk more about this combined analytic/holistic approach in the last chapter.

Complexity and Parts

A commonsense or intuitive notion of complexity concerns the parts of a system and how they work. A system that has many different types of parts might be considered more complex than one with fewer parts. In this sense a brain with more neurons or more different types of neurons could be considered more complex than a brain with fewer total neurons or fewer different kinds of them. But then we must also factor in the number or way in which these parts interact. A brain with one million neurons could be more complex than one with a billion if the interconnectivity between those cells was more extensive, yielding more complicated patterns of communication.

Imagine that we have a system made up of distinct parts. How can we successfully describe the behavior of this system? In physics, there is what is called the n-body problem, where we must describe the subsequent motions of planets given initial information about them such as their positions, masses and velocities. For two objects, one needs to describe how each object acts by itself in isolation. Then we must consider how each body influences the other, what is called the interaction. Finally, we need to consider what happens in the field or environment the objects are in when neither is actually present. So for the two-body case, there are a minimum of four equations: two "isolated" equations, one "interaction" equation and one "field" equation (Weinberg, 1975).

When we increase the number of bodies, there remains a single field equation and a single isolated equation per body. However, the number of interaction equations increases tremendously. For n bodies, there are 2^n relationships. A system with only 10 bodies or parts has 1,024 relations. Most natural complex systems that we would want to study have millions or billions of parts, requiring an astronomical number of relations that would require explanation. Clearly another method of description is needed, one that does not rely on understanding what every part does to every other.

In the case of explaining the motions of the planets in our solar system, it turns out we can ignore the effect of a third body on the interaction of any two. In other words, we can take pairs of bodies, describe how they interact and then simply sum their effects together to understand how the entire system operates.

So, if we had three bodies, A, B, and C, we could describe how all three possible pairs, A-B, B-C, and A-C interact, and then add up each of their effects.

Unfortunately, many systems in nature cannot be solved using summed pair-wise interactions. That is because higher-order interactions do exist. The behavior of one neuron A on another B, is dependent on the activity of a third neuron C. This requires that we factor in the behavior of all "third party" members when considering the behavior between any two. This causes a substantial increase in the number of relations that need describing.

As if this wasn't bad enough, there are yet more complications. If we assume that each member of a pair acts identically on its opposite we can say they have a symmetrical relationship. But in most cases, there is an asymmetry relation between pairs: A acts on B differently than B acts on A. In human relationships this is certainly the case, for a husband may treat his wife differently than she treats him. Asymmetry relations require that we understand pairs twice, once as A acting on B and a second time as B acting on A. Symmetry relations make a homogeneity assumption because they imply that all bodies act identically. Asymmetry relations instead imply heterogeneity, the assumption that not all parts in the system act identically.

To complicate things further (sure, why not?) we must also consider that parts, in addition to affecting others, might also affect themselves. A part might produce a behavior that affects not just another part but itself as well. In this case, we have the equivalent of learning or experience; the part's current actions can alter its future actions. We cannot always assume that a part's actions are static, they are usually dynamical and evolve over time. This means we need to consider the history of a system when describing it. Full analytical explanation of the sort we are talking about here requires knowledge of how a system acted in its past.

Another problem with using parts to define complexity is that we have to stop somewhere. At what level of detail do we wish to stop counting parts? At the cellular level? The molecular level? The atomic level? After all, a neuron by itself can be considered a system made up of interacting elements that are themselves made up of even smaller interacting elements and so on. The same can be said for an upper limit. At what level of global scale does a mind stop? The nervous system? The nervous system plus the endocrine system? The entire body? The body plus the local environment?

In physics, the level of microscopic detail at some point must be ignored. This is known as *coarse graining*. However, we need to stop at a global scale as well, what we might call *fine graining*. Notice that the decision to stop at either end results in the formation of a closed system, since we do not allow influences from smaller or larger scales. There is thus a trade-off involved in graining. By introducing cutoffs we delineate and simplify our system, making it more amenable to investigation. But we also removed it from other causal influences that could have substantial explanatory power.

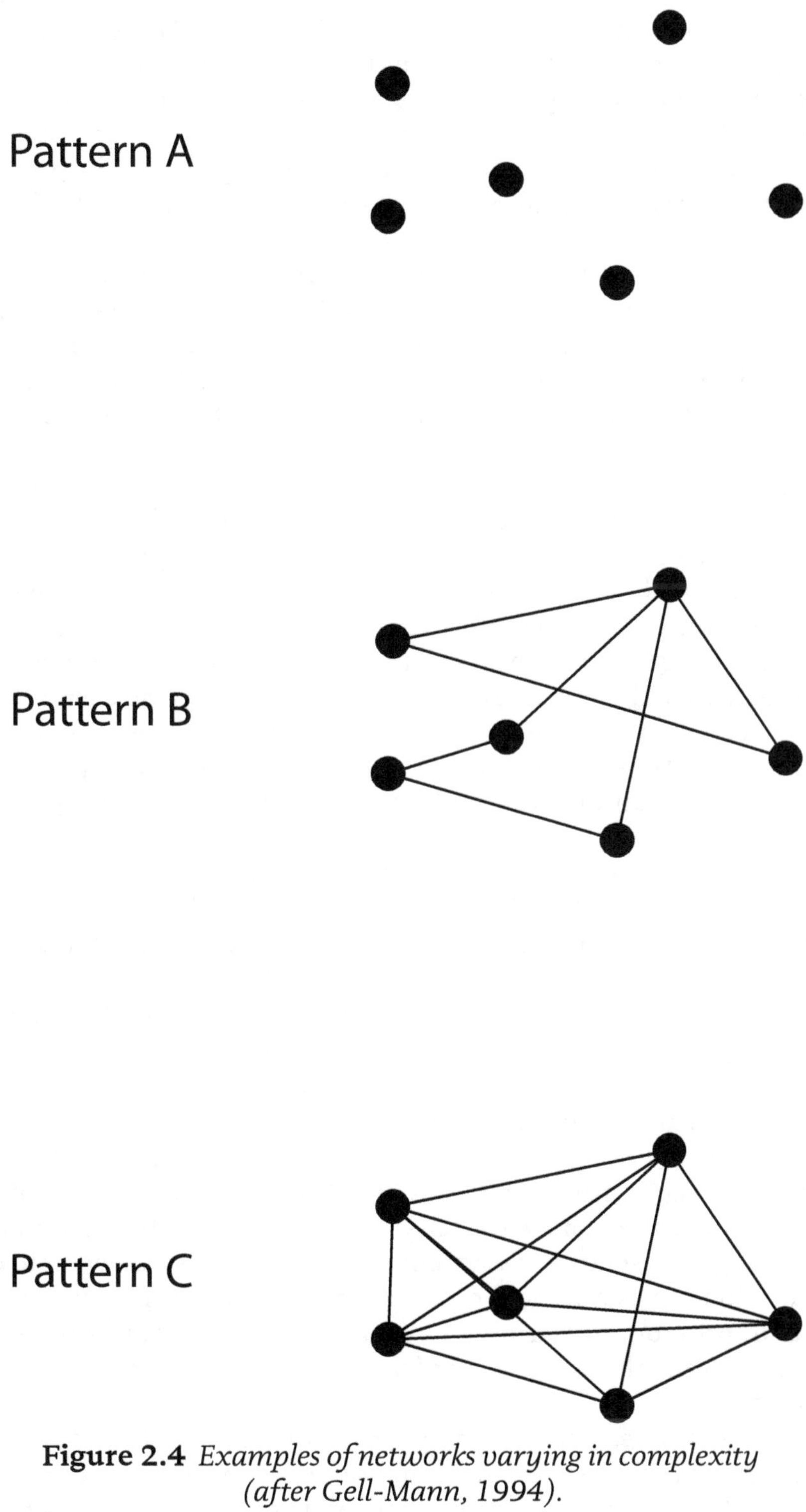

Figure 2.4 *Examples of networks varying in complexity (after Gell-Mann, 1994).*

Information and Complexity

Gell-Mann (1994) raises another important aspect of complexity. He asks us to consider a network such as the one shown in Figure 2.4. This consists of several nodes with varying types of connections between them. We can consider these as simple brains. One configuration has only nodes and no connections. At the other end of the continuum we can connect every node to every other node. If we just compare these two cases, we could say that pattern C is more complex than pattern A because the number of possible circuits that can be formed in it are greater.

But in an informational sense, patterns A and C are equally complex. That is because the length of their descriptions is equal. We could describe pattern A by saying "no dots connected" and pattern C as "all dots connected". Pattern B, intermediate between them in connectivity, would then be more complex because we would need to specify in detail exactly which nodes are connected. This description would be longer than that for patterns A and C. So we see that the informational view of complexity, in which we judge the complexity of a system by the length of its description, is not necessarily the best method.

One reason for this inconsistency is in the language we use. In English, the words "all" and "none" effectively compress more complicated explanations, in effect throwing away information. If we used a different language, one where we had to list each node and all the other nodes to which they were connected, then we would return to our original conclusion. Another problem with description is that even if we agree on the language, there is no guarantee that two observers will use it in the same way. Observers come to a system with different perspectives and background knowledge that will consequently affect their descriptions.

Given what we know so far, we can define a measure of *crude complexity*. This is the length of the shortest message that will describe a system, at a given level of course graining, to someone using language, knowledge, and understanding that both parties share beforehand (Gell-Mann, 1994). This is a crude definition only because the language used and the graining omit detail. The knowledge and understanding of the two parties is also limited.

Algorithmic Information Content

The concept of *algorithmic information content* (AIC) can help us understand complexity better (Chaitin, 1987). Assume that we start with an idealized computer that has infinite storage capacity. We then present the computer with a message string representing the system whose complexity we want to describe. The description consists of the shortest program or algorithm that will cause the computer to print out the string. The length of this program is the AIC of the string. It is basically the shortest set of instructions that can be used to generate or reproduce it.

As an example, take the binary string 00001111. The shortest program that would reproduce this would be something like: "print four zeros, then print four ones". This is a short program, so the string would be considered simple. Next take the string 01001110. Our program would be: "print one zero, a single one, two zeros, three ones and one zero". This program is longer so the string is more complex. The first string is redundant, since its elements repeat more often. The second string has fewer repetitions and so is less redundant, making it more complex.

Redundancy in information theory is encompassed by the notion of *compression*. A highly redundant string is said to be compressible, meaning that we can economically code for it using a short description. In other words, the AIC of compressible strings is low. However, there are many bit or binary strings that are incompressible. These strings have no redundancies and can't be compressed. There is no program shorter than the length of the string itself that will describe it. Random strings, those generated by random processes, have a maximum AIC.

There are problems though in using AIC as a measure of complexity. First, we can never be sure that there isn't some program that could compress a given string further. We may think that we have a random string when in fact there could be some as yet undiscovered rule that would generate the string and make it compressible. Alternatively, we may think we have found the shortest program for a given string when there is in fact another program that would lower the AIC by compressing it more.

There is another issue. Randomness is not the same thing as complexity. As described elsewhere, complex systems have a mix of randomness and order. They have both variability and unpredictability as well as regularity and predictability. The action of air molecules bumping into one another is more random than the collection of molecules that make up a brain or body, but nobody would suggest that the former is more complex.

Computational Complexity

The computational difficulty of a problem depends on the number of possible solutions. Problems that require a large number of steps to solve can be considered difficult. In *combinatorial problems*, where there are a fixed number of possible solutions based on the number of elements in the problem (N), each computational step corresponds to the evaluation of one of these possibilities. In effect, the computer is said to be searching through the space of possible solutions in an attempt to locate the right one. We can measure the difficulty of this type of problem by examining how the number of possible solutions to be searched increases with an increase in the size of the problem (Figure 2.5). The size of the problem here means an increase in the number of elements that can be combined.

There are at least three classes of combinatorial problem difficulty (Bossomaier & Green, 1998). In easy problems, the space of solutions can be searched linearly. Imagine that we have a ball hidden underneath one of three bowls. How do we find out which bowl it is under? Well, we would have to look through all of them to be guaranteed of locating it successfully in every case. If the number of elements is three then the number of searches is also three. So the time needed to search in simple problems is directly proportional to the number of elements.

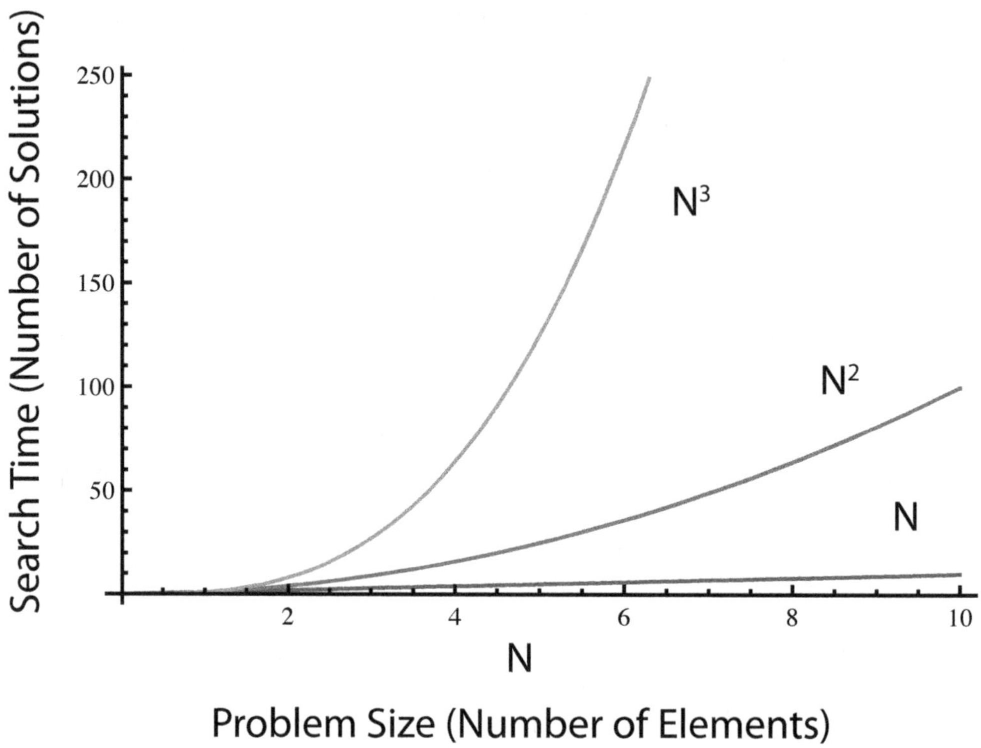

Figure 2.5 *Search time (the number of solutions to be tried) as a function of problem size (number of elements combined) for linear and polynomial cases.*

Problems of moderate difficulty show a greater increase in search time with an increase in elements. Imagine that we have five cities and we want to determine what is the shortest distance between any pair of them. For four cities there are six pairs. Once we pick one city we are left with three others, once we pick one of those we are left with two others and so on. So the total number of pairs is 3 + 2 + 1 = 6. In this class of problem, the search time is more than linear. It is polynomial, meaning it is an algebraic power of the number of elements, such as N^2 or N^3.

Although polynomial problems can be difficult to solve, they are not unreasonably so. It is just a matter of cranking through the various possibilities. Solutions can be found in most cases using the brute computing power of a typical desktop computer. For this reason, polynomial problems are said to be *tractable*, the length of time required to solve them does not become unbounded as the size increases.

Hard problems show an even greater than polynomial increase in the time needed to find a solution. The classic example is the *traveling salesman problem* (TSP). Imagine that you are a salesperson. You have to visit four cities just one time each and you want to do it traveling the shortest distance possible. The problem is to find the shortest possible route from any given starting city. The solution space has (N-1)! / 2 possible options.

It turns out that if we plot the number of routes as a function of the number of cities N, things quickly get out of control. For small numbers of cities things are doable. For example, with five cities, we only need search through twelve possibilities. For twenty-five cities, the number of possible journeys is so large that even a computer searching a million possibilities a second would take 9.8 billion years to look through them all! Problems of this sort are *intractable*, meaning the time required to solve them quickly spirals out of control. Solutions increase exponentially with number of elements. Computers are of little use here. These problems are said to be *Non-deterministic Polynomial complete* or NP complete.

So we now have a measure of computational complexity. A system may be said to be complex if the number of computations it can perform is NP complete. Notice that this complexity is based on a small number of basic elements. This echoes a common theme in complexity science, which is that complexity arises from simplicity. Complex systems have a small number of basic building blocks, but the number of ways these elements can combine or interact is enormous. The complexity emerges not from the elements themselves, but in the intricate way they come together.

Complexity and the Nervous System

The human brain is often touted as the most complex system in the universe, with some ten to one hundred billion cells and as many as ten thousand connections per neuron. But complexity is more than just structure. Koch and Laurent (1999) discuss the complexity of nervous systems in general from four different perspectives. These are teleology or functionality, evolutionary history, anatomical structure, and codes and computation. We briefly summarize their views here.

Perhaps the brain's complexity is linked to its function. After all, we have brains for a reason. They allow us to recognize patterns, learn, remember, use language, and solve problems. These are all specific processes. In a more general sense, the brain keeps us alive in a complex and unpredictable environment and seems to be the organ underlying our capacity to be conscious. There are other complex systems in the natural world that contain an equivalent number of parts, such as galaxies with their numerous stars or sandstorms made up of billions of particles, but these systems don't perform the myriad tasks our brain does. They also don't seem to act in a goal-directed fashion. It seems that complexity is necessary to support complex intentional action.

Nervous systems are the result of a history of evolutionary selection. Evolution, acting on species over generations, can shape biological organs such as eyes and brains. Intuitively, it would seem that the more complex the structure, the longer the evolutionary history required to give rise to it. This notion is supported by the idea of *logical depth* from information theory. According to this idea, the complexity of an object (in this case a bit string) is proportional to the amount of computation required to generate it from a random sequence as a starting point. If we think of evolution as a computational process and the resulting organ or organism as the object, then more evolutionary time ought to be required to produce more complex objects.

From an anatomical perspective, brains and nervous systems are incredibly complex. This becomes even more evident when we examine them at a scale smaller than that of the neuron or dendrite. In the human brain there are numerous neurotransmitter types, including acetylcholine, dopamine, serotonin and others that form functionally specialized pathways. There are also many different types of synaptic receptor, including both voltage- and neurotransmitter-gated ionic channels. Activity in nervous systems is plastic and adaptive, with synaptic modulations based on learning and memory.

Tononi, Sporns and Edelman (1996) derived a theoretical measure of neural complexity that measures the amount and type of statistical correlations within a neural system. Their measure reflects the mutual information shared between subsets of units. Using another measure of matching complexity, they found low complexity with random connectivity between neural units in a simulated cortical area. The complexity increased when the connectivity was modified to amplify intrinsic correlations enhanced by sensory input. Functionally, neural complexity should be low if neurons are all firing independently (randomly) or if they are all firing in a correlated manner, such as when their activity is completely synchronized. Complexity is maximal when there are a large number of subassemblies of different sizes in the population.

One must also consider how neural systems code information when considering complexity. At the individual neuron level, a frequency code is used where variations in firing rate, typically measured as number of impulses per second, can represent information. Pulse trains or bursts of action potentials constitute another type of code. At the global level of neural populations, more possibilities exist. Here, representations can exist in the form of an overall pattern of neural activity, which can be amplitude modulated and distributed over space. Relative timing is also important at the population level. Phase disparities in the arrival of signals between different neural regions can also serve as a code. We talk more about neural codes and patterns of neural activity in the section on neurodynamics.

It would be nice if we had a valid index of neural complexity, a single measure that we could calculate for different species or individuals and which we could then use for purposes of comparison. This type of measure might even correlate well with measure of intelligence such as I.Q. But the task seems hopeless when we try to take into account all four of the different categories outlined above. Any measure of neural complexity must consider that nervous systems are nonlinear and can change with experience. It must also consider that brains are heterogeneous. Different regions have different architectures and different functional characteristics.

Koch and Laurent (1999) conclude that continued use of reductionist methods will probably not lead to greater understanding and that we must take into account the context in which neural activity occurs. A true measure of complexity cannot myopically examine just one neuron or neural population and generalize to the whole brain or nervous system. It must look at how that cell is connected to its immediate neighbors, how activity in one population is affected by activity in other others and how brains operate in the larger context of the body and environment.

Complexity Science

As we have seen in this chapter, it is difficult to come up with a universal definition of complexity. Different researchers focus on different aspects of complexity and several theories about it exist. Because of this, it is better to refer to the study of complexity as *complexity science* rather than complexity theory (Lewin, 1999). We can say a few general things about this discipline. First, we can say that complexity science is the study of complex systems. Although an understanding of exactly what a complex system is can be debated, a good working definition is that a complex system is "one whose properties are not fully explained by an understanding of its component parts" (Gallagher & Appenzeller, 1999).

Another important principle underlying the study of complexity is the notion that complex behavior can arise from the operation of a few very simple rules. It is the goal of complexity science to discover those rules and explain how they give rise to a system's behavior. Complexity science does not restrict itself to one domain, but is truly interdisciplinary. This perspective has been used to explain phenomena in chemistry, biology, ecology, climatology, economics and other fields. Commonalities such as the power law distribution have been found in these disciplines, suggesting that they are governed by identical or similar rules.

Phelan (2001) argues that we should focus not on what complexity science studies, but on how it does so, i.e., its methodology. He contrasts the approach of complexity science with the approach of traditional science. First, traditional science is built on the reductionist premise which argues that a system can be understood by a cataloguing and description of its parts. Once we know how the parts work, an understanding of the whole they constitute will simply fall into place. Whereas the focus in traditional science is on parts, the focus in complexity science is on their interactions, how they work together.

Furthermore, traditional science is linear. It expects behavior to correspond in a simple and direct way to stimulus or causative inputs. The complexity approach views behavior as nonlinear, where outputs can be non-proportional to inputs. In traditional science one goal is the discovery of laws that describe the overall behavior of a system. Instead, complexity science looks for generative rules. A *generative rule* determines how a set of artificial agents will behave in a virtual environment over time (Phelan, 2001). This behavior can include how these agents interact with one another. These agents are the basic component parts of a complex system and can, for example, be cells in biology or human consumers in economics, to use a few real world examples.

Traditional science seeks to explain, predict and control. So does complexity science. However, the focus in complexity studies is not on predicting what a system will do. It is on understanding how the generative rules use feedback and learning algorithms to enable the agent to adapt to its environment over time. It is adaptation rather than prediction that is considered important. A complexity researcher would be happy if she understood how a system adapted to changes in its environment, even if she could not predict its exact behavior.

Finally, traditional science uses experimentation as its primary tool. An independent variable is manipulated in one or more conditions while other factors are held constant or randomized. The outcome is then measured using a dependent variable and compared to a control condition where no manipulation was made. In contrast, the main tool of the complexity researcher is modeling. A model, usually mathematical or computer-based, is constructed. The model has agents that are given a set of rules specifying how to act in response to environmental conditions. The behavior of the agents is then studied with particular attention paid to how their interactions give rise to global action.

A Critique of Complexity Science

Horgan (1995) provides a critique of issues in complexity science. He starts off by saying that the researchers at the Sante Fe Institute, where much work in complexity takes place, tend to "degenerate into computer hacking". In other words, there is too much emphasis on modeling and not enough on experimentation and empirical validation of computer models. What's more, he notes that the investigators there suffer from what he calls the "reminiscence syndrome". They see a result that reminds them of another phenomenon and then apply that older existing model to the current data. In many cases though, the data simply have some accidental features that make it seem as if they are another instance of the older case.

One claim that some complexity scientists make is that it will provide a unified theory for all complex systems. John Holland, one of the outstanding proponents of complexity science, states that we may be able to discover the underlying laws that govern all complex adaptive systems, be they economies, ecologies, or computer networks. Knowledge of these general principles, Holland believes, will allow us to solve the problems that arise in these systems such as financial depressions, species extinction, or computer viruses. This is a bold claim, of course, and other investigators don't share Holland's enthusiasm. To date, there have been only the most general of properties found linking the operations of adaptive systems in different fields.

Another problem concerns definitions. There has been some difficulty distinguishing complexity science from other closely related endeavors, like chaos theory. There is also no commonly agreed-upon definition of the term complexity itself, as discussed earlier. In fact, Seth Lloyd at the Massachusetts Institute of Technology has compiled a list of 31 different definitions of complexity. Some of these definitions draw on somewhat ill-defined concepts themselves, such as entropy, randomness, and information.

Horgan (1995) next points out that complexity scientists believe in a seductive syllogism. This is that simple mathematical rules when followed by a computer give rise to very complicated patterns. Since the world also contains many complicated patterns, these must also be governed by simple rules, ones that can be discovered using computers. This assumption has been criticized. It may be impossible to show that any numerical model of a natural system is valid. This is only possible in formal closed systems such as mathematics and logic. Because natural systems are open, models of them only approximate the real state of affairs.

Self-organized criticality has been touted by its originator Per Bak (1999) as a ubiquitous phenomenon, one that underlies most of the processes in the natural world. We describe this phenomenon in greater detail later. However, the concept is so general and statistical that it does little to explain how specific systems operate. At an abstract level of description, sandpiles, earthquakes and stock markets may all display this behavior, but this doesn't prove that the particular mechanisms that underlie each are the same. In fact, there is no evidence thus far that this is the case.

CHAPTER 3
SELF-ORGANIZATION

Self-Organization

Self-organization is a process by which the internal organization of a system increases in complexity without being guided or managed by an outside source. Self-organizing systems typically display emergent properties. In self-organization, the pattern at the global level emerges solely from interactions among the lower-level components. The rules specifying the interactions are executed using only local information, without reference to the entire pattern (Camazine *et al.* 2001). There are numerous examples of self-organization in the physical, biological, and human realms. These include crystal growth, chemical autocatalytic sets, bird flocking, cellular automata and various phenomena in economics. Table 3.1 shows a more complete listing of the sort of self-organizing processes found just in biological systems.

Features of Self-Organization

Camazine *et al.* (2001) list several important features of self-organizing systems. First, these systems are dynamic. This means they require continual interactions among their components to produce and maintain order. The second feature is obvious: self-organized systems demonstrate emergent properties. These properties as defined earlier mean that certain global characteristics cannot be accounted for solely by a description of the component parts. This does not however, mean that these properties arise by a magical or supernatural force. They can be accounted for entirely by physical mechanisms.

Self-organizing systems also demonstrate stability. It is true that complex systems can transition between states abruptly as when a parameter value is changed producing a bifurcation. Most biological systems though are remarkably stable. The majority of them function in a parameter range distant from bifurcation points and actually resist transitions to alternate states. This makes sense from an evolutionary point of view. Natural selection should favor a single operating state for a system, one that is maximally adaptive to local conditions.

It is probably the case that evolution selects for changes to the processes underlying a given structure rather than producing an entirely new structure (Camazine *et al.*, 2001). If this were true, selection would select for new organizations by tuning an existing system parameter rather than developing entirely new mechanisms. To illustrate, look at the wide variety of different surface patterns on animals ranging from seashells to fish to zebra. Since self-organizing systems are believed to form these patterns, the differences we see between species may be the result of the same mechanism operating under different parameters.

Organism	Behavior
Slime molds and bacteria	Pattern formation
Bark beetles	Feeding aggregation
Fireflies	Synchronized flashing
Fish	Schooling
Honey bees	Nectar source selection
Ants	Trail formation
Army ants	Swarm raids
Honey bees	Colony thermoregulation
Honey bees	Comb patterns
Ants	Wall building
Termites	Mound building
Wasps	Construction algorithms
Paper wasps	Dominance hierarchies

Table 3.1 *Examples of self-organizing processes found in biological systems (from Camazine* et al. *2001).*

The last general feature of self-organizing systems is that complex global behaviors can arise from simple local interactions between parts. This theme turns up again and again. The rules governing cellular automata, we discuss elsewhere, are quite basic. Yet, they give rise to an enormous multiplicity of forms. In the brain, the rules governing neuron function are simple, yet the global result of billions of neurons following these rules is complex beyond comprehension.

Self-Organization and Leadership

A key element of self-organization processes is that they display ordered, coordinated and coherent activity without any external direction. These processes operate as if they were being told what to do by some leader or boss. This is in fact not the case. There is no "blueprint" or instructor providing explicit instructions to each member of the group. They operate only on the basis of local information that is available to them. For example, fish schooling is the result of each fish adjusting their actions so as to not be too far or too close to their nearest neighbors.

Camazine *et al.* (2001) give examples of what self-organization is not. A marching band in which all the members are instructed to step forward a given number of steps and then make a 90° turn is not a self-organizing system. That is because each member was given a set of instructions beforehand and is simply carrying them out. A team of carpenters building a house is also not an instance of self-organization. This is because their actions are guided by a predetermined externally-imposed set of instructions that specify the exact structure of the house.

This view of organization without leadership is a break from the traditional notion in cognitive science of a central planner. This leader is introduced to provide for coordinated action. In computer science, we see a central processing unit (CPU) that performs computations. A prominent theory of working memory postulates a central executive that supervises and controls activities such as retrieval and rehearsal (Baddeley, 1992). The Cartesian theater postulates a center or neural locus where consciousness occurs. Fortunately, these views are gradually being replaced by the distributed processing approach as witnessed by multi-core processors, connectionist models of memory and the multiple drafts model of consciousness (Dennett, 1991).

Self-Organization and Complexity

In physical systems the unit of action is a particle such as an atom, molecule, or grain of sand. Their behavior can be described using only the laws of physics. In biological systems, the unit of action is capable of much more complex behavior. It can be a cell like a neuron or an entire organism such as an ant or a termite. Here, the laws of physics alone are insufficient to explain the unit's behavior. We need to derive new rules. Genetics determines these rules, but they are also shaped by an organism's unique interaction with the environment.

One way to conceptualize this is to think of a unit in a system as an *agent*, an entity capable of taking in information about its environment, processing it in some way and using the results of this computation to perform an action. We talk more about agents and agent-based modeling later.

The simplest "agents" are those that all follow the same rules, those rules being the laws of physics. The instructions each unit follows would then be a simple mapping of a stimulus input to a response output. A sand particle that is bumped by another sand particle would move based on such factors as wind speed and gravity. In this case, it would move a given distance in a given direction based on an equation taking these factors into account. For particles, there really is no internal information processing taking place. The action is more like a reflex that directly connects the impinging stimulus to a behavior with no intervening interior processing.

When we go up one level of complexity, to that of a neuron, we see much greater variability of behavior. The instructions or code that we put into our unit still map inputs to outputs, but does so using more complicated intervening rules that are affected by a much larger set of factors. A neuron will fire only if it is depolarized past threshold. That is a simple enough rule, but the processes underlying it are affected by more possibilities, including the type of input, excitatory or inhibitory, the number of inputs of each type, their distance from the axon hillock, etc.

If we step up one level more to collections of neurons or nerve nets as can be found in simple organisms (e.g., in phylum porifera, a marine sponge), then even more factors come into play. Here we must take into account all of the dynamics that can occur between cells that are connected to each other in intricate ways. The study of intermediary systems like nerve nets can serve as a link between individual neurons and entire organisms, but the linkage is not easy or straightforward, as there is a tremendous jump in complexity in going from one level to the other.

For each level we go up, we find an increase in the variety of behaviors that can be generated, the rules that underlie them, and the factors that affect those rules. If our agent is now an entire organism or person, then stimulus inputs feed into a complex body and brain. That brain performs extensive computations on the information that can yield a behavior. The rules underlying cognitive processes are governed by nature and nurture. Some are hardwired, such as reflexes. Others are learned. Many are influenced by both nature and nurture.

So we see that there is a continuum of self-organizing systems going from simple to complex. As we increase complexity, we notice a number of changes. The first is an increase in complexity itself in terms of the number of parts, and the number and type of interactions. Part of this is due to increased nesting or hierarchical organization where agents are made up of other agents. For example, cells are made of molecules. These cells then make up organs that make up organisms and so on. Generally speaking, in the natural world, more complex systems are made up of more levels or hierarchies of interacting agents.

Increased complexity is also correlated with increased internalized control over behavior. For atoms and particles, there is little to no internal control of behavior. The actions of these units can be explained entirely by the influence of external forces that act on them.

As we move up in complexity, we see an increase in the length of the rules that describe behavior. The length of the programs these agents follow gets longer the more complex they are. The environment thus has less and less of an effect on them. At sufficiently high levels of complexity it makes less sense to even say that a stimulus causes an action. The agent can in fact serve as its own stimulus and can generate its own driving behaviors. In philosophy, this continuum of complexity is equivalent to a determinism-free will dimension, with simpler agents being determined and more complex agents acting of their own will.

The Process of Self-Organization

Feedback

The key process that drives self-organization is feedback (Camazine *et al.* 2001). There are two types. In *negative feedback* a change in a process produces a signal that serves to reduce the change. A thermostat in a house is a good example. If the temperature in a house drops, a sensor detects the change. It is then compared against an internal set point. If the temperature is below this set point, another process is engaged that returns the temperature back to its original value. The second process in this case would be turning on a heater. Notice that the direction of the change is irrelevant. It could be raising a temperature that drops or lowering a temperature that goes up, as would be true for a thermostat connected to an air conditioner. Negative feedback keeps conditions cycling slightly above and below a constant set point. Its function is therefore to maintain homeostasis.

While negative feedback keeps things the same, *positive feedback* does just the opposite. It promotes change by causing a system to respond to a perturbation in the same direction as the perturbation. The result of positive feedback is an amplification, where a small initial change can cause large dramatic change. The rapid growth rate seen in the human population during the past few centuries is a good example. In each current generation, more people are born than in the previous generation, causing an exponential or even more rapid increase in the number of people.

Both negative and positive feedback are present in neural processing. If a neuron population transmits excitatory signals that are converted to inhibitory signals that feed back onto itself, the rate of firing in the population will remain constant. Negative feedback here serves to maintain a baseline firing rate. If these excitatory signals instead feedback positively, the population's rate of firing will go up. One possible example of a feedback process in the early visual system is contrast gain control where the output strength of a neural signal is adjusted relative to the locally prevalent contrast (Bonin, Mante & Carandini, 2006).

These two types of feedback usually work together. If there were only positive feedback in a system, its behavior would spiral out of control. There would be too much change. A process, once initiated, would continue to either increase or decrease explosively. But, if negative feedback is added, this action can be limited. In contrast, if a system had only negative feedback, there would be no behavior. The system would simply stay as it is. The two forms working together provide the right balance of change. Examples of self-organizing systems that use positive feedback coupled with negative feedback include ripples forming across the surface of wind-blown sand, pigmentation patterns on fish and the formation of ocular dominance columns in the visual cortex (Anderson, 1990; Forrest & Haff, 1992; Murray, 1981; Swindale, 1980).

Information Transfer

The organization present in self-organizing systems comes from the interactions between their units. When examining biological systems, we see that for coordination to occur, information must somehow be passed between members. There are two ways for this information transfer to occur (Lloyd, 1983; Seeley, 1989). *Signals* are stimuli shaped by natural selection whose function is to convey information. *Cues* are stimuli that convey information in an incidental fashion. Pheromones released by female moths to attract males are signals. They are coded for genetically to facilitate mate localization. The path an animal such as a deer leaves by walking through the woods is a cue. It occurs as an accidental by-product of walking but can also be used to locate mates.

In the brain, tracts are bundles of axons that serve to transmit information over relatively large distances, that is, from one part of the brain to another. They are coded for genetically, but their formation is also probably influenced by emergent factors such as chemical gradients during nervous system development. Information transmitted along tracts are the neural equivalent of signals. They play a very specific adaptive function and so have been selected for by evolution. The corpus callosum in humans for example, serves to send signals between the two cortical hemispheres. This pathway may enhance problem solving and other cognitive skills.

Pathways can also be formed in the brain through experience. Repeated stimulation in the presence of a stimulus can strengthen synapses so that they are more likely to be activated when that same stimulus is presented again in the future. This is the neural basis of learning and of plasticity. These sorts of pathways are equivalent to cues. A retrieval cue in memory is in fact activating a learned pathway that associates the cue with some other piece of stored information.

The phenomenon of retinal waves may be thought of as a hybrid form of information transfer because it uses genetic and emergent patterning to "train" the visual system early in development. These pathways are restructured and organized by spontaneous rhythmic bursting activity generated by the retina before it becomes light sensitive, i.e., even before vision happens (Wong, 1999). This activity spreads across the retina as a wave causing synchronous firing in neighboring cells and contributes to the connectivity patterns of ganglion cells found in the mature adult.

In most biological instances of self-organization, information transfer is local. Each "unit" in the collective is only processing signals and cues that are close by. Fish when schooling follow their neighbors, turning when they do, matching their speed and heading. So each individual fish may have little idea where the school is traveling, yet it can travel along and get there just the same. This is also true for neurons. Most connectivity between neurons is local. The vast majority of cortical neurons for instance, communicate only with their neighbors. Very few have long-range connections. Those that do are typically sensory or motor, in which case their job is to take information from senses to the brain or from the brain out to muscles and glands. So neurons also rarely see the big picture, but in performing their function they are always part of it.

Global information in self-organizing systems is often difficult or impossible to acquire. It would take too long to propagate through the system. Local information however is easily accessed and transmitted. Because local information only has to travel a short distance, it can be passed on quickly and efficiently. So we see that there is no need for individual units to be aware of what the whole is doing. They serve as carriers of information and contribute to global action without knowing what those actions are.

Stigmergy

Information in self-organizing systems can be obtained from two primary sources. The first, as we have already described, is from other members. The second is from the environment. Much of the cooperative action among social insects is based on traces of previous work in the environment. In some ant species if a worker locates a food source, it will walk back to the colony laying down a pheromone trail. Any other workers who stumble across this trail will head out from it to help carry the food back. This is a very simple but effective system that does not rely on complex communication. The ants don't have to master a language or have direct links to one another.

The process by which individuals in a self-organizing system communicate with one another indirectly through modification of their local environment is called *stigmergy* (Grasse, 1959). It was first noted in nest-building behavior among termites. Worker termites at the beginning of nest construction will scoop up a small ball of mud, invest it with pheromone and deposit it on the ground more or less at random. Other workers are attracted to the pheromones and over time are more likely to place their balls next to other balls. What starts off as a random distribution of mud balls turns into piles that form the basis of arches, pillars, chambers, and tunnels. The stigmergy effectively turns disorder into order. It converts disorganized and haphazard activity into organized and cooperative work. The fact that it does this with no central coordinator and without direct communication between members is one of the wonders of self-organizing systems.

A process akin to stigmergy takes place during the development of the nervous system. During this period, an axon will grow but needs to connect with a target such as a muscle fiber. Axons that make successful contacts with their targets cause them to release a *neurotrophin.* This chemical promotes the survival and activity of the neuron so that its connection remains. Axons that fail to connect properly don't receive adequate neurotrophin and ultimately die off. In this process the neurotrophin is like the ant or termite pheromone. It serves to organize the system, converting an initial disorganized growth of pathways into a selective and ordered one where each axon connects to its proper site.

It is important to note that the issuing of a command does not create order here. The nervous system does not give explicit instructions to the axons telling them where to go. It simply waits for them to arrive and keeps them alive if they make it. Similarly, the ants don't know where to go to find food; they simply wander around and when they find it, provide a means for others to do so. Termites start off in a disorganized state of laying down nest material and end up with structure through chance clusters that aggregate based on attractive factors. In each case, the system started off disorganized but then through simple feedback from the environment moved toward order.

Alternatives to Self-Organization

Although self-organization is one route to creating order in nature, it is not the only path. Camazine *et al.* (2001) list four alternatives. In order, these are well-informed leaders, blueprints, recipes, and templates. We describe these and the problems associated with them in what follows.

The first alternative to self-organization involves the use of a *well-informed leader.* As mentioned above, this leader has knowledge about the process to be performed and provides explicit instructions to the members of the collective on how to execute them. For example, a foreman can tell the members of a construction crew what to do when assembling a house or building. In the brain, the frontal lobes may issue orders to other brain regions, initiating, terminating and monitoring ongoing processes. But in these cases, orders from the top are not the only ordering force. Feedback between the workers or between the neural subroutines also guides action.

Nature probably selects for self-organization over leadership because too much is expected of a leader. The leader must have an extensive understanding of the emerging structure, must be able to maintain a birds-eye view of it and be able to communicate instructions, in some cases instantaneously, to all group members. A huge command and control burden thus rests on the shoulders of the leader. This also creates vulnerability. If the leader perishes, the entire system comes to a grinding halt.

A second alternative is the use of a *blueprint*. This is a compact representation of the system's structure. Construction workers can consult a blueprint to aid them in the building of a house. In cognitive science, a visual image is a good example of a blueprint. The image captures the basic structural relationships among the parts. If we were imagining what a bicycle looks like, the picture in our mind's eye would show us the tires, handlebars, seat, and other parts in relation to one another. Images serve as useful representations of objects upon which transformations such as scanning, zooming and rotation, can be performed (Kosslyn, 1980).

The problem with blueprints is that they can contain a lot of information. Imagine how much information is needed to code for a blueprint of the entire human brain or body. The human genome cannot possibly contain a complete description of the organism. It would be far too costly and the risk of errors or mutations increases with information load. Instead, genes appear to regulate or modify self-ordering processes during development. Much less information is required to do this.

Another problem with blueprints is that they specify *what* is to be built, not *how* (Camazine *et al.*, 2001). What is needed then is an instruction manual that not only lists the parts and their arrangement, but also describes how to put them together. For this, we need a *recipe*. A recipe provides a step-by-step set of instructions for how to build a pattern. A novice chef who does not know how to make German chocolate cake can do a reasonably good job by following the recipe closely. A good example of a recipe in cognitive science is a *script* (Schank & Abelson, 1977). Scripts are mental representations that tell us what to do in a particular situation. A script for a restaurant would tell us that we first sit down, then order, then eat, then pay the bill, etc.

But recipes have their difficulties. A recipe is static and inflexible. It does not say what to do if a problem arises. A restaurant script does not tell you what to do if you lost your wallet and cannot pay the bill or if you spill tomato sauce on your shirt. Real-life situations demand that we deviate from learned behavioral norms. These cannot always be specified in scripts, which are generic and detail performance under ideal conditions only.

The fourth alternative to self-organization is the use of a *template*. This is a guide or mold that specifies the final pattern. A cookie cutter is a good example of a mold because it specifies the shape of a cookie. The template-matching model of visual pattern recognition posits the use of a template drawn from memory that is matched to an image derived from perception. The greater the degree of overlap between the template and the image, the greater the probability of recognition. This model has difficulty accounting for stimulus variability, since different templates would be needed for stimuli that vary in size, orientation, shape and location. However, a template may be used for simple perceptual processes such as matching, comparison, or copying.

A problem with templates is that they may not always be available (Camazine *et al.*, 2001). In nature an animal's body may serve as a template for the construction of a nest or burrow, indicating how large a chamber has to be. But templates seem to be more often used as tools by humans. Early humans may have used the rules of symmetry in conjunction with templates to produce tools such as stone axes (Wynn, 1999). For instance, one half of the head of a double-headed stone axe could serve as the template for the production of the other half with a corresponding mirror-image reflection.

All four of the alternatives to self-organization are instances of top-down management. Leaders, blueprints, recipes and templates all require the presence of information outside of or extrinsic to a system. That information is then disseminated to the system's members to enable them to perform their specified functions. In each of these four cases, information is also localized. It is concentrated in one place. Self-organization in contrast, is an example of bottom-up organization. Here, information is inside or intrinsic to the system. The information guiding behavior does not need to be sent from one to multiple locations because it is already distributed throughout the system. Each member has access to limited but locally accessible information that is sufficient to guide behavior.

In the natural world, self-organization is the rule, not the exception. But in the examples given above, we saw that cognitive processing is also governed by centralized, top-down mechanisms. Which cognitive/neural processes are self-organizing and which are governed by alternatives? They are probably both at work in most cases. For example, feature extraction and grouping in object recognition may be more self-organized, while verbal identification and labeling, which involve comparison against stored representations, may be more top-down. Self-organization may also be more predominant in development. It may lay down the foundation for top-down processes that evolve later in response to environmental input.

Self-Organized Criticality

Self-organized criticality (SOC) is a characteristic of dynamical systems that have a critical point as an attractor. In other words, these systems tend toward a point similar to what we see in a phase transition, where there is a sudden and dramatic reorganization from one state to another. Most phase transitions in nature occur only under certain conditions. Liquid water for instance, will turn to steam only at a specific temperature and pressure. What makes SOC interesting is its robustness. Systems can display this behavior under a wide variety of conditions. It suggests that certain natural systems reach criticality spontaneously, that they may in fact self-organize themselves to this state without any external manipulation (Bak, Tang & Wiesenfeld, 1987).

The classic example of SOC is the sandpile. Imagine that we have a hand dropping grains of sand one at a time onto a table. Eventually a pile of sand will develop. As the pile grows, its slope increases until it reaches a critical angle. At this angle avalanches happen. Every so often, sand will cascade down the side of the pile. How big are these avalanches? It turns out that there will be very many small avalanches with just a few grains of sand. Large avalanches on the other hand will occur but they are rare.

If we measure the size of avalanches while we continue to add individual grains of sand we see a relationship between avalanche size and frequency. Frequency of occurrence decreases with increasing size of avalanche. If we next transform the data by taking the log of frequency and size and then plot the result, we have a straight line that fits the points (Figure 3.1). *A power law* describes this line. The presence of a power law is significant. Unlike the normal curve that has the highest concentrations of elements at a given scale, a power-law distribution is "scale free". No single representative scale dominates.

Fractal structures show similarity at different scales. They also have no single dominant scale. If we could understand the mechanism that gives rise to sandpile behavior, then this type of mechanism might underlie natural phenomena that show fractal characteristics and SOC. The physicist Per Bak has come up with such a model (Bak, 1996) It is a computer model that simulates sandpile avalanche behavior. This model is a cellular automata (CA). It operates based on a few simple rules but results in complex behavior.

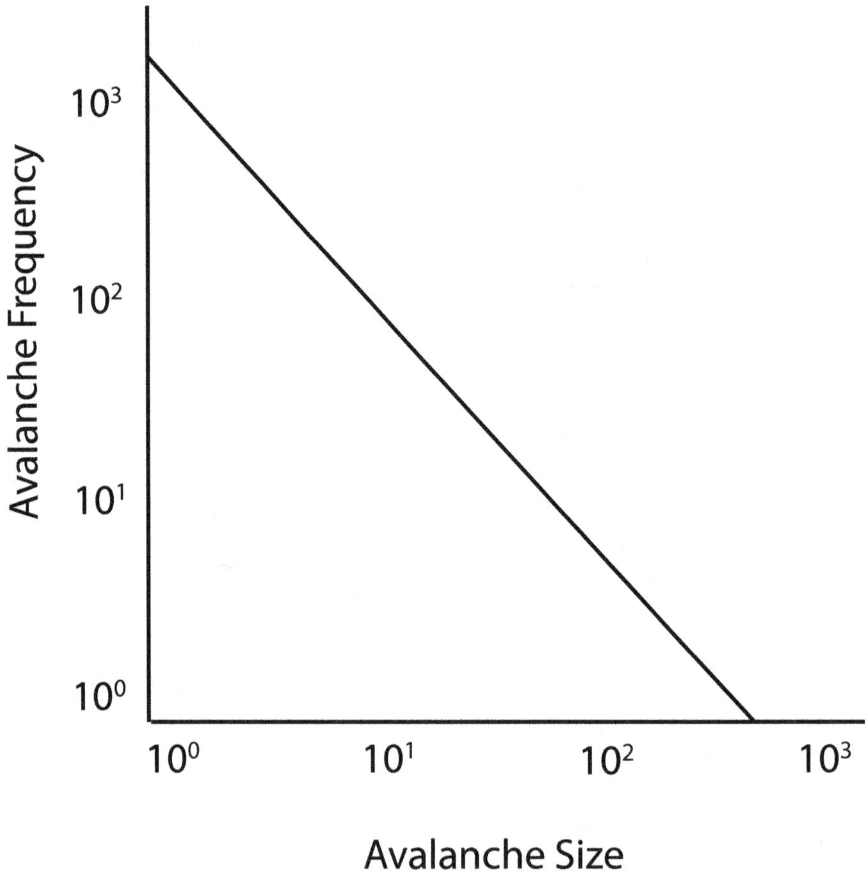

Figure 3.1 *A power law describes the relationship between frequency of occurrence and avalanche size. The larger an avalanche is, the less likely it will occur.*

Here is how Bak's model works. We start off with a lattice. Each location in the lattice represents the location of a pile of sand with a given height. The height of a pile is just the number of sand grains on top of one another at that location. There is a critical height for each of these local piles, say four grains high. We drop each new grain of sand onto one of these lattice points picked at random. If the grain drops onto a local pile that with this new addition is less than the critical height nothing happens. But if the added grain drops onto a pile and makes it four grains high, an avalanche process is triggered.

The four grains in this pile then fall and distribute themselves onto the four nearest lattice points. If any of these grains happen to create another local pile with four grains, then it too will collapse and distribute itself. In this way a chain reaction can be initiated. On some occasions the process may not go very far since none of the nearby piles exceeds the critical height. On other occasions, depending on the distribution of heights, the collapse of the initial pile will continue to collapse additional piles and the process may spread throughout the entire pile causing a widespread avalanche of sand.

This model suggests that simple rules may be the driving force behind dynamical systems that are scale-free. If this were true, then it could explain a wide variety of phenomena, because power law distributions are ubiquitous throughout the natural and artifactual world. They are found in the distribution of earthquakes, river floods, forest fires, landslides, epidemics, the internet and financial markets. Investigators are now turning to CA models to help explain these processes.

Mental phenomena may also be governed by these kinds of mechanisms. The sandpile model provides an interesting explanation for creativity. We can think of sand grains as simple ideas and the configuration of these grains in piles as complex ideas. If new ideas are added to the system, as occurs during learning, a critical point may be reached where they spontaneously reorganize themselves into a new configuration yielding novelty. This corresponds to our intuitions about creativity. It is difficult to come up with new ideas if we know too little about a topic. As we gradually acquire more information on this subject though, we can combine what we know to yield new and interesting thoughts. In this analogy, "brainstorming" or creative reasoning is equivalent to the avalanches, but like the sand grains, relies on a critical number of pre-existing ideas if it is to occur.

CHAPTER 4
DYNAMICAL SYSTEMS

The Dynamical Systems Approach
Nonlinearity

Chaos theory or *nonlinear dynamical systems theory* is the study of how systems change over time. A system in the most general sense is a collection of interacting parts. Change is a characteristic of all systems. The weather, the stock market, governments, animal populations and of course the human body and brain all change with time and have been studied using this approach. Perhaps the easiest way to see change is to use a graph with time on the x-axis and some measure of interest on the y-axis.

The case where search time is equivalent to problem size (i.e., where it is N) in Figure 2.5 depicts a *linear relationship*. It shows a continuous steady increase that allows for prediction. In linear relationships, the output is proportionate to the input. A linear equation is of the form: $y=c+bx$ where x is the input variable, c is a constant and b is the slope or rate of change. For any value of x we can predict what y will be. This equation, when fit to real world data, enables us to predict with a reasonable amount of accuracy. There are many examples of linear relationships in psychology. These include mental rotation, visual search, and the scanning of items in short-term memory.

In contrast the logistic equation in Figure 4.1 shows a *nonlinear relationship*. These are much more complex and difficult to predict. The output in these systems is not directly proportional to the input. In other words, a small change in input values can produce either small or large changes in output. Nonlinear relationships are characterized by fluctuations and in some cases, sudden dramatic qualitative changes. Nonlinearity is probably the norm for psychological phenomena and recent studies bring to light numerous examples in psychophysics, perception, learning, memory, cognition, human development, motivation, and problem solving to name just a few areas (Guastello, 2000).

In recursive equations, the input value at a given time determines an output value. This output value is then fed back into the equation to produce the next output value and so on. In nonlinear relationships the value at any time is thus dependent in part on the value obtained the previous time. An example of a nonlinear equation is: $x_{t+1} = c + x^2_t$. In this case the value of x at time t is squared and then added to the constant c to yield the output x_{t+1}. This process is then iterated.

Predictability

One of the interesting things about chaotic systems is that they are not *random*. A random process is one that is based strictly on chance, where the likelihood of any possible event that could occur is equal. Random events are disorganized

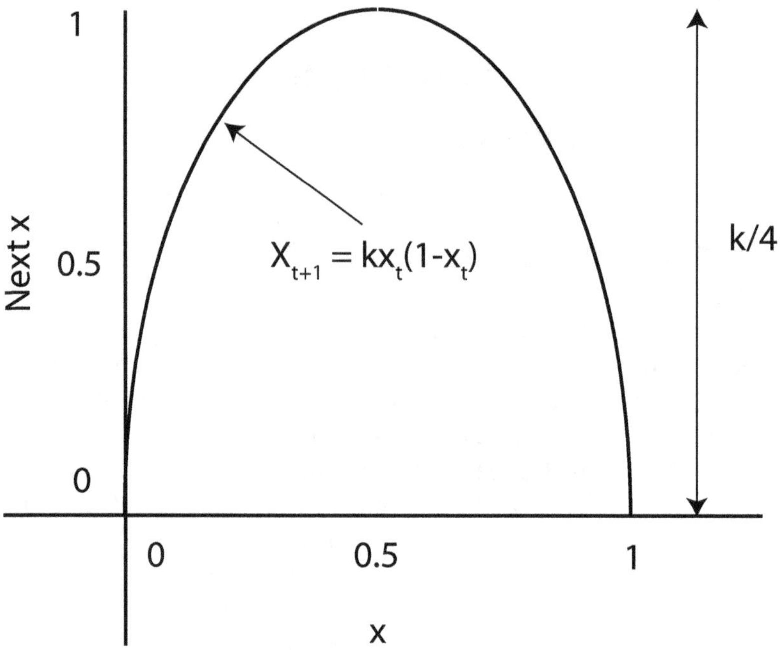

Figure 4.1 *The logistic equation is nonlinear. It takes the shape of a parabola.*

and haphazard. They cannot be predicted with any reasonable degree of certainty. Although chaotic processes look random they are in fact governed by knowable rules. Their processes can be exactly specified by equations. A *deterministic* process is therefore one that is completely and to within some degree of precision, exactly specified. Chaotic systems are deterministic but only predictable to a limited degree.

A hallmark of chaos is *sensitivity to initial conditions*. What this means is that two slightly different starting conditions in a chaotic system will over time diverge to produce two very different outcomes. Imagine two pool tables where we have arranged the colored balls in exactly the same way. We then use a pool stick to hit the white cue ball into the colored balls on one table, scattering them across the surface. Theoretically, if we could hit the cue ball in precisely the same way in the second table, the balls would move and come to rest in the same locations. That is because known physical laws such as friction and inertia govern the ball's behavior. In other words, the pool table system is deterministic.

However, it is not possible in reality to hit the cue ball in exactly the same way twice. No matter how hard we try, there will be minute differences in the angle of the stick, the force exerted, the location with which it contacts the cue ball, etc. These differences, although small, ultimately lead to different patterns of action in the system. Sensitivity to initial conditions is also referred to as the butterfly effect whereby a butterfly flapping its wings in Brazil could conceivably lead to a tornado in Texas.

We can quantify this sensitivity by taking a nonlinear equation and letting it iterate with two different starting values. When the difference in starting values is small, their trajectories start off identical but then begin to diverge and follow different paths. When the difference is larger, this divergence occurs much sooner. Figure 4.2 shows both small and large differences in initial values for a nonlinear equation.

What this means is that in chaotic systems, we have short-term predictability only. We can extend the length of a prediction by increasing the precision of our input values, but we can never be so precise that we obtain complete long-term predictability. In processes that are truly random, we cannot predict at all, even in the short term. We can only statistically describe what might hap-

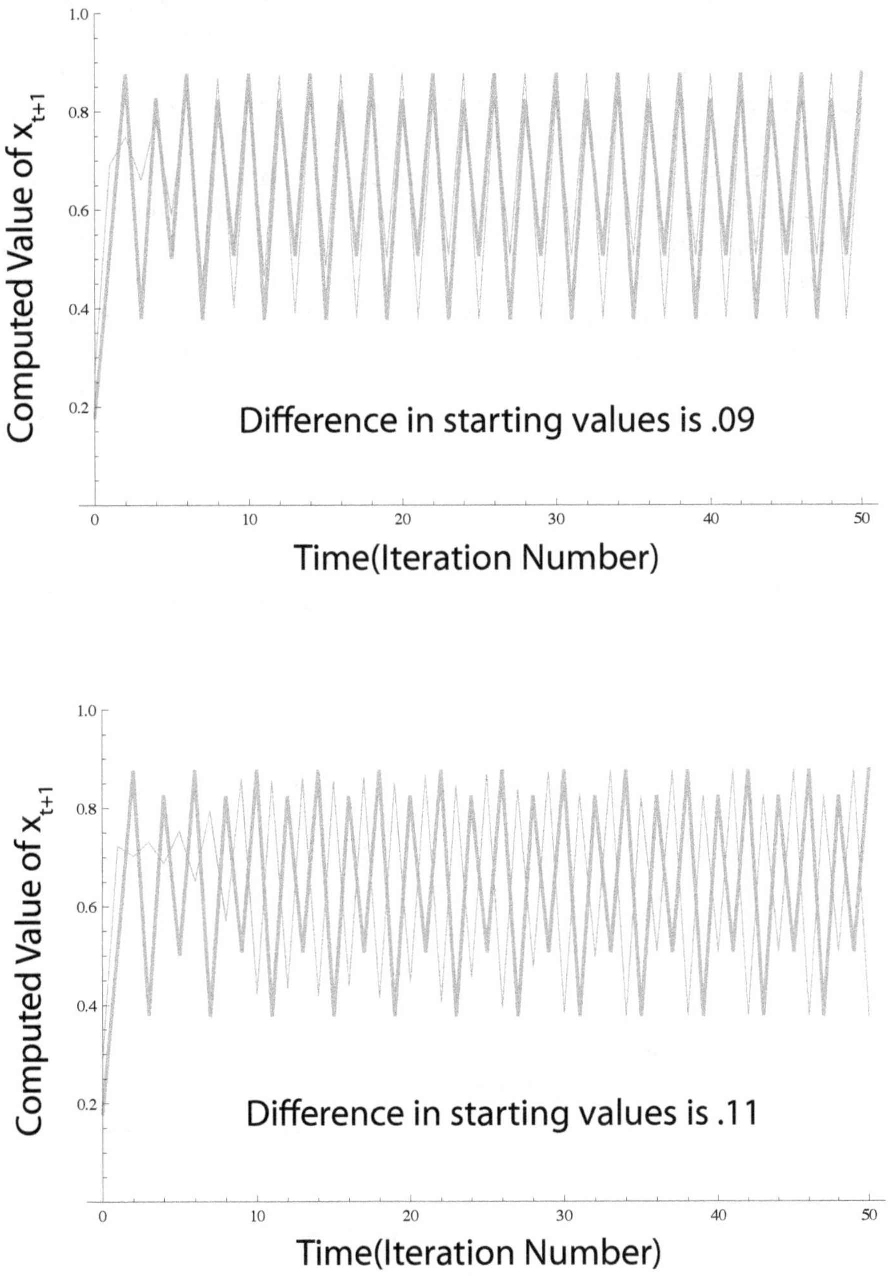

Figure 4.2 *Divergence in the logistic equation (k=3.5) based on small differences in starting points (initial value x_0=0.18).*

pen. So chaotic behavior, although not as predictable as simple linear systems, is more predictable than truly random processes.

State Space and Trajectories

Earlier, we saw that one way to depict chaos was to plot how a single variable changes over time. But what happens when we have a chaotic system made up of more than one variable? In this case, we plot the activity of the system in a space consisting of two or more variables. The space depicted in this type of graph is known as *state space* or sometimes phase space. Each axis of this graph represents a dimension for a single variable. A point within this space represents the state of a system at one point in time. A sequence of points represents how the system changes over time and is called its *trajectory*.

Figure 4.3 shows a state space representing change in two dimensions. The first dimension is extroversion and the second is confidence. Both are personality traits that can be measured and expressed on a scale that in this case runs from one to ten. The point t_1 represents an individual at the start of an assertiveness training program. The successive points t_2–t_7 represent the person's score on the two traits at the end of each month of training. The line linking

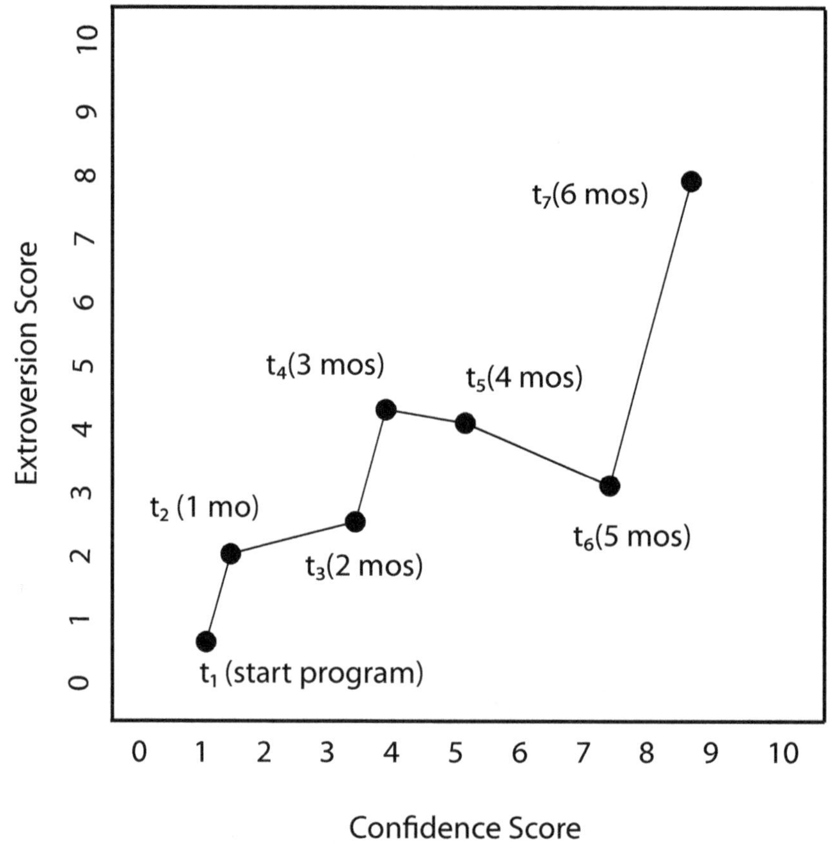

Figure 4.3 *A state space consisting of two personality dimensions (confidence and extroversion) and the trajectory showing how a person might change over time on them.*

each of these points is the trajectory the individual travels during the six-month training period. As you can see, she started off scoring low on both traits but increased with some degree of fluctuation over time.

We are by no means limited to studying or plotting just two variables. Represent a state space with three dimensions. We could include the two personality traits in our previous example and adds a third, a measure of family relations. Low scores on this dimension reflect that the individual is not getting along well with her family. If the family relationships measure decreased while extroversion and confidence increased we would now have a more complex relationship between the three outcomes. The path the trajectory travels in this higher dimensional state space would correspondingly be more complex.

You might be wondering what kind of paths trajectories follow. Do they wander aimlessly around their state space? The answer in most cases is no. Trajectories typically stay within certain regions of their total state space. For example, we could represent space on and above Earth's surface as a three-dimensional state space, with latitude, longitude and altitude. If we plotted the positions you occupied in this space, the trajectory would show how you move around over time. This trajectory would be confined to a very small region of the total space. It would stay mostly in the city you live, with occasional forays farther afield when you go on vacation or travel. Most of the space, such as all the points above 10,000 feet, or the areas over the oceans, would never be traversed.

Trajectories usually also exhibit regular behavior. In this example, your location at night would be constant since you would be in bed and not moving. Here, the trajectory becomes a steady state since it is fixed at one location for a period of time. During the day however, there would be a recurrent pattern between your home and office, reflecting your commute between these two locations. In this case, the trajectory would form a loop, moving back and forth between your house and office. In both cases, the behavior of the system is regular and ordered, at least for part of the time. One study tracked the location of about 100,000 individuals through their mobile phones and found that people move less than 10 kilometers on a regular basis and tend move around in a series of very predictable patterns (Gonzalez, Hidalgo, & Barabasi, 2008).

Attractors

The behavior of a trajectory can be captured by the notion of an attractor in the state space. An *attractor* is the point or points that over the course of time attract all trajectories emanating from some range of starting conditions. In other words, an attractor is a place that the system likes to stay. Trajectories that are near enough to an attractor that they eventually go to it are said to be in the attractor's *basin of attraction.*

Another analogy is useful here. Imagine that a valley is the basin of attraction, the valley floor is an attractor and a ball is the state of the system. If the ball is inside the valley, it will eventually roll down the sides until it comes to rest at the bottom. Once a system comes to rest at an attractor, it doesn't mean that it always has to stay there. An input of energy, such as the wind blowing or a person kicking, can move the ball to a new location.

There are several different types of attractor. Generally, we can classify them into two categories. A *nonchaotic attractor* has a regular geometry such as a point, loop, or smooth surface. Errors or minor perturbations usually don't have significant long-term effects and neighboring trajectories tend to stay close to one another. There is some meaningful prediction despite some fluctuation for these attractors. Chaotic attractors have characteristics opposite of these. The point, periodic, and toroidal attractors described below are examples of nonchaotic attractors.

A *point attractor* is a single fixed point in state space. It represents those systems that come to rest with the passage of time. These systems maintain a steady state without further change. Examples include a ball that stops bouncing and a train that chugs along at a constant speed. There are various ways a system can approach a point attractor. A trajectory can converge directly to the attractor from a single direction, it can move toward it in an alternating fashion from either side, or it can spiral down toward it. Figure 4.4 shows how a perturbed pendulum follows a trajectory toward a point attractor in two-dimensional space.

Next we come to *periodic attractors* that are also referred to as *limit cycles*. These consist of two or more values that keep repeating in the same order. Many periodic attractors map as simple loops in state space that circle back onto themselves. The classic example is the cyclical predator prey relationship between sharks and shrimp shown in Figure 4.5. Initially we can conceive of the shrimp population as increasing. But this causes the sharks that feed on them to multiply as well. As predation increases, the shrimp population falls. That then

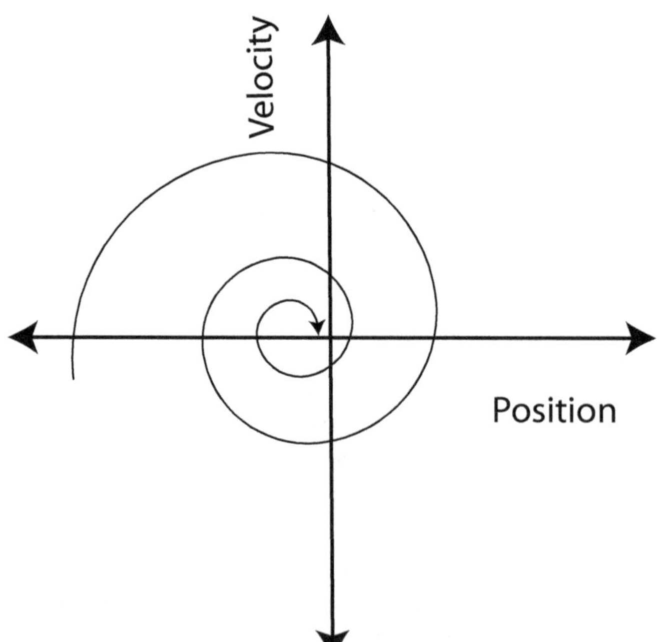

Figure 4.4 *A pendulum evolves to a point attractor. After the pendulum is moved, its velocity and position simultaneously trend toward a single value of the resting position (based on Crutchfield* et al., *1986).*

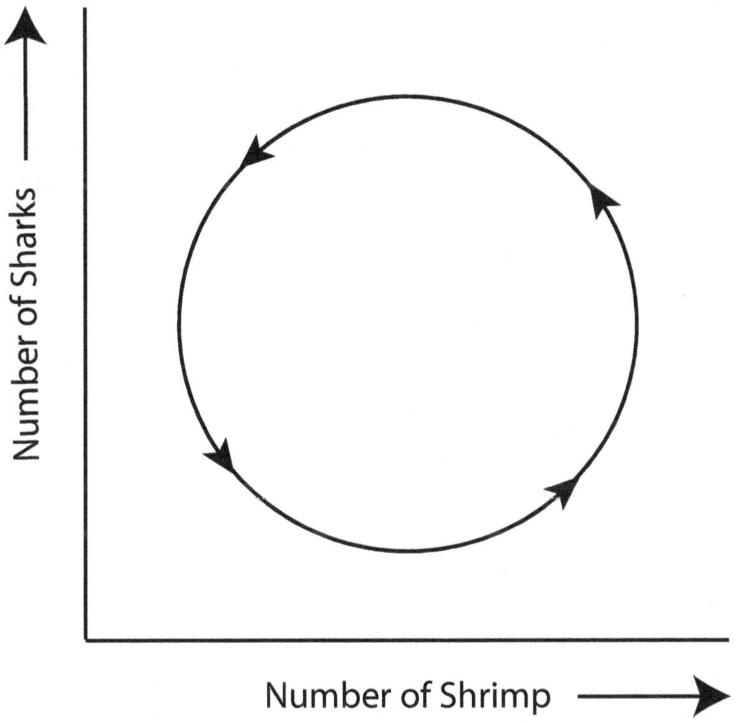

Figure 4.5 *A periodic attractor or limit cycle showing the cyclical relationship between shrimp and shark populations (after Stewart, 1989).*

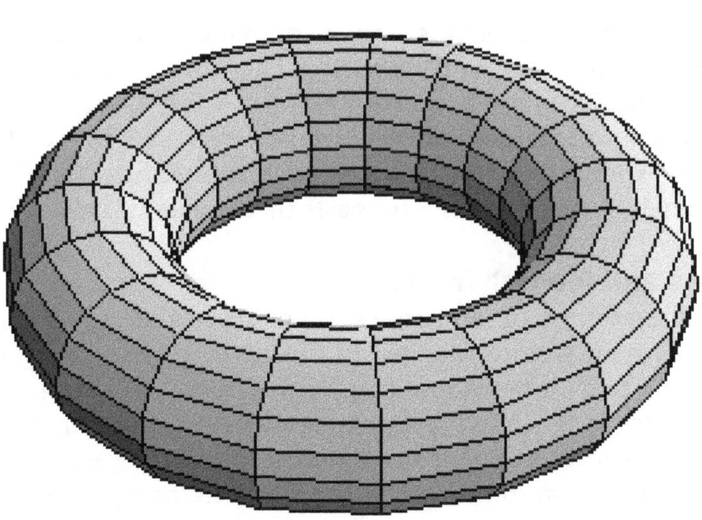

Figure 4.6 *A toroidal attractor has the shape of a torus.*

causes the number of sharks to decrease because their food supply drops. The decrease in sharks then spurs a second growth in the shrimp population, repeating the cycle.

There are also *toroidal attractors* that can be considered as the combination of two or more limit cycles. The shape of this attractor is a torus which resembles an inner tube or doughnut (Figure 4.6). Any point on the torus is defined by four variables. The state of the system over time is traced out by a continuous curved path that moves across the surface of the torus, usually winding around in a corkscrew type pattern. There are two type of toroidal attractor. Those that wind around a given number of times and come back upon themselves are periodic. Those that don't repeat exactly are quasiperiodic.

A *chaotic attractor* or as it is usually referred to, a *strange attractor,* has a much more complex geometry than the nonchaotic types just listed (Figure 4.7). It can travel erratically over a wide region of its state space. A minor perturbation can send the system flying off into a very different path. Two points that start near one another can diverge substantially due to sensitivity to initial conditions. Long-term prediction, and even short-term predictability can be impossible in some strange attractors.

Dimensionality and Lyapunov Exponents

The *dimensionality* (*D*) of a dynamical system is a measure that reflects the complexity of its behavior. A point attractor in a finite dimensional space has a dimensionality equal to zero (*D*=0). A limit cycle attractor inside such a space has a dimensionality equal to the number of independent directions defining the cyclic activity (Heath, 2000). A circle would then have a dimension of one (*D*=1) while a torus in three-dimensional space would have a dimension of two (*D*=2).

The *Lyapunov exponent* is a quantity that characterizes the rate of separation of infinitesimally close trajectories. There are a whole spectrum of Lyapunov exponents. They are in fact equal to the number of dimensions of the state space. Usually, the largest one is what is referred to. This is called the maximal Lyapunov exponent (MLE), because it determines how predictable the dynamical system is. A positive MLE suggests that a system is chaotic.

Landscapes and Mental Trajectories

We can use the notions of landscapes, trajectories and attractors to visualize how a mental system's state changes with time by creating a *vector landscape.* This is usually a two-dimensional map with arrows depicting what direction the system moves in. Each axis refers to a relevant dimension for the system, such as activity in groups of neurons. Figure 4.8 shows a vector landscape with two attractor basins. A system starting anywhere in the figure would, following the arrows, be eventually drawn into the bottom of one of the basins. However, as we noted earlier, the system does not have to come to a complete stop at either of these two points. External forces could move the system out of a basin. Also, the landscape itself is dynamic and could change with time, meaning the

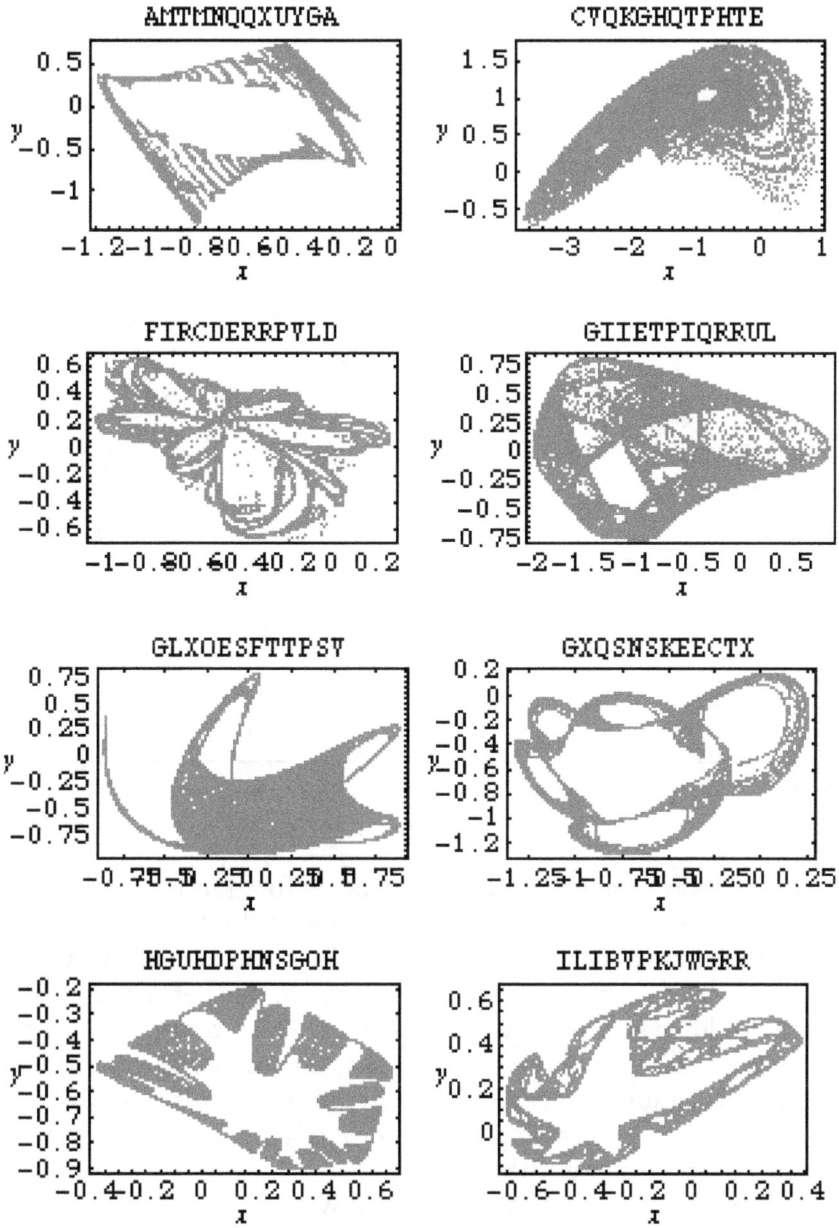

Figure 4.7 *Examples of different strange attractors. By Eric W. Weisstein. ©2005 Wolfram Research, Inc.*

region containing one or both of the attractors could alter, pushing the system to another place.

The two basins in this landscape could represent alternate perceptions of an ambiguous figure. For example, you might see a woman's face in the dark. The details are unclear, but the face appears to be either your mother or your aunt. As more evidence accrues over time supporting the grandmother outcome, the system would be pushed closer to the corresponding "grandmother" basin. If it came to rest at the bottom of that basin or even reasonably close, let's say by

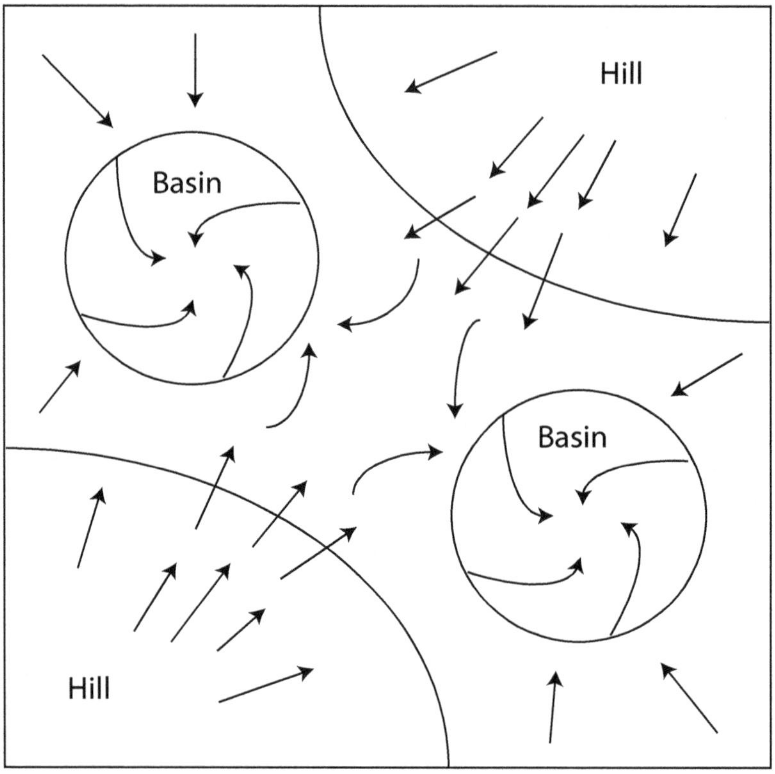

Figure 4.8 *A vector landscape with two attractor basins. Arrows indicate the direction of trajectories.*

orbiting slowly around the rim, then recognition for grandmother would occur. If there were more evidence that the face you were looking at resembled your aunt, the system would travel in the opposite direction, moving toward the aunt basin. In the case of ambiguous perception where neither percept was strong, the system would be somewhere in-between the two basins. Other outcomes are possible. For example, if the stimulus was an ambiguous figure such as the Necker cube or the rabbit/duck face, a viewer might oscillate back and forth between the two basins, indicating they alternately perceive the figure one way and then the other (DeMaris, 2000).

An alternate way to conceptualize movement in a dynamical system is to use an *energy landscape*. This is a three-dimensional plot. The x-axis and y-axis represent two dimensions of interest, again perhaps neural activation. The z-axis or height of the plot stands for energy. The system naturally wants to move to a lower energy level. The system at any given time can be represented as a ball that is rolling around on the surface of the plot. If we place the ball at any arbitrary point on the landscape in the absence of external forces or further deformations in the landscape, it will roll down toward one of the basins. Just as a valley or basin serves as an attractor, causing the ball to roll toward it, we can also consider a mountain or high point in the landscape as a *repellor*, pushing the ball away from it.

It is worth digressing for just a moment to talk more about the dimensions in these landscape plots. The primary dimensions that make up these spaces stand for neural activity. If we were to be completely accurate, each dimension would represent the firing rate of a single neuron, averaged over some temporal window of a given duration. Since there are some one hundred billion neurons in the human brain, the actual state space would contain this many dimensions! It is impossible to visualize or even plot such a high dimensional space, so typically we must content ourselves to considering just a few dimensions at a time.

To complicate matters even further, firing rate is not the only relevant characteristic of a neuron. We could look at many other parameters such as changes in firing rate, bursts of impulses, synchronization, amount and type of neurotransmitter released and so on. To add in these other dimensions further complicates the space. Researchers need to agree on what the relevant dimensions are that they want to measure. Some dimensions will certainly be more relevant for investigating particular phenomenon and will consequently reveal different dynamics.

Ghosts and Rivers

Landscape models provide an intuitive framework to help us think about the temporal aspects of thought and consciousness. Consciousness has often been likened to a "ghost in the machine". The ghost in this metaphor corresponds to the ball in the landscape, while the machine would be the landscape and other forces that drive the system. In the brain, the ghost would be the state of activation in one or more neural population codes. The ghost's movement would be governed primarily by the steepness of the landscape(s), i.e., its potential energy. In other words our thoughts and perhaps conscious experience are much like the water that flows through a river basin. When the incline of the riverbed is steep, the water flows rapidly. When it is shallow, it flows slowly. When the riverbed swings to the left or right, the water again follows suit.

The direction and speed of the ghost or water in these analogies is affected by two primary influences. The first are external factors outside the brain. These include the constantly changing flux of stimulus input from the environment. The second are internal factors endemic to the brain itself including past experience that has shaped our predilection to respond in certain ways. Imagine a man who in a relaxed state of mind, daydreaming or just gazing out at the world allowing his thoughts to be governed mostly by perceptual input that will travel down pre-existing landscapes carved from prior experience. In philosophical parlance, this individual's thoughts are more determined by the environment. In the language of cognitive development, he is assimilating, fitting information into already existing categories. The transition between one of his thoughts and the next is like free association between ideas and more akin to subconscious or effortless automatic processing.

In contrast, imagine a person who is studying for an upcoming exam. She must focus her attention on the terms she must memorize. In this case, her attention may be like will that serves to forge new neural connections, the equivalent of deforming the topography of the landscape in state space. Her thoughts are flowing down landscapes as they are being formed. This takes mental energy and effort and is a more difficult way of thinking. These alterations to the landscape could take many forms such as adding in new attractor basins, building mountains, carving rivers and leveling fields. From a cognitive developmental point of view, she is accommodating, changing her existing concepts to fit new information. This mode of thought is effortful and is akin to conscious reasoning.

Control and Order Parameters

The variables influencing the behavior of a dynamical system are called *control parameters*. Changes in the values of a control parameter cause the system to shift between qualitatively different states. At some range of values, the system may exhibit stable behavior. Then, at some critical value, the system may shift and display radically different behavior. One can think of a control parameter roughly as an independent variable that drives a complex dynamical system since in experiments, it varied in order to see what effects occur. Finding a system's control parameters is a first key step in understanding it.

Many dynamical systems, although complex, still show some degree of regularity or order when measured at a global or large scale. Variables that capture this kind of regularity are known as *order parameters*. The different states that a system can assume can be considered as values of one of its order parameters. Not all global or emergent properties of a system are order parameters. Only those that reflect key aspects of its macroscopic state and which demonstrate qualitative shifts in behavior qualify. An order parameter can be thought of roughly as a dependent variable, since it changes and undergoes different types of behavior in response to alterations of the control parameter.

Let's illustrate with a familiar example. The change in temperature applied to a collection of H_2O molecules is a control parameter because it produces qualitative shifts in behavior. The different states that the molecules assume, such as ice, water, and gas constitute the system's order parameter. The distribution of attitudes discovered in social psychology is governed by several control parameters, including how close the individuals are to one another. Varying this proximity by altering the geometry of the space within which people interact can produce very different outcomes of attitude distribution. It can produce complete agreement in opinion, clusters of minority opinions embedded in a majority group and other patterns of fluctuating opinion. These different distribution states would be considered values of the order parameter. We discuss this example in much greater depth in the section on social psychology.

The Logistic Equation and Bifurcation

A popular equation for studying chaos is the *logistic equation*. It was originally designed to model changes in the population of animal species. The formula is: $x_{t+1}=kx_t(1-x_t)$, where x_{t+1} is the animal population in the current year, k is a control parameter representing the net birth or death rate and x_t is the population in the previous year. Population in this example is the order parameter. Typically, x_t takes on values between 0 and 1 while k ranges between 0 and 4.

Figure 4.1 shows the logistic function that takes the shape of a parabola. Notice that when x_t is less than 0.5 the population always grows. When x_t is less than 1 but greater than 0.5, the population dwindles. As we saw previously in the shark and shrimp example, this type of relation makes sense, since when animal populations get too large their predators increase and resources become scarce, reducing their numbers.

Now let's set k equal to a particular constant, vary x_t and plot the change in population following examples given by Williams (1997). We will do this by inputting a starting value for x_t plugging it into the logistic equation, determining a value for x_{t+1} and then putting this new value into the equation again. In other words, we will iterate the logistic with our starting point and plot the resulting change. If we do this with k set to 0.95 and put in any value for x_t the resulting plot always converges to zero (see Figure 4.9). What this means is that no matter how large the starting population, with a growth rate of 0.95 (in fact any $k<1$), it will always decrease to zero, becoming extinct. In this example, zero is a point attractor and the starting population values are in the basin of attraction. No matter where the starting point is in this basin the trajectory will always move to the attractor. The trajectory in this case is the changing values of x_{t+1} that occur during the iteration.

If we repeat this process for k=1.4 and different, non-zero values of x_t we find that all iterations converge on a different point attractor whose value is slightly less than three (x^*=0.286). Figure 4.9 shows this process. Similarly, when k = 2.8 for any value of x_t between zero and one, the attractor is now x^*=0.643 as shown in Figure 4.9. Notice that the approach to the attractor though is different here. Instead of directly converging, the trajectory alternates above and below the attractor by smaller and smaller amounts.

A surprising thing happens if we repeat this experiment for k values greater than three. When k=3.4, the trajectory now moves toward and then alternates back and forth between two attractors at x^*_1=0.452 and x^*_2=0.842 (see Figure 4.9). Our animal population now fluctuates between a low of 45.2% of its maximum in one year and a high of 84.2% the next year, since our time interval for each iteration in this example is one year. The system is now characterized by the presence of a periodic attractor with two points.

Continuing along, when k=3.5, any value of x_t produces now four attractors at x^*_1=0.875, x^*_2=0.383, x^*_3=0.827, and x^*_4=0.501. The animal population now has a periodic attractor with four points. As you can imagine, this process continues. Further increases in k produce periodic attractors with 8, 16, 32, 64, points and so on. This two-fold increase in attractor points is called *period-doubling*. More generally, we refer to any sudden qualitative change in a system's behavior as a *bifurcation*. With the logistic equation, bifurcations occur at specific values of the control parameter. This process is shown graphically in Figure 4.10 that plots the attractor points versus k.

As k increases, period-doubling happens faster and faster, with only small increases in the control parameter. When k is about 3.57, the number of points in the attractor becomes infinitely large. At this stage we have transitioned into a realm of chaos and the path of the trajectory is very long and complex. In

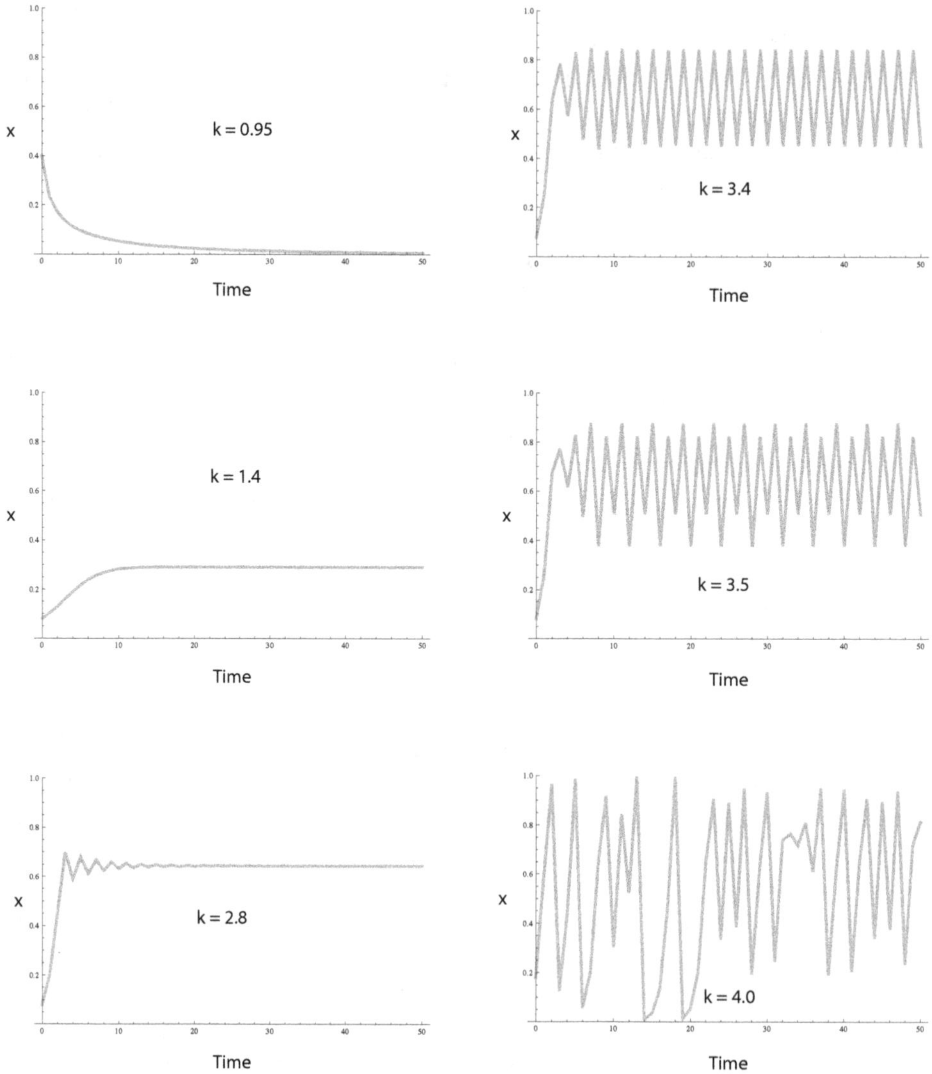

Figure 4.9 *Iterated values of the logistic equation plotted for different values of the k control parameter. The system demonstrates convergence to a fixed attractor, period doubling, and then chaos.*

fact, it becomes difficult in some cases to distinguish this chaotic behavior from that which is purely random.

To summarize, as we increase the control parameter *k* in the logistic equation, we obtain different patterns of behavior. For different ranges of *k*, there are distinct regimes. When $k \leq 1$, the attractor is a fixed point at zero. When *k* is between one but less than three, the attractor is fixed and has a value greater than zero but less than 0.667. When $k \geq 3$ but less than 3.57 we see period doubling. Finally, for *k* values greater than 3.57 but ≤ 4, we move into the chaotic regime.

Order out of Chaos

Although the behavior of a chaotic system seems random and lacking in any type of order, this is not the case. As we pointed out earlier, chaotic behavior is not the same as random behavior. Chaotic systems do demonstrate order. There are attractors in these systems, but they are chaotic and can follow very complex routes. There are also *windows*, zones of *k* values in the logistic example, where any starting value produces a simple periodic attractor. This region of chaos is depicted on the right side of Figure 4.10. The windows, although not present in this figure, can be seen as vertical white bands in this region.

The order in chaotic systems develops spontaneously, without any external cause. In other words, it is an inherent property of the dynamical system itself. There are numerous occurrences of self-organization in nature. Examples include tornados, certain chemical reactions, and the flocking of birds. Of course, mind and consciousness may also be an instance of self-organization that arises from the brain's dynamical behavior. We address this topic in greater detail elsewhere in the book.

Hysteresis

Hysteresis is a feature found in some nonlinear systems (Hock *et al.*, 1993). It refers to two distinct ways a variable *B* can act in regards to the values of another variable *B*. The value of *B* that is observed crucially depends on its history. If *B*

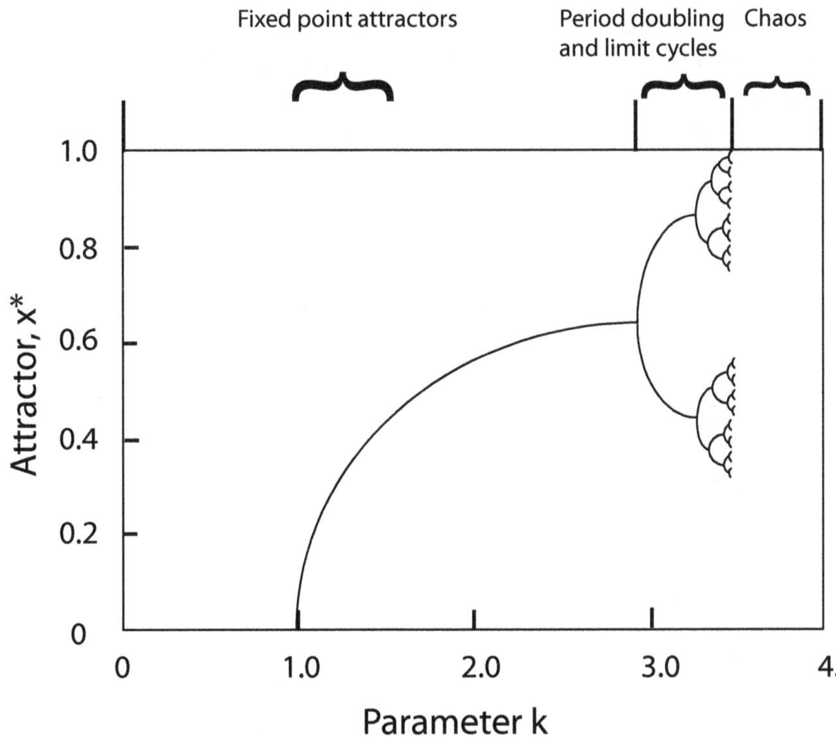

Figure 4.10 *Variation in the attractor X* with control parameter k for the logistic equation. Each fork indicates a bifurcation.*

has a history of low values, it will suddenly increase dramatically past some crucial higher value of *A*. On the other hand, if *B* has a history of high values, it will suddenly drop in value past some crucial low value of *A*. This is shown in Figure 4.11. Hysteresis can alternatively be considered as the phenomenon of path dependence, where the path from state *A* to state *B* is different than the path from state *B* to state *A*.

A system with hysteresis shows both gradual continuous change and at the critical values, abrupt sudden change. The range of values of variable *A* for which *B* can exist in those two different states is known as the hysteresis region and is also depicted in Figure 4.11. Typically, systems that are more stable demonstrate greater hysteresis. Examples of hysteresis include transitions between alternate perceptions of apparent motion, between in-phase and anti-phase motor actions, in social behavior and in other psychological phenomenon described throughout this book.

Catastrophe Theory

Many natural phenomena seem to exhibit slow gradual change. However, as we have seen, under certain values of a control parameter there can be sudden and dramatic changes, demonstrated as a bifurcation to a new value in an order parameter. *Catastrophe theory* can be considered as a branch of dynamical systems in that it studies and classifies phenomena characterized by sudden shifts in behavior arising from small changes in circumstances (Saunders, 1980).

In catastrophe theory we usually start with a two-dimensional space that defines the possible combinations of two variables that alter the system. This is known as the control surface. We then map onto this a two-dimensional behavior surface. The behavior surface shows how the behavior of the system reacts to the paired values of the control surface. What we are interested in is the

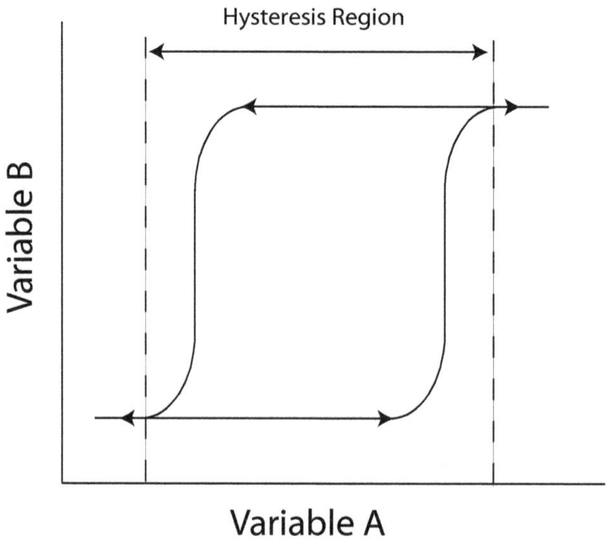

Figure 4.11 *A system demonstrating hysteresis (after Nowak & Vallacher, 1998).*

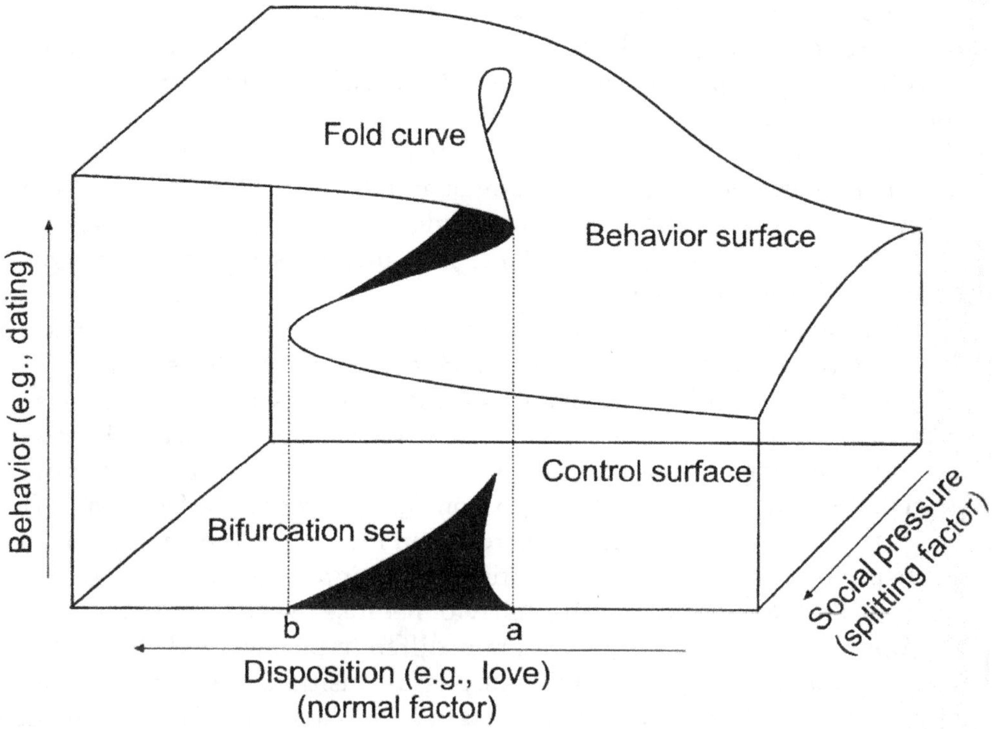

Figure 4.12 *A cusp catastrophe model of dating behavior. From Tessor, A., Human Relations (Volume Number 33) pp. 393-407, Copyright © 1980 by Sage Publications. Reprinted by Permission of Sage.*

relationship between these two surfaces, especially those areas where dramatic and sudden (i.e., catastrophic) change in the system occurs.

It turns out that for a number of natural systems, there are several geometric possibilities for the behavior surface. Figure 4.12 shows one that turns up time and time again, known as the cusp geometry. The cusp or fold curve is a place where the response surface folds over on itself. On either side of this fold there are stable points. Changes in the control surface only produce small changes in response in these regions. But the cusp itself consists of unstable points. It is here where dramatic change happens.

If we approached the cusp from the left in Figure 4.12, we essentially "fall off" the point where the cusp folds over. We would drop, or shift to a different part of the response surface. This is what it means to be in the region of unstable points. Small changes of input values on the control space here correspond to larger jumps in the output behavior space. The area occupied by the cusp, if we projected it onto the floor of the control space, would resemble a triangle with two curved sides. Any area inside this triangular region would cause a jump to a new value. The cusp is just one of a number of different surface geometries. Other more complex ones include the swallowtail and the butterfly.

Let us illustrate the cusp geometry using an example from social psychology. Tessor (1980) has formulated a cusp catastrophe model of dating behavior. In his model, strength of dating or mating behavior represents commitment to a romantic partner and is the dependent measure. It is shown as the behavior surface in Figure 4.12. This is influenced by two control factors. They are social pressures against dating and the emotional involvement or love of the couple for one another. These represent the two dimensions of the control space shown at the bottom of Figure 4.12.

In the model, when there is weak social pressure against dating, the relationship between emotional involvement and dating is linear: the greater the emotional involvement, the greater the dating behavior. This can be seen in the back region of the surface in Figure 4.12. The surface rises going from right to left in a gradual continuous way.

For high social pressure against dating, seen here as the front portion of Figure 4.12, the pattern is discontinuous. Here there is very little dating until a moderately high level of emotional involvement. At this critical value we see a large sudden increase in dating. This is because the surface takes a rapid upswing at the cusp traveling in that direction. The reverse is true when traveling in the opposite direction. In this case, as we go from left to right on the front part of the behavior surface, we see that small decreases in love under high social splitting pressure produce small decreases in dating. At the intermediary value, there is a sudden and dramatic drop in dating behavior because the cusp is crossed again. Catastrophe theory has been used in many different physical and social science disciplines ranging from thermodynamic phase transitions to embryological development to brain functioning. It also has been applied to the fields of business and economics to explain stock price stability, market planning, and business cycle activities. Specific psychological applications include psychophysical judgments, clinical decision-making, perception of optical illusions, attitude change, and cognitive development to name just a few (reviewed in Guastello, 1987).

CHAPTER 5
NETWORKS

A *network* in the most general sense refers to any interconnected group or system. Usually information of some sort is shared through the network. Examples of networks in nature include food webs in ecologies and immune, circulatory and digestive systems in organisms. Technological examples include electrical power, transportation, and computer networks like the Internet and the World Wide Web. Networks can be schematically represented using nodes and links, where a circle or dot stands for a node and lines stand for the connections between them. In some networks computation or processing occurs at the nodes. The result of that processing is then transmitted to other nodes along the links. Obviously, the brain is a network, with neurons as nodes and dendritic and axonal connections linking them together.

In this chapter we will not discuss artificial neural networks that for some time now have been used to simulate and replicate mental functioning. A vast literature on these networks already exists and information on them is readily available elsewhere. Instead, we focus on other types of networks that are usually discussed as part of complex systems and network science. These are Boolean networks, random-biased networks, hierarchical networks, and small-world networks.

Boolean Networks

Kaufmann (1995) describes a very simple network consisting of just three nodes. Every node is connected to every other. Each node can only assume one of two possible states, either on or off, denoted as a 1 or 0. If a node is on it sends a signal to its two neighbors. If it is off, it does not. We can imagine that each of these nodes is a neuron, the links between them are synaptic connections and a signal is an action potential. Figure 5.1 shows this basic arrangement. Because these kinds of networks follow simple logical rules, they are called *Boolean networks*, named after George Boole, the founder of mathematical logic.

The "neurons" in this net will fire or be active based on the type of input they receive from the others. They do this by following logical rules. If a neuron follows an AND rule, it will only fire if it receives input from both of the other neurons. So neuron A, if it followed the AND rule, would only fire if it received a signal from neuron B *and* C. On the other hand, neuron A, if it followed the OR rule, would fire if it received a signal from B (but not C), from C (but not B), or from both B and C together. The AND rule is more restrictive, since both other neurons must be active before A turns on. The OR rule is more lenient, since it only needs inputs from either or both of the other neurons to cause A to turn on.

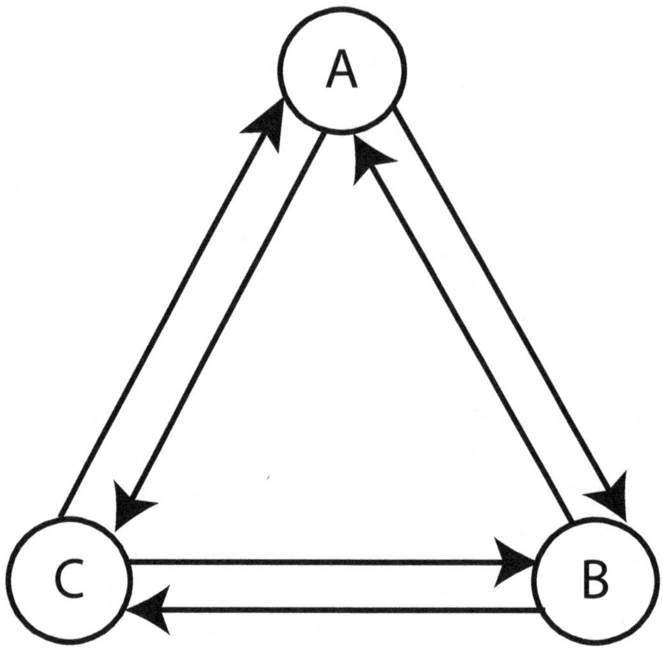

Figure 5.1 *A Boolean network consisting of three nodes.*

Since there are two rules each neuron could follow, there are twelve possible ways we could configure the network. Each possible combination consists of assigning one rule to a given neuron with the remaining possible assignments going to the other two. These twelve rule assignments are shown in Table 5.1. To simplify things we are going to examine the activity of the network using just one of these configurations, where neuron A is assigned the AND rule and neurons B and C are assigned the OR rule.

There are eight possible states of the network under these conditions. These are shown in Table 5.2. If the network is in this state at time 1 (*T*), the resulting pattern of activity that occurs following the application of the rules at time 2 (*T* + 1) is shown in the right half of the table. What this means is that given any particular starting configuration, we can specify exactly what the resulting pattern of activity in the network will be like.

We can analyze this network's activity using dynamical concepts. Each pattern of activity can be considered as a trajectory in the state space of possible configurations and can converge toward point and periodic attractors. The collection of trajectories that flow into an attractor would be its basin of attraction. For example, if the network starts off in the state (000) where none of the neurons are firing, it will remain frozen in that state with no more activity happening. Similarly if all the neurons are firing (111), the network will also remain that way. In these two cases, the network would never change and would remain fixed in only one of its possible eight states.

Other trajectories are more interesting and pass through more steps. If the net started in state (110) it would pass next to (011) and then to (111). Here, we have a three -step cycle that converges to constancy with (111) being a point

Rule		OR				AND				OR	
Node	A	B	C		B	C	A		A	C	B
Activity States	0	0	0		0	0	0		0	0	0
	0	1	1		0	1	0		0	1	1
	1	0	1		1	0	0		1	0	1
	1	1	1		1	1	1		1	1	1

Table 5.1 *The twelve different possible ways two nodes can activate a third node for the network depicted in Figure 5.1 (after Kauffman, 1995).*

Time		*T*				*T*+1	
Node	A	B	C		A	B	C
Activity States	0	0	0		0	0	0
	0	0	1		0	1	0
	0	1	0		0	0	1
	0	1	1		1	1	1
	1	0	0		0	1	1
	1	0	1		0	1	1
	1	1	0		0	1	1
	1	1	1		1	1	1

Table 5.2 *The different possible states of the Boolean network at time T and T+1 when node A follows the AND rule and nodes B and C follow the OR rule (after Kauffman, 1995).*

attractor and the other two states being part of its basin. If the network started in either (001) or (010) it would oscillate back and forth between these two states and be caught in a periodic attractor.

Figure 5.2 shows all possible trajectories a network with this rule assignment could take. Arrows indicate the transition from one state to the next. Of course, the pattern of trajectories would change if we altered the rule assignment. The more AND rules there are, the less activity there will be, since each neuron requires inputs from both of its neighbors. The more OR rules, the greater the possible number of trajectories, since only one a single input triggers activity. This network is deterministic, meaning the rules that govern it are completely known. Yet, it can exhibit a surprising variety of behavior. With only three neurons the number and variety of such states is somewhat limited. But the complexity of this kind of network can increase dramatically simply by adding in

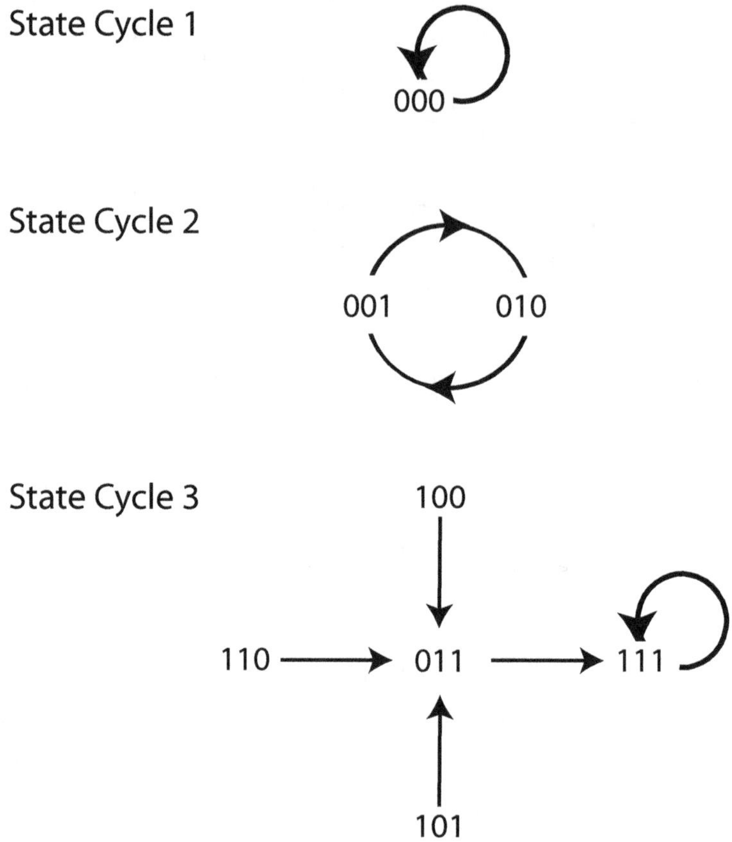

Figure 5.2 *The different possible trajectories the Boolean network can take. Notice there are point, periodic, and other types of attractors (after Kauffman, 1995).*

additional neurons and altering the rule assignment. Imagine a Boolean net with ten nodes. It could have 2^{10} possible states. That is 1,024 states.

Now imagine the state space of the human brain, which contains some 100 billion neurons with some 10,000 connections each. To complicate matters further, real neurons can assume many more states than just on or off. They can fire faster or slower, in which case each firing rate or speed could constitute a distinct state. The total number of states in this kind of network is astronomically large, as would be the variety of different trajectories, basins, and attractors.

Biological Meaningfulness and Boolean Networks

Many of the functions or rules governing Boolean networks are not biologically meaningful (Raeymaekers, 2002). For example, although logically possible according to a Boolean rule, it is unlikely that an element receiving two activating inputs (out of two in total) would fail to turn on. Whereas an OR rule stating that an element can turn on if one or another or both of its inputs are activating is more likely. If all possible Boolean rules are used, the number of states the network can cycle through increases dramatically. This can be measured as a rapid increase in the mean number of cycles the network runs through as a function of the number of elements, *N*, in the network.

Large numbers of states and long cycle lengths are symptomatic of a network that is disorderly and therefore unlikely to be a candidate for a biological network such as a genetic regulatory network or a neural network. Raeymakers (2002) created Boolean networks with two, three or four inputs per element (*K*=2, *K*=3, and *K*=4). He then ran these networks for different numbers of elements starting at ten (*N*=10) and going up to one hundred (*N*=100). He compared networks with all theoretically possible Boolean functions to those with only biologically meaningful functions. The results showed that the meaningful Boolean nets were much more orderly. They showed only a very gradual increase in mean number of cycles as network size *N*, was increased.

Raeymakers (2002) also varied the ratio of activators to inhibitors. In a purely random network this expected ratio is about 50%, with equal numbers of activating and inhibiting connections. However, in biological networks there are usually many more of one type than another. In the brain, the number of activating connections far exceeds inhibitory ones. The percentage of inhibitors was varied from 0 to 100% in steps of 10%. Biologically meaningful rules were sampled randomly depending upon the conditions. The results showed a drop in cycle length when the number of activators or inhibitory reached about 70%. The effect was much more pronounced and occurred earlier for a majority of activators.

It makes sense that natural biological networks should be in a regime that is neither too orderly nor too disorderly, as we have frequently seen throughout this text. If there are too few states the network can be in, it will be unable to operate. For instance, if the network is to hold memories and each cycle or attractor corresponds to to a memory, then too few cycles means an inability to store a sufficient number of memories. If, on the other hand, there are a vast number of states the network can be in, then there may be interference effects where similar memories form adjacent attractors that aren't distinct enough or where the cycle length needed to retrieve or activate a memory becomes hopelessly long.

Network Science

Recent years have seen the development of a new area of study called *network science* (Watts, 2003). Scientists, mathematicians and others in this area have begun to explore the way complex networks work. A network in this framework is considered as any collection of interacting parts. Typically, the parts are represented as nodes and the relationships or actions between the parts as links or connections. Like cognitive science, which is made up of investigators from multiple disciplines, network science is also interdisciplinary. Investigators come from such diverse areas as mathematics, physics, finance, management and sociology. This is because networks are everywhere, as electromagnetic fields, stock markets, companies and societies.

Graph theory, a branch of mathematics that studies networks, has been around since 1736. So what differentiates the new science of networks from this well-established field? Graph theory focuses on the structure of networks and assumes their properties are fixed in time. Contemporary network scientists additionally consider networks as dynamical systems that are *doing* things. Power grids distribute power, companies manufacture products, and brains think. All of these are networks in action. The structure of networks can also change over time, new links can develop and new organizations emerge. The new science of networks studies not just their structure but their function, how they operate and how their operations evolve and change.

Network scientists can consider networks as abstract structures, disregarding their specific context. Whether the network is a brain, a traffic system, or the Internet in some sense doesn't matter. Researchers have elucidated commonalities these networks share with one another and derived from them universal mechanisms of action. For example, all networks exhibit a critical point, where activity propagates suddenly through the entire system. It seems amazing that cars on the road and employees in a company should act the same, but because they are both elements in interconnected relationships, they do.

Having said this, however, it is important to note that although networks do share common operating characteristics, they don't act exactly alike. In some cases, the units or nodes that make up the network do make a difference. A neuron in the brain acts differently from a consumer in a society. So we need to acknowledge that there are both universal and particular features to networks. This is why network science must proceed as an interdisciplinary endeavor. The specialists in each field have the training to understand the particulars of specific networks and inform us about these differences.

Network science in its current incarnation is more emergent than reductionist.

Network researchers currently focus more on the overall structure and function of a system and how this affects the flow of information or action in it. The hope is that the organization of the system, rather than the things that make it up, will tell us how it works. This approach has been very fruitful but if we are to understand how networks differ then we also need to practice reduction. This requires peering inside the nodes of a system and understanding how these nodes behave in the context of their relationships to other nodes. Ultimately, network science will only be able to advance by comprehending the mechanisms of the parts that make up a network and how these interact with the mechanisms of the network as a whole.

Random Graphs

Most networks in real life are ordered. That is, there is a pattern or structured way in which the elements of the network are "wired up" to each other. Take the personnel in a corporation. The people that know each other usually work in the same department and do similar things. A secretary in the marketing department cannot be expected to know a manager in the quality control department. In a similar fashion, it is unlikely that a janitor will work closely with the chief financial officer.

But if we are to understand network function in a basic sense we need make some simplifying assumptions. One of these is that each node in a net has an equal likelihood of being related (connected) to any other. This way, we can ignore all the possible ways that networks can be ordered, since this ordering affects how they function. Networks of this sort are called *random graphs* (Erdos & Reni (1959). One way to think of a random graph is to imagine dumping a group of buttons on the floor. Pick any two of the buttons at random and then tie them together with a piece of thread. The buttons represent nodes and the threads stand for the links that connect them.

Since this is a random process, sometimes you will pick a button twice, or even three times or more and continue tying it to others. So in some cases a string will only connect two buttons. But some strings of course will connect a much larger group of buttons. Now pick any button at random and pull it up off the floor. The number of buttons that are attached to it are called its *connected component*. Clearly the size of these components will depend on the number of threads. The more threads we use, the larger the average size of the components.

What is the size of the largest connected or giant component? This depends, as we said, on the number of threads. If we only use one thread, it will be two buttons. If we use so many threads that every button is connected to every other, then it will be equal to the total number of buttons. What happens in between these two extremes is interesting. Let's plot the fraction of all nodes in the largest component versus the average number of links per node. This is shown in Figure 5.3. As you can see, when the average number of links is less than one, the number of nodes in a component is small, close to zero. But as this average

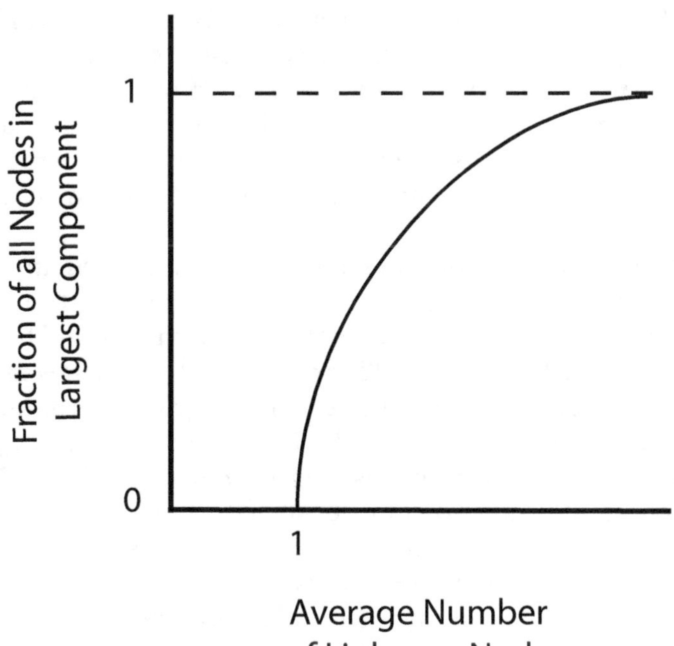

Figure 5.3 *The fraction of all nodes in the largest connected component in a random graph as a function of the average number of links per node. The point at which there is a large increase is a critical point (after Watts, 2003).*

exceeds one, we get a sudden and dramatic increase in the size of the largest component.

This threshold is known as a *critical point*, because it separates a state of disconnectedness from a state of connectedness. It is as if suddenly the whole system changes from isolated buttons like tiny islands in an ocean, to groups of increasingly large clusters of buttons connected to a continent or landmass. Below the threshold we have only small groups of buttons. Above it, they unite to form a giant component. Why is this important? Because below the critical point, the chances are that no single button will be connected to any other. In this state, the system is fragmented and all action is local. Above the critical point most buttons form part of the giant component. Most any two picked at random are now linked and can communicate with one another. So the network at this stage acts as a coordinated whole. Practically all nodes in the system are in touch with each other. Action is now global and coherent. The critical point thus represents a fundamental shift in a system from fragmentation to integration, from local piecewise action to global coherence. What is really fascinating is that nearly all networks show these kinds of transitory states.

Random-Biased Networks

In the real world networks are rarely random. Rather, they have organization, with certain nodes being more likely to be linked than others. Perhaps the most obvious way to show this is in the social world. Here, individuals that share a friend are more likely to become friends themselves. We can represent this in a network as a triangular configuration with A first knowing both B and C which over time will cause B to know C. This principle is known as *triadic closure* and is depicted in Figure 5.4 (Rapoport, 1957). Triadic closure is a natural consequence of a social network. So even if we start off with a net that is unbiased, that is, has random connections of equal likelihood, it can naturally evolve into a net where some connections will be forged by non-random processes. These kinds of nets that contain both random and non-random connections are known as *random-biased nets*.

This type of closure holds in neural networks as well. Imagine two cells that are both receiving activation from a third. Let's call this third cell a *facilitator neuron*. If these two already have a connection, then it will be strengthened by their repeated activity. Hebb (2002) first postulated in 1949 that the synapse between two neurons that are simultaneously active will become strengthened. This principle later came to be called the *Hebb rule*. The physiological process underlying the Hebb rule is known as *long-term potentiation* (LTP). According to LTP rapid repeated stimulation induces structural changes in a synapse. These changes can include an increase in the number of post-synaptic receptors or increased neurotransmitter release (Edwards, 1995).

Once triads are present in a network, they themselves can serve as the foundation for further connections. Figure 5.4 shows how one triad can lead to the formation of another. There are more complex forms of emergent structure in these kinds of nets. Figure 5.4 shows *quartic* closure is also possible. All that is needed to complete a link is the presence of a single facilitator neuron with

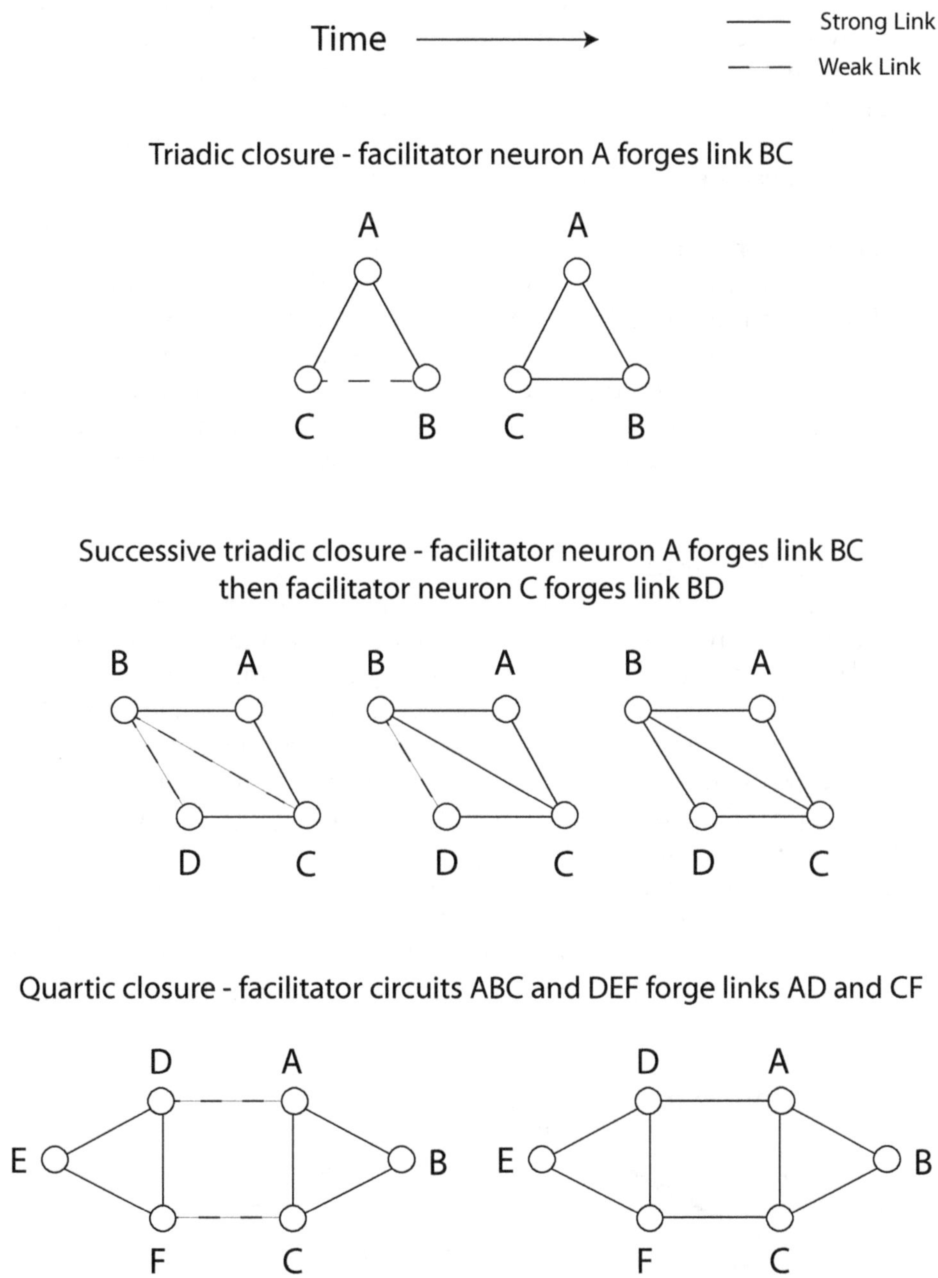

Figure 5.4 *If one neuron has connections to two others, it increases the likelihood that those two will also form a connection. This is an example of triadic closure. In successive closure, the formation of one triad contributes to the formation of a second triad. Quartic closure is also possible in which case two circuits contribute to the strengthening of links.*

links to both endpoints of the new link. In this way we can build shapes of any number of sides. Notice that links can be forged by facilitator circuits as well.

These examples demonstrate that once organization is introduced into networks with these dynamics, it leads inevitably to new organization. The structure created by one process triggers the formation of new, in some cases, more complex structure. Organization thus does not need to be designed or "built into" a network, it emerges in a self-organizing way from it. The starting point is the geometry of the network itself and the rules governing activity within it.

The network's initial geometry is set in place during development when neurons are migrating into their locations within the growing brain. This geometry is the net's anatomy. The dynamics operating in the network, including processes like LTP, are its physiology. The emergent structure then forms based on the anatomy and physiology. But since links are being forged, the result is a new anatomy for further processes to operate on. What we see here then is that structure not only serves as the basis for function, but that function can also give rise to structure.

Hierarchical Networks

If random nets are completely random and random-biased nets are partially random, then we can think of hierarchical networks as having next to no randomness. The connections in these networks are highly organized. Most workplace environments demonstrate this type of structure, which is characterized by the presence of different levels. The highest density of connections is within the levels with fewer connections between them. The number of nodes decreases as we travel upwards with each group of nodes converging onto another group above it with fewer elements. This is shown in Figure 5.5.

Let's illustrate what a hierarchical net is like in the world of academia. Students form the bottom of the hierarchy, with a given number of them in each class. Professors form the next highest level, with each professor teaching a given number of classes. Department chairs manage the professors in their de-

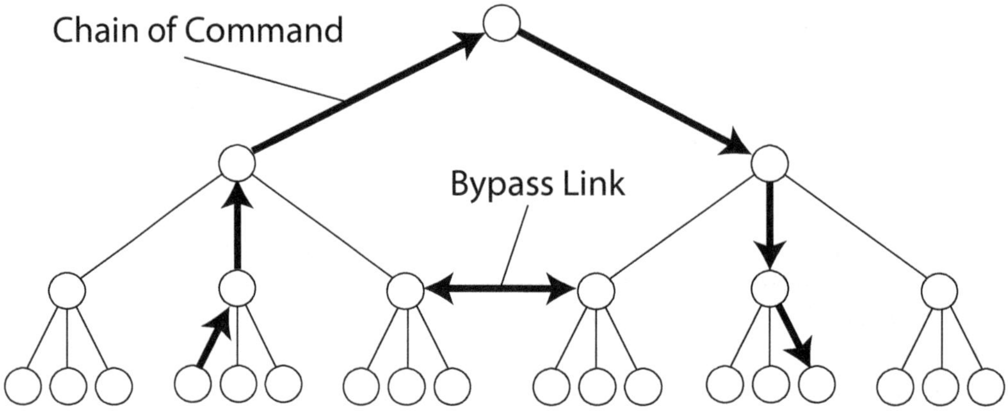

Figure 5.5 *A hierarchical network with a chain of command and bypass link indicated.*

partments. A Dean typically supervises the chairs, and a Provost supervises the Deans.

Hierarchical nets are good at performing a variety of tasks, among these being distribution of goods or services. For instance, if a company needed to deliver goods to stores located all over the country they might start with a single central distribution center. This would ship the product to a smaller number of regional facilities that can in turn ship them to more local districts servicing ever smaller areas.

However, one of the things hierarchies can't do well is transmit information (Figure 5.5). Suppose that a person in a company wanted to do something, but always needed approval from their immediate supervisor. The path for approval for the lowest-ranking official would have to travel all the way up to the highest level and then back down again before anything could be done. Similarly, if the highest-level supervisor wanted to give an order involving the lowest subordinate, the path would have to travel downwards all the way from the top to the bottom. These pathways are known as the *chain of command.* These problems can be avoided of course by giving local autonomy to members farther down the hierarchy, which effectively shortens the chain of command.

Passing a message from any one person to another in a hierarchy can also follow a long path, especially if the person is in a department far removed from the sender. The granting of local autonomy cannot shorten these paths. This problem can however, be solved by adding *bypass links* that act as shortcuts in the network. This is akin to the addition of random links in a highly clustered/ locally connective network. The links create long-distance shortcuts and are the basis for the small world phenomenon.

There are at least two ways to add bypass links. We can add them in randomly as is the case in some small world networks, or we can use another procedure called a *locally optimal algorithm.* In this technique only nodes that send many messages back and forth to each other are linked. Other less popular message routes are not. A local bypass effectively relieves those nodes that are relaying the most traffic, freeing them to up to pass on other messages and thereby increasing overall routing efficiency. The local optimal algorithm is much better than the random method. Far fewer of these kinds of links are needed to reduce the burden of the most congested nodes.

Centrality

One of the most important questions concerning networks is how activity in them is coordinated. A network might have one, or perhaps several centers that perform this function. This is known as the issue of *centrality.* If there is a "leader" or part of the network that receives information, evaluates it and issues commands, then centrality is accounted for. Many examples of such systems exist, such as computers with a central processing unit, political organizations run by dictatorships and the ever-present military beauracracy. The more interesting case is how networks without any such center achieve this activity.

This question has particular relevance for the human mind. For many years, it was believed that the brain must have a central coordinating agency, a center where everything comes together. Descartes, for example, believed it was the pineal gland, a locus whereby the immaterial mind controlled the physical brain. This center begs the question though, because if we allow for a homonuculus, then we must explain how it coordinates. If we cannot, we are stuck with the familiar infinite regress of little men inside of little men.

When we introspect it seems that we have a single unified self. This seems to run contrary to the known physical evidence that activity is occurring throughout many separated brain areas simultaneously. Some have suggested that consciousness therefore must be the mechanism by which this distributed activity is unified into a single form of awareness. If we could figure out the centrality issue we might also determine the answer to the mystery of consciousness.

A variety of explanations have been given. Dennett (1991) postulates that consciousness is made up of multiple drafts or streams that are constantly edited and tapped into. Neural synchronization is another more neural based account (Singer, 1996). It postulates that a pacemaker, perhaps located in the frontal lobe, coordinates widespread neural activity. This pacemaker forces neurons to fire in phase with one another, linking them together into a single unitary awareness.

In some networks, coordinated global activity happens simply as a function of spreading activation that disperses throughout the system quickly but which can arise from any part of it. Take for example the clapping of an audience after a show. If one group of people happen to clap in synchrony their clapping summates and is louder than the clapping in their immediate surround. This causes the audience nearby to clap in synchrony with them, which raises the group's volume even more and perpetuates its spread until the entire audience is clapping at the same speed. What makes this interesting is that it can originate in any part of the audience. All that is required is a minimum number of initial starting clappers as a seed to start the propagation.

Hierarchies and Control

There are at least three models by which hierarchical networks can exert control (Watts, 2003). The first and most obvious is where there is a central leader or coordinator (Figure 5.6). This is a single person or group of individuals at the top of the hierarchy who do the decision-making. They constitute the core, whereas the rest of the network makes up the periphery, so we can refer to this as a core-periphery architecture. This system works best when messages need to be passed globally between distant parts of the network.

In contrast, when message passing is more local, occurring mostly between individuals within a level, a different architecture works best. Here, we allow local members to communicate by forming into teams at every level (figure 5.6). Information processing in this team architecture is distributed and occurs throughout the entire system, not just at the top.

But what about the case where information flow is neither global or local, but some mixture between the two? This corresponds to the case confronted by most businesses. In this situation, individuals at the bottom of the network

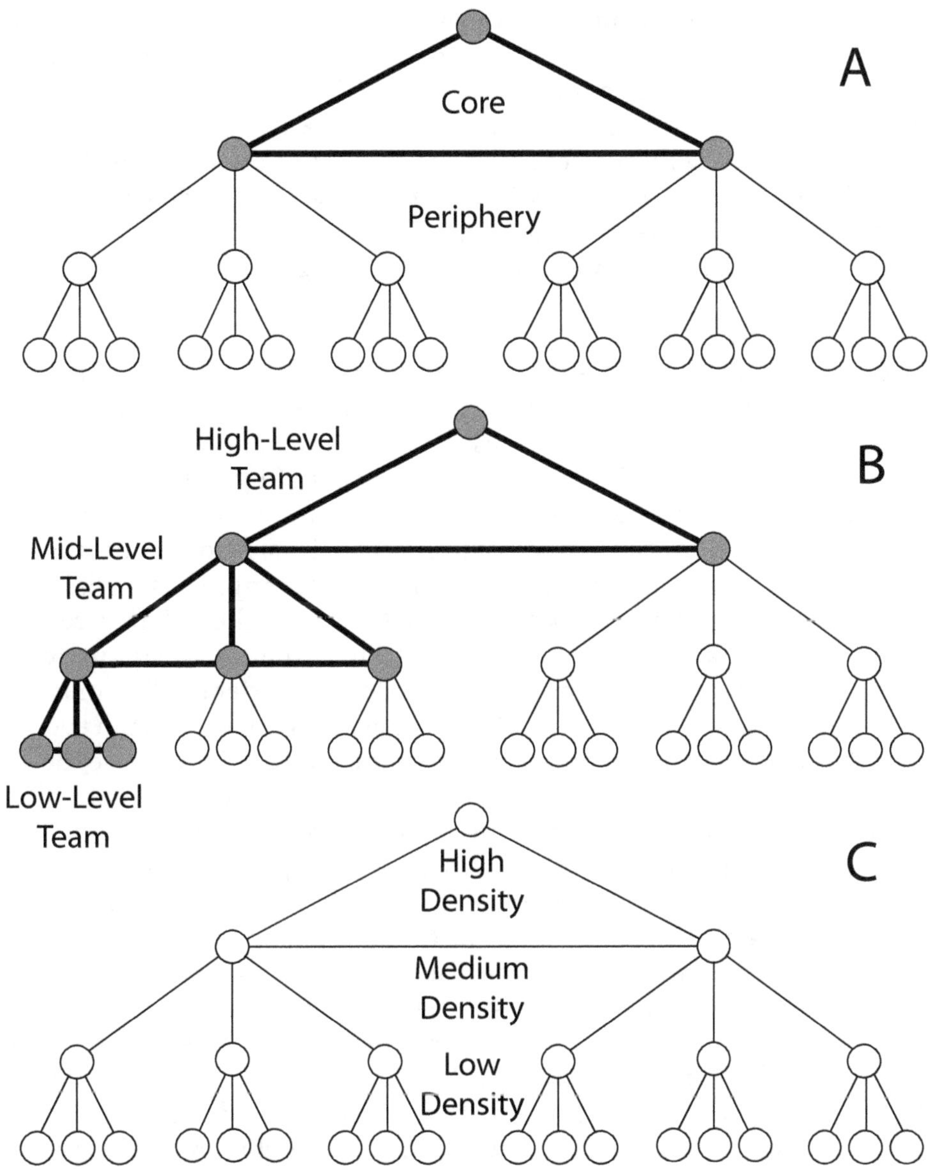

Figure 5.6 *Three different forms of control in a hierarchical network. In (A) a central coordinator or executive team passes messages downward. In (B) local teams are formed that communicate with each other at each level. In (C) we have multiscale connectivity, with meta teams having progressively fewer connections farther down in the network.*

mostly pass messages upwards, while those near the top of the hierarchy must be able to transmit signals to any other portion of the network.

The solution in this case is to employ meta teams that span larger chunks of the network. These teams have stronger ties at the top and then progressively weaker density of ties at we move downwards. So the meta team at the top-most portion would be very tightly integrated. The team down one level would be slightly less integrated, and so on, with the least integration at the bottom levels (Figure 5.6). This sort of network is said to have *multiscale connectivity*.

Core-periphery hierarchies are vulnerable to attack. If the core is damaged, the rest of the network will be unable to coordinate itself. Distributed team-based hierarchies are somewhat more resilient because decision-making is spread to the teams at each level. But these architectures are also at risk. If one team is removed, it will interfere with information flow through the system. Multiscale connected hierarchies with meta teams are the most resilient to damage. They can still operate when almost any portion of the network is removed. The messages are simply rerouted around the damaged region.

Hierarchies and the Brain

We have already discussed the role of centrality and how it may be instantiated in the brain, but there is more to say on this topic. Hierarchical organization is perhaps most evident in the visual system (Hubel & Wiesel, 1979). Here neurons at the bottom of the hierarchy code for spots of light. Connections from these neurons then converge to form representations of features such as lines at a particular orientation at a specific location in the visual field. Connections from these *simple cells* converge on neurons in the next layer up. These cells also code for the combined input of the lower features that feed to them. If they receive inputs from multiple adjacent oriented line detectors, they would in turn code for the sum of these representations, which would be an oriented line moving in a specific direction. These *complex cells* next converge to neurons at an even higher level that would again stand for their combined inputs. In this case these new *end-stopped cells* would represent an angle made of two appropriately-oriented line segments that are moving conjointly in the same direction (Figure 5.7).

If we extrapolate up the hierarchy, we end up with cells in the highest layers that code for large complex objects. These have been nicknamed "grandmother cells" because there may be some that fire only in response to the stimulus of your grandmother's face. So the visual system uses hierarchical networks to good advantage. It uses them to successively integrate aspects of the visual field and build a percept. In effect, the hierarchy allows the visual system to employ a "divide and conquer" strategy where it breaks down the complex visual image into microscopic features and then assembles these features into parts and then wholes that can be recognized. Similar hierarchies may exist that allow concepts to be formed from component characteristics.

This points out an important feature of hierarchies. Traveling up through them allows for convergence and integration, that is, the putting of pieces together. Traveling downwards through them allows for divergence and separation, a splitting apart or distribution of elements.

Communication between levels in hierarchies can also allow for the resolution of ambiguity in visual perception. In the *word superiority effect*, it is easier to determine the identity of a feature, say the letter "A" when it is presented as part of a meaningful word ("CAT") than when it is presented by itself as an isolated letter (Reicher, 1969). The surrounding letters in the word provide a context that helps disambiguate the letter's identity. But this could only happen if information about the word, processed at a higher level, was compared against information about the letter processed at a lower level. This shows that percep-

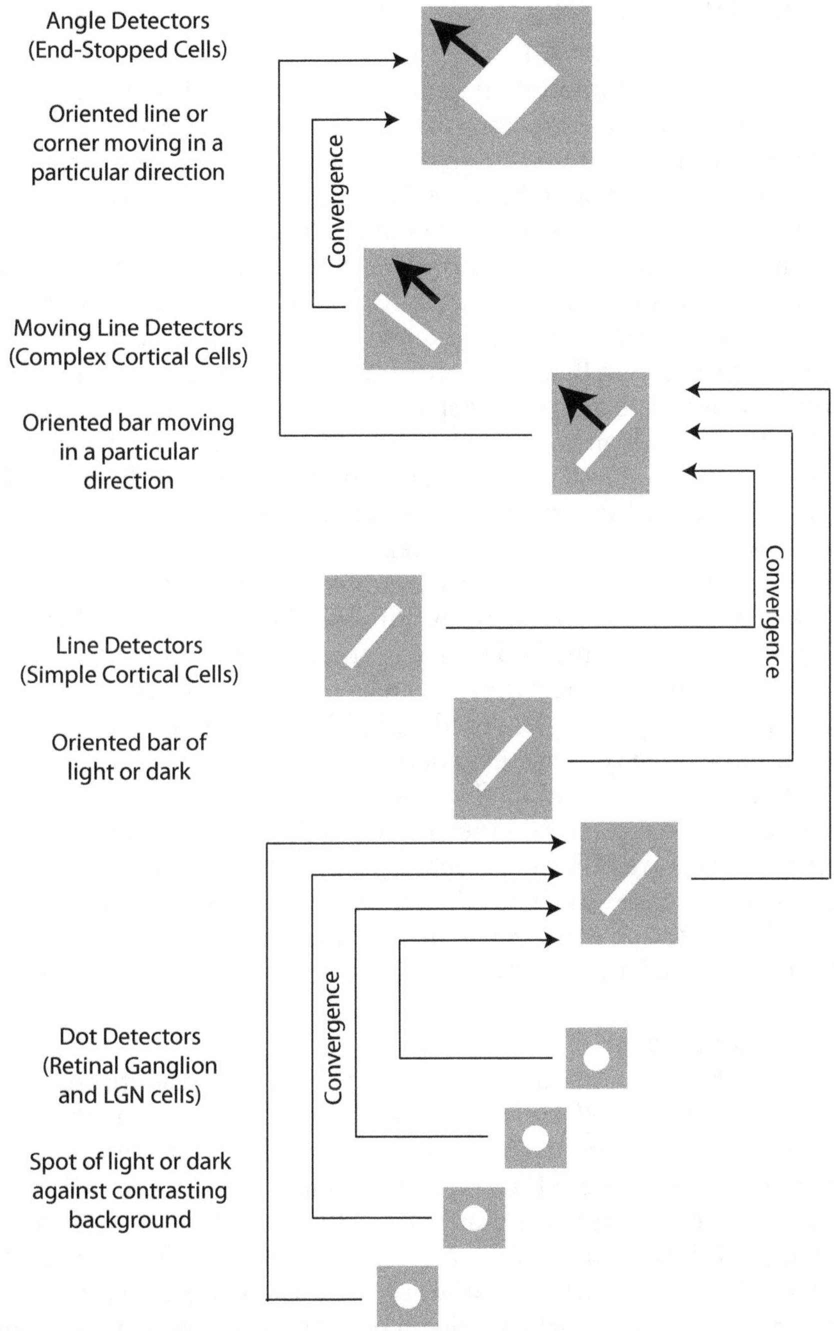

Figure 5.7 *Hierarchical connectivity in the visual system. Multiple neurons coding for a simple feature at a lower level converge to code for more complex features at higher levels.*

tion is not purely "bottom up", with only parts forming wholes. It must also be "top down", with wholes additionally helping to form the developing percept. So information in these hierarchies travels in both directions, both feedforward and feedback, as part of the assembly process.

Small-World Networks

It's a Small World After All

In the 1990s several undergraduates at Albright College in Reading, Pennsylvania stumbled upon an amazing finding. They realized that every American movie actor could be connected to the actor Kevin Bacon in four steps or less. The links are through actors who appeared in the same film together. Let's take Will Smith for example. He was in the movie *Independence Day* (1996) with Harry Connick Jr. who was in the film *My Dog Skip* (2000) with Kevin Bacon. So only two steps separate these actors. The public received this idea with wild enthusiasm. It garnered attention on television and the Internet. One website, called the Oracle of Kevin Bacon, received 20,000 hits a day at its peak of popularity and was ranked as one of the top ten websites of 1996.

The finding is legitimate. Not only Kevin Bacon, but any two American actors can be connected by a maximum of four links. In fact, the principle holds not just for film actors but for most social relationship networks. The psychologist Stanley Milgram first noted this when he sent out 160 letters to people addressed to a stockbroker. He asked them to pass on the letter to someone who they thought would know the man. A number of the letters got through, most in about six steps (Milgram, 1967). This "small world" phenomenon, it turns out, is actually far more general than even social relationships. It has turned up in the U.S. electrical power grid, road and railway webs that spread out across continents, and the nervous systems of many animals.

What is the explanation for this phenomenon? How could the units of such networks be so easily connected when the networks themselves are so vast? After all there are billions of web pages and billions of people on earth. It seems like it would take a huge number of connections to link any two elements together. There must be some intrinsic organization or structure to these networks that allows such ease of connectivity. Let's examine what this is.

Ordered and Random Connections

The answer to the question posed above has to do with network structure. Two mathematicians at Cornell University were the first to reveal this (Watts & Strogatz, 1998). They simulated a continuum of network types. At one end, they created a highly ordered network with entirely local connections. Here, a node was only connected to other nodes in its immediate neighborhood. At the other end of the continuum they had a completely random network where a node could be linked with any other node in the net with equal likelihood. Random networks of this sort contain an equal mix of link distances both local and global.

Let's start with a network of six billion nodes and imagine them arranged in a circle (Strogatz, 2003). The population of the world was six billion several years ago, so we can imagine each node in this example as standing for a person. In the ordered case, each node would be connected with 1,000 others, 500 to the left and 500 to the right. Each connection can be represented as a line drawn between nodes that are themselves denoted as smaller circles (Figure 5.8). In this ordered network, it would take three million steps on average to link any

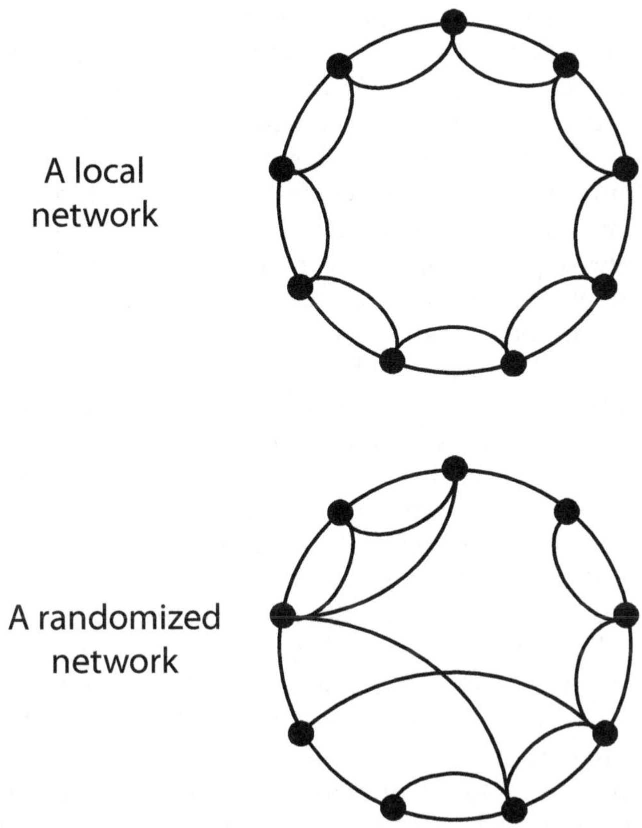

Figure 5.8 *In an ordered local network all nodes are connected to their nearest neighbors. The number of jumps needed to go between any two nodes is large. In a randomized network some of the local links have been severed and replaced with longer distance ones. In this case the number of jumps to get between any two nodes is far fewer.*

one node to any other. The farthest connection is a node diametrically opposite its counterpart on the circle, three billion smallest steps distant. To reach it would require hopping around the circle in steps of 500. It would thus take three billion divided by 500, or six million steps, to reach. Since the closest node is just one step away, the average number of steps separating any two nodes would be midway between these one and six million, a whopping three million.

Now let's take the case of the completely random network. Here it only takes four steps to get between any pair of nodes on average. If every person knows 1,000 other people and each of those knows 1,000 additional people, then there are one million people ($1{,}000^2$) within two steps of each other, one billion within three ($1{,}000^3$) and one trillion ($1{,}000^4$) within four. So any two people in this system are separated only by between three and four degrees of separation.

In addition to these two extreme cases, Watts and Strogatz (1998) were able to generate networks at any point in-between the two by varying the degree of random connectivity. They did this by deleting a local connection from the ordered net and replacing it with a random one. The greater the number of connections for which this was done, the less ordered the network and the more global long-distance connections it contained. What resulted was a unitary dimension of networks varying in their degree of global connectedness.

To evaluate these nets, the authors collected two measures. The first was *average path length.* It was determined by taking any pair of nodes, calculating the number of links in the smallest chain between them, repeating this for all node pairs and averaging the result. Average path length reflects the intuitive notion of "degrees of separation", how many steps it typically takes to get from one node to another. The second statistic was *clustering.* This was defined as the probability that two nodes linked to a third would also be linked to each other. In human terms, it is the likelihood that friends of a friend are also friends of each other. Whereas average path link indicates global structure, clustering reflects the interconnectedness of neighborhoods, a local measure.

In their computer simulation Watts and Strogatz (1998) used a network with 1,000 nodes and ten connections per node. They measured average path length and clustering for this network as they varied its degree of randomness. When the network was made only slightly random by altering just one percent of its links, the path length plummeted by 85%. Further increases in randomness had little additional effect on path length, it had bottomed out at a low level. In contrast, a one percent increase in randomness had barely any impact on clustering, it dropped by only three percent. Only at high values of randomness did clustering begin to decrease.

Here's the interpretation of this result. Just a slight bit of rewiring had introduced long-range connections into the network. These connections brought huge regions of nodes that were previously far away into closer proximity. It only took a few such connections to bridge the gap between what used to be widely separated areas. At the same time, the strong local connectivity was preserved and only began to fall apart when a large proportion of the connections were altered. In effect, a local network was converted into a global network with the introduction of just a few long long-distance bridges. This or any other type of network where the number of links between any pair of nodes are few are called *small-world networks.*

Egalitarians and Aristocrats

It turns out there are at least two kinds of small-world networks. The first is of the type discovered by Watts and Strogatz, where a few random connections are thrown in among mostly local neighborhood links. Buchanan (2002) calls these *egalitarian networks* because the links are distributed more or less equally. But there are also what he calls *aristocratic networks.* Here, the long-distance connections result from the presence of a small number of hubs, nodes that have many more links than the rest of the network's nodes. As is the case with random egalitarian networks, there are also numerous examples of hub-based aristocratic networks, both in the natural and technological realm. Examples include the Internet, the spread of sexual disease between people, scientific papers linked by citations, and the probability of words occurring next to one another in English sentences. The World Wide Web is currently aristocratic in nature but likely to become less so with the recent decision by the Internet Corporation for Assigned Names and Numbers (ICANN) to allow more top-level domains.

The aristocratic organization seems to emerge through a very simple process, one Buchanan (2002) refers to as "the rich getting richer" or more properly as *preferential attachment.* When a network is born, there may be slight differences in the number of links each node has. Those with more links will gradually be linked to more often because they serve as good portals to gain access to large numbers of other nodes. If you created a website on the World Wide Web and wanted lots of people to access it, you would link to other popular sites such as yahoo.com or google.com because you know lots of other people are already visiting those pages. Hubs therefore turn the network into a small world one where information flow between widespread points is facilitated. The fact that networks naturally and unintentionally develop hubs as they grow suggests that it is an emergent property, a solution nature has found to solving the problem of rapid and efficient communication.

The two types of small-world network both accomplish the same function, but in different ways. As a result, they each have different weaknesses (Baran, 1964). Egalitarian nets are more robust to nonspecific damage. They can stand up better to the destruction of randomly selected nodes. As long as there are a few long distance connections intact, messages can get rerouted along them. Aristocratic nets are more sensitive to selective or intentional attack where the hubs are targeted and destroyed. Because the hubs carry the lion's share of links, taking out only a few effectively shuts down the network. Since the current Internet and World Wide Web are aristocratic, this means special attention should be given to securing those hubs and shielding them from attack. Figure 5.9 shows examples of both type of network.

The Power and Development of Hub-based Webs

In 1999, three computer scientists who all happened to be brothers, Michalis, Petros, and Christos Faloutsos, conducted a study of the structure of the Internet. Keep in mind that the Internet is not the same as the World Wide Web. The Internet is the hardware the World Wide Web and other network software applications like Email run on. The Internet consists of routers that are effectively links that transfer data to many other locations in the network. Their study found that just a small number of nodes contained many router links. When they plotted the distribution of routers versus nodes, they found a power-law curve with a high peak for small numbers of nodes and a gradually tapering tail with increasing nodes. This distribution reflects a hub structure, where a few nodes have most of the links and a decreasing number of nodes have fewer and fewer links. A "fat tail" is a hallmark of these curves, it tails off toward zero much more slowly than the normal curve. A similar finding has been obtained for web pages and the links between them (Barabasi, 2002).

Another interesting feature of aristocratic webs is what we might call *elasticity*, meaning they have the capacity to handle a tremendous increase in traffic as they grow. Albert, Jeong, and Barabasi (1999) estimated the diameter of the World Wide Web, essentially the number of clicks it would take to travel

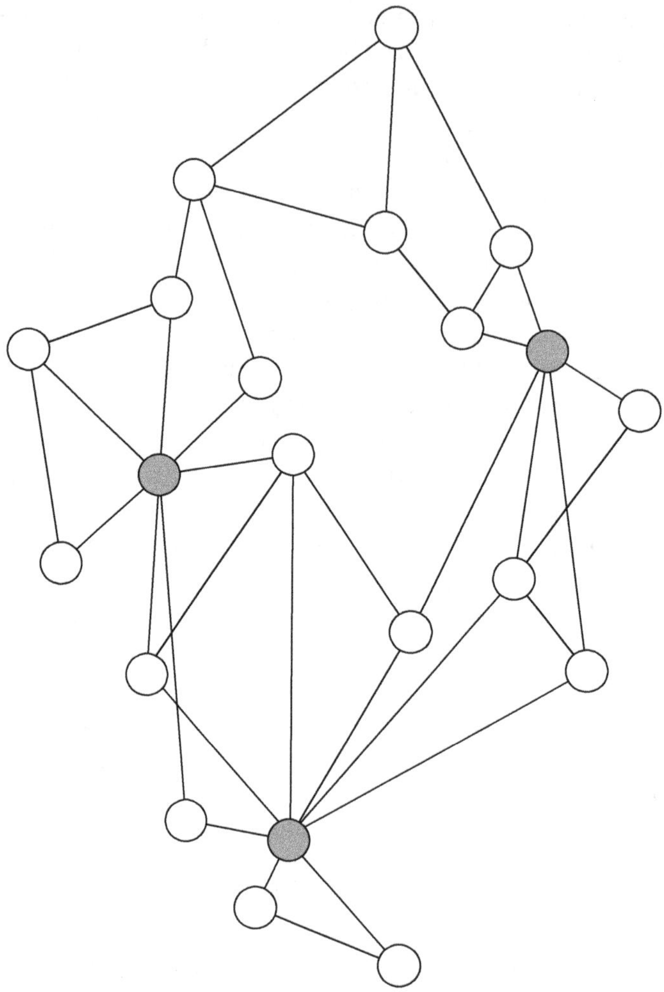

Figure 5.9 *A hub-based network with the hubs indicated. Because aristocratic networks are hub-based, they are more susceptible to attack.*

between any two documents. The number they discovered was 19. They then determined how this diameter would change with an increase in the total number of web documents. Surprisingly, the diameter changed very little, with a 1,000% expected increase over the next few years altering the diameter from 19 to only 21. This means that hub-based webs have the built-in characteristic of being able to accommodate increased usage.

However, there is also an upper limit on the growth of hubs. Hubs with too many links can become overburdened with excessive traffic. Delays and blockages can then occur, slowing travel time. In these cases, rerouting around the hubs may be necessary. Alternatively, new hubs with fewer links may form. A case of this is the U. S. airport system. Despite the presence of hubs, analysis reveals that air routes actually conform more closely to an egalitarian pattern. Another case is electrical power grid nodes. Linking new transmission lines to these stations requires the addition of large and costly equipment that limits their size.

Neuroscience and Networks

What about the brain? Is it a small-world network? Scannell (1997) studied the cortex of cats and monkeys. He located 55 distinct cortical regions in the cat, each associated with a particular function. In macaque monkeys there are 69 such regions. For both of these animals about 400 to 500 links connect these areas. These links are single axons but also nerve fibers made up of numerous axons. Latora and Marchiori (2001) analyzed Scannell's data, and determined that there are only between two and three steps separating neurons. They also found these neurons were highly clustered, with many shared local connections.

So it appears that mammalian brains are small-world networks after all. This makes sense. If an animal is to respond rapidly to a stimulus, say to the sight of a predator or prey, the information can't pass through too many synapses. Each synaptic link would add additional response time, slowing down the animal and putting it at risk.

Epilepsy is a brain disorder characterized by uncontrolled neuronal firing. Much like a forest fire that can start in one part of the forest and then spread, epileptic seizures can also multiply. Unfortunately, the brain's long-range connections allow a seizure in one part of the brain to leap to another that may be far separated. It is not uncommon for a seizure to spread from say one hemisphere to another via the corpus callosum or from one cortical area in one hemisphere to another on the same side.

In an ischemic stroke, a vessel supplying blood to a brain area bursts, killing the neurons it formerly supplied by denying them oxygen and nutrients. The deficits observed indicate what brain area was destroyed. For example, if damage occurred to the left temporal lobe near the centers implicated in language, the patient would develop an aphasia, or language deficit. If the damage was in the motor cortex, the patient would have motor difficulties.

Many stroke patients can recover all or most of their former ability through therapy. What is happening is that their brains are reconfiguring, in some cases forming new pathways that avoid or flow around the area of damage. The ability of the brain to recover from injury in this fashion is called *plasticity*. It is the dense weave of local connections that allows this reconfiguration to occur. If one node or set of nodes is knocked out by the stroke, the signal passing through there can reroute along alternate intact pathways. Clustering thus ensures that the brain doesn't fragment into isolated islands.

Small-World Networks and Synchrony

Elsewhere in this book we discuss the importance of neural synchrony, which generally refers to correlated activity between neurons. It turns out that small-world networks seem to be necessary for synchrony. Kuramoto (1984) found that synchrony between coupled oscillators with distributed natural frequencies (each firing at a variety of different rates) failed to occur in ordered networks, where each oscillator was only connected to its immediate neighbors. However, taking this network and randomizing it just a little bit, by one or two percent, produced global synchrony where all the oscillators beat in unison (Strogatz, 2003). The explanation is intuitive: just a few long-distance connections transmit timing information to more distant parts of the network, allowing mutual influence to spread quickly. The fact that small-world networks and synchrony are both properties of brains is no accident. The former is a prerequisite for the latter.

Lago-Fernandez *et al.* (2000) report on a number of small-world network simulations. In one, they created a simulation of the olfactory antennal lobe in the locust. They connected the neurons in their model together in different ways and measured how the network responded to simulated olfactory input representing smells. When the network was ordered with only local connectivity the simulated neurons synchronized. However, they took a long time to do so. This sluggish response was ostensibly due to the time it took for the timing signals to propagate throughout the system. Next, they wired up the network at random. In this case, the signal spread rapidly through the system, but synchrony didn't happen. The randomly wired nets apparently lacked sufficient order to allow the network to produce an organized response.

Finally, they tried ordered networks with just a slight amount of randomization. Now the system synchronized and did so quickly. But Lago-Fernandez and his colleagues discovered another interesting finding in addition to this. In real world locusts, some of the neurons either slow down or speed up compared to the group frequency. This frequency deviation probably codes for the intensity of an odor or for the difference between different types of odors. Only the small-world wiring pattern was able to replicate this effect.

Small-World Networks and Collective Computation

A well-known problem in computer science is called the *density classification problem for one-dimensional binary automata.* This is a fancy title for a relatively simple situation. Imagine a network consisting of 1,000 lightbulbs. They are arranged as an ordered network, each lined up next to one another forming a giant circle. The connections are entirely local, with each bulb connected to just six of its nearest neighbors.

Like the nodes in a Boolean network, each lightbulb can be either on or off. The job of this network is to "decide" whether or not most of the bulbs in the circle are initially on or off. If the majority to start with are on, the network needs to signal this by ending in a state where all the lights are on. If the majority of the lights to begin with are off, it should ultimately determine this too, by ending in a state where all the lights go off.

Each bulb can only "view" a total of six lightbulbs, three on its right and three to its left. The trick is the come up with some sort of rule each bulb must follow that will produce the correct solution by the whole. The entire system is allowed to iterate for twice as many steps as there are nodes. So if there are 1,000 bulbs, the system is only allowed to compute for at most, 2,000 steps.

This type of problem is one of *collective computation* where a system as a whole can be said to possess intelligence and to solve problems, even though none of the individual system's elements know what is going on. The elements, by following a simple local rule, work together to produce a global solution.

The first rule that seems as if it might work is to simply have each bulb look at its neighbors and turn on if the majority of those six are on or to turn off if the majority of them are off. Unfortunately, this majority rule doesn't work. It causes the ring to freeze into a striated pattern with stripes of bulbs that alternate between the on and off state. For a while, the best rule for this problem would only produce correct solutions 82 percent of the time.

Duncan Watts and Steven Strogatz decided to see if turning the bulb into a small-world network would help. They randomized some of the connections and applied the majority rule. Lo and behold, the network now solved the problem correctly 88 percent of the time. This is an interesting result, because they got better performance not by changing the rule as everyone else had up to that point, but by altering the architecture on which the rule operated. By creating a small-world network, they had increased the problem-solving capacity of the system. In some sense, the structure of the system is as "intelligent" as the functions that operate in it. This is an important lesson for psychology, where intelligence is viewed primarily as a "software" or computational process.

Percolation

In our discussion of hierarchies we saw that passing a message from one point to another could be facilitated by introducing a bypass or by creating new connections equivalent to forming teams. In the section on small world networks, we saw that we could connect any part of the network more easily by introducing a few random long-distance connections. Both of these situations are really equivalent in the sense that the architecture of the network can be altered to change its connectivity. Connectivity in a broad sense refers to how easy it is for information or activity to spread through the network.

A good metaphor for understanding spread in networks is disease. A disease starts off by infecting one or more individuals who then come into contact with others. These new people become infected and infect yet additional people. A number of factors affect how far a disease will spread in a social network. The interconnectedness of the network is important, as is the susceptibility to infection. An elementary school teacher can contribute greatly to the spread of a disease because she comes into contact with many children who may all be highly susceptible to infection. A loner who works at home most of the time on the other hand may not infect anyone at all.

So the spread of disease can be understood using the science of networks. A number of disease models have been proposed. We discuss one of them here, called the percolation model, where *percolation* refers to the propagation of the disease through a social network.

We start with a large population of people, or sites, connected by bonds along which a disease can spread. Each site is either susceptible or not according to an occupation probability. Each bond is either open or closed. Open bonds allow the disease to be transmitted. Closed bonds do not. A disease starts at a source and flows along any open bonds, traveling from one susceptible site to another. It stops when it fails to encounter any new open bonds. The portion of the network that becomes infected from a randomly chosen starting point is called a cluster. Once a disease gets into a cluster, it will infect all the bonds within it.

Obviously, when the occupation probability is high and most bonds are open, the disease will spread to fill up most of the net. In this case, we see a *percolating cluster*, a single giant group of susceptible sites connected by open bonds that occupies the majority of the network. If the occupation probability is low and/or most bonds are closed, the disease will be contained. In this situation, the clusters of diseased individuals will be small and scattered.

The presence of a percolating cluster determines whether an epidemic turns into a pandemic. If a disease originates in the percolating cluster, all individuals within it will ultimately be infected. Under these conditions, it spreads very rapidly and is difficult to contain. If the disease starts outside the percolating cluster, it will eventually die out after having infected only a proportionately small number of individuals.

Percolation and Psychology

The percolation model gives us a theoretical perspective with which to understand the spread of activation in neural networks. If sites are neurons and bonds are synaptic connections, then we can determine how easily messages between neurons will spread. Some brain areas may have the equivalent of a percolating cluster. Neural activity that makes contact with the cluster will diffuse throughout it unimpeded. There are several psychological conditions that suggest the presence of percolating clusters in the human brain. These are seizures encountered in epilepsy, disorganized thinking in schizophrenic patients, and divergent thinking. We describe each of these next.

There are many different types of epilepsy, but they are all characterized by seizures. A *seizure* is an uncontrolled surge of synchronous electrical activity in the brain. For causes that are not well known, large populations of neurons in an epileptic's brain begin to fire at the same time. In some cases, the seizure originates in one part of the brain, such as the temporal lobe or hippocampus, and spreads to other previously unaffected areas.

Seizures result when there is either too much activity in excitatory neurons and/or too little activity in inhibitory ones. Medications used to treat epilepsy work by decreasing activity in excitatory synapses using the neurotransmitter glutamate. Others work by increasing the level of inhibitory transmitters such as GABA. The neurons participating in a seizure are analogous to the sites in a percolating cluster. Once these sites are triggered, activation spreads rapidly through the entire cluster, much the same way a disease would spread throughout a population.

One of the symptoms of schizophrenia is *disorganized thinking*. It is characterized by a loose disassociation between ideas and manifests itself in bizarre speech. Susan Sheehan (1982: 25) recorded what Maxine, a young female schizophrenic patient, said to herself one day:

This morning, when I was at Hillside [Hospital], I was making a movie. I was surrounded by movie stars. The X-ray technician was Peter Lawford. The security guard was Don Knotts. That Indian doctor in Building 40 was Lou Costello. I'm Mary Poppins. Is this room painted blue to get me upset? My grandmother died four weeks after my eighteenth birthday.

The train of thought in disorganized thinking is illogical and jumps from one unrelated topic to another. It is as if the patient enters a neural percolating cluster that "takes over" their thought processes, causing them to leap along open bonds between disparate sites.

Brain imaging reveals that schizophrenics have excessive activity in dopaminergic synapses. Dopamine is a neurotransmitter implicated in abilities such as attention and motor behavior. It could be that disorganized thought and other symptoms occur when there is hyperactivity in these dopamine pathways. Each topic" that comes to mind in disorganized speech might then correspond to a localized area of neural tissue made up of a cluster of sites. The jumps in thought from one topic to the next seem to be the long-distance bonds linking these clusters together.

Creative individuals tend to score high on tests measuring *divergent thinking*. The goal of divergent thinking is to generate many different ideas about a topic in a short period of time. It typically occurs in a spontaneous, free-flowing manner, such that the ideas are generated in a random, unorganized fashion. One question measuring divergent thinking on a creativity test asks: "How many uses can you find for a brick?" A highly creative person would produce more responses, such as "paperweight", "bookend", etc. than a less creative individual. Creative people are usually also better at generating ideas during free association, which involves stating whatever thoughts come to mind.

Divergent thinking seems to be a milder version of disorganized thinking since it is controlled and more focused. During divergent thinking or free association, a person is allowing neural activation to flow along without inhibition. Ideas within a mental percolating cluster that aren't normally linked or are only weakly linked can become related during this state. Attention seems to be implicated in both disorganized and divergent thinking. During normal thinking we may consciously and effortfully direct the flow of thought to stay on topic. When this attentional focus is released, it may become easier to enter a percolating cluster of ideas.

The Future of Network Science

We have seen that network science is a powerful framework from which to study the behavior of complex systems. It has uncovered basic architectural and functional features that underlie many different systems whether those are social relationships, brains, transportation networks, the spread of infectious disease or biological ecosystems. Network theory can provide us with explanations for emergent behavior and self-organization and may be the cornerstone of a real understanding of mind and consciousness.

But we have a long way to go. The next step is to combine reductionism and network theory together. We need a better understanding of the nature of the parts in networks not in terms of how they act in isolation, but of how they act in concert. The interface between the part and the whole, of how the part's function affects the whole, is sorely lacking. Many current network models assume that all nodes in the network are identical. This is clearly not the case. In many systems, the parts can differ dramatically. Surely not every individual in a society thinks or acts alike. So network theory needs to take this specialization or knowledge of individual differences into account. The same signal sent to two nodes will not always produce the same response if the nature of the nodes differs.

One way to reconcile this part-whole dilemma is to apply network theory inside and outside of parts. For instance, one could model the behavior of individual neurons or of neural circuits or of the entire brain as a network dynamic. In this approach, each network would be nested inside another at a different level of spatial scale. Alternatively, there could be networks that operate at different time scales, some slow, others fast. Getting these networks to mesh or interface with one another would inform us as to how the whole manages to operate.

CHAPTER 6
NEURODYNAMICS

Brains are incredibly complex networks and all sorts of intricate patterned activity can occur within them. We refer to this activity as *neurodynamics*. In this chapter we start off by discussing neural oscillation, which is the basis for nervous system communication. We then examine how the timing of oscillations in single- and multiple-neurons serves as a form of information code. This leads to the concept of neural synchrony and coupling by which neurons coordinate their action. We then investigate some of the work on metastability, a state similar to that between order and chaos where there is sufficient plasticity to encode and represent new patterns, but sufficient stability to preserve old ones. The chapter concludes with a summary of the research on nonlinear dynamics and data obtained from brain wave recordings from different brain states.

Neural Oscillation

Neurons obviously exhibit oscillatory behavior. Each action potential is a waveform event with an amplitude, frequency and phase that can be described using a sinusoid function. A neuron's frequency is modulated by the activity of its neighbors. Excitatory inputs increase the firing rate while inhibitory inputs decrease it. The collective activity of a neural population is in essence a group of coupled oscillators that mutually influence one another. Oscillatory models that link oscillators together can therefore serve as a powerful tool for interpreting neural network dynamics (Koch & Segev, 1998).

A number of investigations show that oscillatory neural dynamics in conjunction with synchronization play an important role in how the brain represents and computes information. These processes appear to underlie cognitive processes such as feature binding, attention and associative memory (Borisyuk & Hoppensteadt, 1998; Borisyuk *et al.*, 2000; von der Malsburg, 1999). Different oscillatory patterns that can appear in network models include regular oscillations, chaotic oscillations, and quasiperiodic or envelope oscillations with multiple frequencies (Borisyuk, Borisyuk, & Kazanovich, 2001).

Alexander and Globus (1996) postulate that coherent oscillations are the ordered, non-strange attractor of the brain's edge-of-chaos dynamics. They provide abundant evidence of oscillatory behavior throughout every level of brain organization from neuron to the whole brain. Rapp *et al.* (1985) found oscillations within single neuron membrane potentials. The bursting pattern of some neurons can be thought of as a higher-order type of oscillation (Chung, Raymond & Lettvin, 1970). Rhythm generation can be the result of neuron bursting patterns, a neural circuit or an interaction between the two (Smith & Feldman, 1990). A circuit may be able to modulate the natural oscillation of a bursting cell.

Neural oscillation has been found in many different brain areas. Pettigrew (1984) discovered this behavior in the inferior olive. What is more, there is evidence that neural oscillations are fractal and exist at different time scales. Tech (1989) found fractal spike trains in cat primary auditory fibers. Chung, Raymond & Lettvin (1970) also found nested spike train activity in frog dimming fibers. The smallest scale were the individual spike trains or action potentials. These were organized into bursts and these bursts were ordered into different patterned super-bursts, so there were at least three levels of temporal organization present.

For an introduction to oscillation and a description of several neural oscillatory models, please see the models chapter. In that chapter, we discuss the mathematical basis of oscillations, a general oscillatory model, and two specific models, one for hippocampal function and memory, the other for attention. In that section, we also derive several conclusions regarding oscillatory neural models.

Neural Coding

Single-Neuron Codes

There are a number of different ways neurons can code for information. We briefly review each of these because they are relevant to dynamic neural activity. Keep in mind that neurons usually have a baseline rate of firing. Information is therefore not signaled by a neuron going from "on" to "off" or conversely, from "off" to "on". Instead, the change in the activity from baseline is the better basis for information. Neurons receiving excitatory input will fire faster, while neurons receiving inhibitory input will fire more slowly. One exception to this rule are non-spiking neurons such as photoreceptors that signal information only in the presence of a particular stimulus or input.

In *rate coding,* the average speed of a neuron is used to represent information. A rate is measured as the mean number of impulses per second over some time interval. If a neuron were firing at 300 Hz for some fraction of a second and then switched to 400 Hz for some other time period, its rate over the total interval would be 350 Hz. Because the receiving neuron or neurons must integrate frequency over time to determine the rate of an input neuron, this cannot be the basis for responding to fast changes.

In *temporal coding*, the timing of impulses carries information. Typically, one can measure the inter-stimulus interval (ISI) between impulses over time and analyze them to look for patterns or trends. *Burst coding* is another form. Neurons can fire in bursts of activity, where they exhibit high frequency activity alternating with lower or baseline level activity. The timing and frequency of these bursts and of the intervals between them can also represent information. Figure 6.1 shows examples of rate, temporal and burst coding.

Multiple-Neuron Codes

The three codes just mentioned are ways that individual cells can represent information. It is also the case that cell assemblies or cell populations, large number of cells, can represent information. In *specificity coding* individual neurons

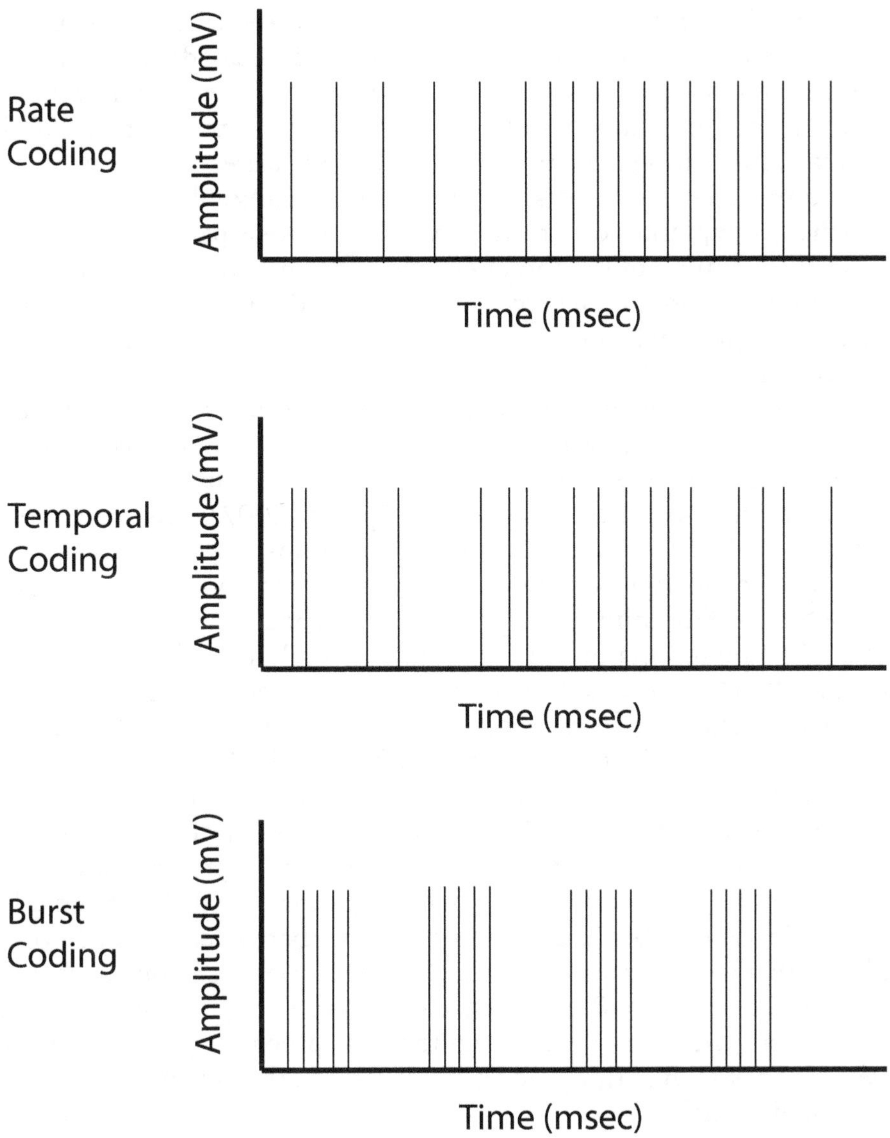

Figure 6.1 *Examples of rate, temporal and burst coding in neurons.*

are specialized to respond to some stimulus attribute. For example, in area V1 of the primary visual cortex, there are cells that respond selectively to lines at a given orientation (Hubel, 1982). It is unlikely that one or a small number of neurons would code for a single object or feature, because if those cells died, the representation would die with it. For this reason, the brain appears to employ large numbers of single cells to code for something, even if they all code for the same attribute.

It is unlikely that specificity coding is used to represent complex stimuli such as faces. That is because cells that respond to specific faces also respond to more than one face (Goldstein, 2002). A distributed coding seems more likely. In a *distributed coding*, a complex stimulus such as a face is represented by the pattern of activity across a group of neurons (Ishai *et al.*, 1999). If there were three cells in a distributed code for faces, then neuron A firing fast, neuron B firing at a moderate rate, and neuron C firing slowly could code for your mother's face. A different set of frequencies in each would then code for your father's face, etc. One of the advantages of a distributed code is that a small number of neurons can represent a large number of different features or objects.

Another advantage of distributed coding is being able to represent something such as your grandmother's face even if her face is partly occluded. In this case, a smaller number of the total neurons in the population would be active, but there might be enough that recognition could still occur. This account of distributed representation is prominent in connectionist, parallel-distributed processing account of mind.

If only 75% of the neurons in the population code for your grandmother were active, this would constitute in essence, a "vote of confidence" toward that particular recognition. We would be somewhat certain, but not completely confident, that we had seen her. According to Spivey (2007), this is true for most of our thoughts. Rarely do we activate all the neurons in a population code for a perceptual object or linguistic word. Instead we activate a certain number based on the condition of the stimulus input, which is usually never the same and never "ideal". The representation then is fundamentally probabilistic, rather than categorically "on" or "off". In fact, we could simultaneously entertain a number of "guesses" (partially constructed representations) as to what the stimulus is. We might be 75% certain it is our grandmother, but 25% certain it is our aunt.

In addition, if enough neurons are activated, they could send activation back and forth between them and reconstruct a representation in a *pattern completion* process (Grossberg, 1980). In this fashion, our representation would get built up over time with successive saccades or glimpses of the face from different viewing angles. A number of experiments show that we can identify occluded stimuli and that we actively reconstruct our memories based on subsequent information (Biederman, 1987; Loftus, 1975). If the information over time was more consistent with our grandmother then the probability of that representation would increase over time and the strength of other alternative representations would diminish.

Synchrony and Coding

We now consider the case where one neural population sends outputs to another. In these situations there are two primary ways that information can be conveyed. In a *synchronous code* the timing of impulses between the two regions is the same. Synchronous coding can be oscillatory. If it is, then the frequency of the two populations can be the same or they may coincide only at certain intervals. In this later case, we could for instance have one region firing at half the rate of the second so that they would synchronize only every other impulse. Synchronous codes however need not be oscillatory. They can synchronize by regular periodic bursting, in which case the bursts rather than the individual spikes are coincident.

There are also *asynchronous codes* where there is a correlation of activity between two neuron populations but without strict spike train temporal alignment. Here, the phase disparity or difference between frequencies carries information. The auditory system relies on phase differences or lag times between the arrival of signals from either ear to localize a sound source (Gaik, 1993). Asynchronous codes allow for dynamic correlations between neuron populations. A dynamic *neural* correlation is one where the pattern of activity in the two regions changes consistently over some time interval (Friston, 2001).

Vaadia *et al.* (1995) found the degree of firing between two neurons in monkey cortex following a behaviorally salient event changed systematically over the order of about one second. An example of a dynamic correlation of this sort would be two neurons that reduce their firing rate for 400 ms, then increase for 600 ms, and then decrease again for 200 ms. If this pattern was consistently observed each time an identical stimulus was presented, we could conclude the dynamic correlation in the pattern of their firing was coding for some aspect of the stimulus. This would be the case even if the frequencies of the two neurons were different.

Neural Synchrony

When you look at a car, disparate parts of your visual cortex become active. Area V1 processes contour, area V4 processes color, and area MT processes motion. But we don't perceive the shape, color, or motion of the car as separate features. Instead, we see the car as a unitary entity with all of these features attached to a single object that is distinct from the background and other objects around it. It is not entirely clear how the visual system combines all these disparate pieces of information. This is referred to as *the binding problem* in vision.

One suggestion for how the brain solves the binding problem is through *neural synchrony* (Malsburg, 1999). This is the coordinated firing of neurons in different regions. In the case of visual object perception, the neurons in the contour, color, and motion areas would all oscillate or fire at the same time for a fraction of a second. Their shared activity would be the brain's code for joining them together such that what is experienced would be all three qualities simultaneously.

A group of German researchers presented an image of a moving bar to a cat while recording activity in its visual cortex (Gray, Korny & Singer, 1989). They found that the neurons fired rhythmic discharges at 30-60 cycles per second. The activity was highly synchronized but short-lived, lasting only about one-third of a second. When they deleted the middle of the bar, the same neurons still fired, but out of synchrony, suggesting the cat no longer perceived the bar as a single unitary object.

However, it is important to distinguish between temporal synchrony and temporal structure. The former refers to the precise timing of changes, the latter to the pattern of changes over time. Guttman, Gilroy and Blake (2007) found that observers preferred temporal structure for determining perceptual organization. Their results show that the visual system groups elements that change globally according to the same pattern even if those changes don't occur at exactly the same time. It seems the brain is capable of integrating similar temporal activity in different neurons across a time lag. That is, it can link neurons that are acting the same even if they are spatially and temporally separated from each other.

Fell *et al.* (2001) found that neural synchrony may underlie memory formation. In their study, they had people memorize a list of words while recording directly from neurons in the hippocampus and rhinal cortex. Later, the participants were tested to see which of the words they could recall. The participants showed synchronous neural firing between these regions only for certain words but not others. It turned out they remembered the words for which synchrony occurred but not those words for which there was no synchrony. This implies that synchrony is involved in the formation of memories.

In addition to perception and memory, neural synchrony may be the fundamental mechanism underlying consciousness experience. Koch and Crick (1990) suggest that consciousness is caused by the synchronized firing of neurons, whereas uncorrelated firing could influence behavior but not produce the subjective feeling of conscious awareness. Many questions remain concerning this hypothesis. For example: Are there specialized consciousness neurons that must synchronize to produce the experience? Is there a minimum number of neurons that must participate? If consciousness requires the rapid synchronous activation of many neurons in far-flung locations throughout the brain, then it seems to require a small-world architecture.

Neural Coupling

Friston (2001) regards the brain as an ensemble of dynamical systems but believes we can understand this complex activity by analyzing it during set time periods. He proposes the notion of a *neural transient*. This is the activity of a particular neuron or neuron population at a given moment as well as just prior to that moment, i.e., its recent history. In other words, a neural transient is what a cell or collection of cells in the brain is doing at a given moment understood in light of the just prior inputs to that cell or population from other brain areas. We could for instance determine the behavior of a single cell in the visual cortex as a function of the activity of all the cells that connect to it over the previous 10 milliseconds.

In a series of studies, Friston (2000) examined the correlated action between different human brain areas. Using MEG he recorded activity in the left parietal and left prefrontal cortical regions of a normal subject over a period of up to two minutes. The subject was instructed during this time to move a joystick to the left about every two seconds or so. Friston then plotted the cross-correlations between the two regions. This plot shows the frequency relationships between the two areas. He found 20 Hz activity in the prefrontal region was correlated with activity just below 20 Hz in the parietal region. There was also 34 Hz parietal activity that was associated with activity at 25 Hz, 35 Hz, and 55 Hz in the prefrontal area.

This technique can be used to assess the degree and type of neural coupling between different brain regions. As such, it is a welcome new tool for studying nonlinear dynamics. However, it only tells us that there is a frequency relationship between these areas, not what the actual mechanism of the interaction is. A mechanism account would detail the physical activity between cells within and between each area. Ultimately, what will also be needed is an understanding of higher-order neural transients, those between three or more areas. In this case we would look at the correlated activity across longer time spans and including more brain areas. We could examine the correlation between say area A and B in one time step and then between B and C in a second, succeeding interval. Finally, these correlations need to be related to the more abstract cognitive functions that emerge from them.

Metastability

One of the problems in neurodynamics is explaining the coordination that occurs between local brain areas that are distant from each other. Most cognitive and other brain states produce global activity, where far-flung regions work cooperatively. What is this activity like? How does global action moderate local activity and vice versa? How does the brain transition from state to another?

Varela (1995) was among the first to outline a systematic hypothesis to deal with these questions. He believed that there was a single large cell assembly that underlies the emergence and operation of every cognitive act. According to him, this pattern of neural activation has three characteristics. First, it is dynamically unstable, meaning that it constantly gives rise to new patterns. This is the notion of *metastability*. A neural pattern can be said to be metastable if it persists for a short time but then collapses and gives way to a new pattern. Each brief pattern corresponds to a different cognitive state that is preceded and followed by other states.

Second, this neural pattern modulates local brain activity and constrains its action. For example, small-scale activity in the visual system, such as that found in the occipital or parietal lobes, may be shaped by inputs and feedback from frontal areas governing attention. It is likely that this influence is two-way and that activity at the local level can also influence what is happening at the global level. The local regions may be considered quasi-autonomous in that they can operate without global influence, but their behavior can be changed by it.

Third, these transient global patterns emerge very quickly in approximately 100 to 300 msec. Connections between cortical areas and between the cortex and the thalamus act as the "highways" that carry messages back and forth between regions. The activity is characterized as re-entrant, recursive, and parallel. This connectivity and rapidity is what allows the brain to quickly switch from one state to another. It also allows for dynamic reorganization, whereby different patterns of brain activity can be added to or deleted from the ongoing global pattern.

Tononi and Edelman (1998) agree with Varela's assessment and call this global pattern *the dynamic core*. The dynamic core can be thought of as a meta-

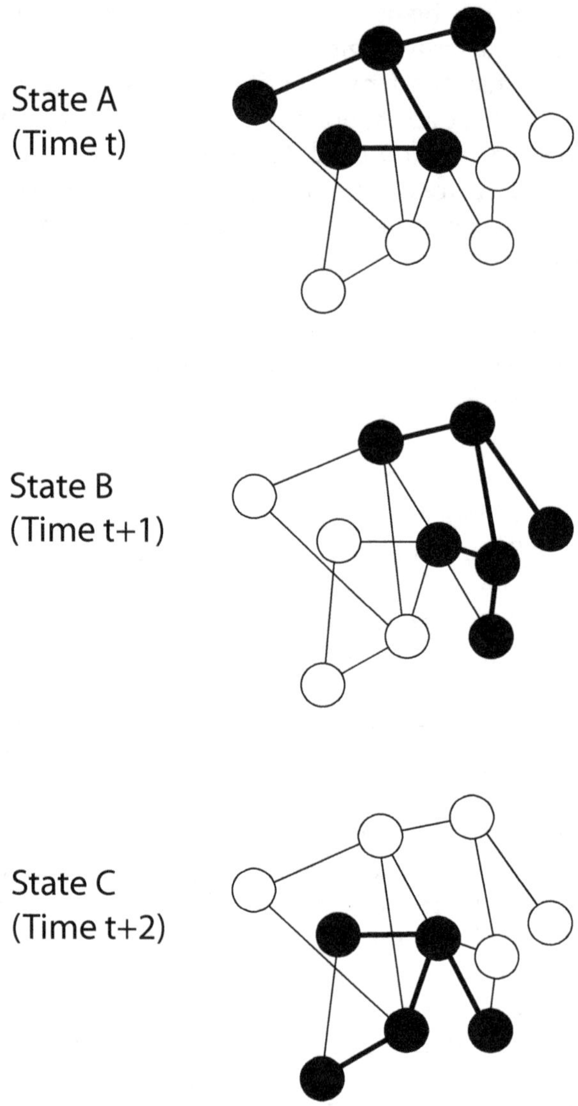

Figure 6.2 *The dynamic core is a metastable pattern of neural activity distributed over different brain areas (Tononi & Edelman, 1998). Here we see the dynamic core represented as activity in a neural system that changes over time.*

stable spatio-temporal pattern of neural activity distributed over many distinct brain regions (see Figure 6.2). There is evidence to support its existence. Brain imaging studies reveal the presence of large-scale cortical dynamics. What is not well understood is how the coupling between regions, which is mostly nonlinear, gives rise to the vast repertoire of cognitive and mental states we are capable of experiencing.

Dynamics and Metastability

Le Van Quyen (2003) summarizes the theoretical and practical work on studying the dynamic core. He also connects its features to principles in dynamical systems theory. Each state represents an elementary behavior or neural pattern.

Transitions between states can occur through bifurcations, where the system undergoes a sudden change. He states that the dynamic core can be viewed as a trajectory in state space. There are no stable regions in the phase space, because the system never stays in the same configuration for long. The system behavior is best thought of a sequence of transient visits to weakly attracting regions. Also, the behavior is modulated by external influences such as environmental stimuli that can shift the pattern into different trajectories.

Friston (1997) postulates that a system in the transient metastable regime is operating over a surface of connected *submanifolds*, local dips in the surface that temporarily trap the system state. Each visit to a submanifold manifests as a transient stable pattern. Metastable dynamics in this perspective are not a collection of separate attractors. Neither do they constitute a single low-dimensional attractor. They are instead a group of many distinct attractors loosely coupled together.

Tsuda (2001) formulates a similar dynamical interpretation of neural events. In his account, perceptions, memories, and other cognitive states are *exotic attractors*. An exotic attractor is one that exists in a high-dimensional state space, as opposed to the low-dimensional attractors considered so far which include fixed points, limit cycles, tori, and strange attractors. A cognitive state in his view is a multi-stable system, in that it can very easily be pushed out of its current attractors and into new ones. This can be triggered by external events or by the presence of internal noise.

If there is a strong instability, then there is little trace of the original attractors. Many chaotic modes exist and the system moves toward a turbulent, noisy macroscopic state. If the instability is weak though, there is an intermediate state between order and disorder. During this state, the system is pushed out of its attractors, but a trace of them remains. Tsuda (2001) calls this trace an *attractor ruin*. The system now consists of a collection of attractor ruins and itinerant orbits connecting them. This behavior he refers to as *chaotic itinerancy*. It is chaotic itinerancy that allows the system to transition flexibly between states, much the same way an itinerant worker might wander from one job to the next.

Connectivity and Metastability

Friston (1997) differentiates between *functional segregation* and *functional integration*. Two neural populations are functionally segregated if their activity is autonomous and uncorrelated. The dynamics of each region proceed along their own courses and are not influenced by what is happening in the other region. This occurs in the brain in areas that are not connected by tracts. In patients who have had their corpus callosum severed and who are presented visual stimuli independently to each visual field, the two hemispheres may be said to operate this way.

Many brain regions though, are connected via one or several synapses. In these cases functional integration can occur. The activity in one region then affects what happens in the other. When having a catch with a baseball, the visual regions are functionally linked with motor regions, since one must know where the ball is in order to catch it. Perceptual-motor coupling takes place whenever we interact with the world in activities such as walking up or down stairs, writing, etc.

The degree of connectivity is clearly related to the degree of functional integration. If two neural regions had no connections, they would not be able to influence one another at all. If they were extensively connected, such that every cell in region A was connected to every cell in region B and vice versa, they would be highly integrated. Any activity in A would have some effect in B.

Friston (1997) created an artificial neural network and varied the density of connections between three groups in it. He found that as connection density increased, the dynamics of the model passed through three phases. At low densities, the system was in a state of stable incoherence where each population preserved its own unique oscillatory dynamics. At intermediate levels of connectivity between the populations there was a regime of metastability. In this state, the system was dynamically unstable, passing transiently from one pattern to another. At high densities there was stable coherence with phase locking and complete entrainment.

So moderate interconnectivity is synonymous with metastable dynamics. This level of connectivity seems to correspond with the small number of random long distance connections we saw in small world networks. Of course connectivity can be across any distance long or short. The neuroanatomical evidence suggests local dendritic connections between cells within a brain region such as the hippocampus are more dense, while longer distance ones are less so. This may be the explanation for why at smaller scales we see more oscillatory and periodic activity, while metastable dynamics should predominate at larger scales.

As the next step in his analysis, Friston (1997) calculated the *correlation dimension* (D_2) of the simulation. This measure of dimensional complexity indicates the space-filling nature of the trajectory. It was estimated using Lyapunov exponents that show the degree of divergence between nearby trajectories. A high dimensionality score indicates that the trajectories wander and that the system is more chaotic. A low dimensionality score indicates that trajectories stay more in one locale and that the system is more ordered.

The results showed that as connectivity between the groups in the model was increased, dimensionality dropped monotonically. With low connectivity, each isolated group contributes its own dimensions to the overall dimension raising the value of D_2. With higher connectivity, each of these groups ceased acting independently and began to act as a single coherent system with a lower corresponding D_2 value. It is important to note that metastability and dimensionality are two different manifestations of complexity. While the first shows an inverted U-shape as a function of connectivity, the latter shows a continuous decrease. Whereas dimensionality reflects the spacing-filling nature of the manifold, metastability relates to its actual shape.

Connectivity between neural groups is a physical aspect of a neural system since an actual physical connection in the form of a synapse is necessary. But the phenomenon of neural plasticity shows that connectivity can also be soft wired. The strength of existing connections can be varied with increased use as seen in Hebbian learning where simultaneous pre and postsynaptic activity produce changes such as increased neurotransmitter release or receptor density. This is the equivalent of increasing the weight in a neural network simulation. This suggests that the brain may be capable of altering connectivity, and hence metastability, on its own. With practice, learning, or attentional focus, neural regions that were weakly connected and mostly autonomous can become linked and capable of maintaining transient dynamics.

Nonlinear Dynamics and EEG

Is the Brain Chaotic?

There is considerable controversy in the research community over whether the brain exhibits chaotic dynamics (Valdes *et al.*, 1999). It could be that what seems to be chaotic is merely stochastic. The difference is important. If chaotic behavior could be verified, then the implication is that it is being driven by deterministic factors. If however, the dynamics were found to be stochastic, then it could be caused by random factors and may be the equivalent of noise. Most of this work is based on brain activity measured by an *electroencephalogram* (EEG). A time series output of the EEG is recorded and analyzed to determine the presence of chaos, for instance by calculating dimensionality or Lyapunov exponents.

Heath (2000) summarizes a number of the problems in this area of research. It is difficult to obtain the large runs of data necessary to compute accurate and reliable measures like dimensionality and Lyapunov exponents. The data are often contaminated by noise. The presence of noise may produce false high dimensionality (D_2) estimates, so that many studies may be incorrect in interpreting high dimensionality in their data. However, relative values of dimensionality estimates may be interpretable when comparing such values across experiments, for example, when comparing healthy individuals to those with psychological disorders or when comparing different stages of sleep in normal individuals. This is just the sort of literature we summarize below.

It seems that many early tests that claimed the presence of chaotic brain dynamics only discriminated nonlinear from linear activity, not chaotic from stochastic activity. In order to avoid these kinds of mistakes in the future, Valdes *et al.* (1999) make a number of suggestions. First, researchers need to estimate the dynamical equations themselves rather than just compute invariants such as the correlation dimension, since these are only a limited description of the global dynamics. Second, neural models should be based on stochastic as well as deterministic dynamics since it seems likely that the brain is based on both types of processes. Third, neural models must be checked against the EEG data. Finally, neural systems are not closed and so models must take into account exogenous inputs, such as environmental stimuli that can impact on dynamical activity.

Neural Dynamics During Different Brain States

Stam (2005) provides a comprehensive overview of the literature on nonlinear dynamical analysis of EEG. In this section we only summarize a select number of studies to illustrate what neural dynamics are like for certain brain states. We review the broad findings from research investigating dynamics during resting wakefulness and perceptual, emotional, and cognitive states, including memory. We also look at what dynamics are like during sleep, different stages of development, and various pathological conditions such as epilepsy and dementia.

Normal Resting-States

Early researchers believed EEG patterns of healthy awake subjects reflected low-dimensional chaotic activity. Palus (1996) found evidence for a nonlinear component in the EEG recordings of normal healthy human volunteers and reports that a number of similar findings have been published. However, none of these findings were of low-dimensional chaos. Kulkarni, Parikh and Pratap (1999) confirm this. They analyzed the dynamics of EEG signal from normal individuals with their eyes closed during audio and mental activity and found the signal was deterministic and chaotic, but not of low-dimensionality.

It is now generally acknowledged that there is no substantial evidence to back up the low dimensionality claim (Palus, 1996). The observed EEG activity in healthy awake adults instead seems to indicate weak nonlinear activity (Stepian, 2002). It seems likely that nonlinear coupling between different brain regions occurs during this state (Breakspear, 2002; Breakspear & Terry, 2002).

What can we conclude regarding the brain "at rest"? If there were a default state to brain activity, it seems to be more characteristic of ongoing activity rather than rest. Stam (2005) states that the awake adult brain has relatively high dimensional complexity. It is not random noise, but has weak nonlinear properties. He additionally describes this activity as having relatively weak levels of synchronization. The synchronization, rather than being constant, fluctuates over time in a scale-free manner. This implies an ever-changing collection of neural populations that couple and decouple with one another over time.

Perceptual States

Kondakor *et al.* (1997) obtained higher global dimensional complexity in an eyes-open condition relative to an eyes-closed group. Aftanas *et al.* (1994) found a dimensional increase in frontal central brain regions in participants who were imagining compared to those in a perceiving condition. Lutzenberger *et al.* (1992) also found more complex brain dynamics during imagery in comparison to perception.

Emotional States

Aftanas *et al.* (1997) had participants view movies with negative, positive, or no (neutral) emotional content. Kolmogorov entropy and the largest Lyapunov exponent were higher for those watching the emotional films. In a similar study, there was a decrease in coupling for left frontal areas in negative emotion subjects whereas there was an increase in coupling in posterior central areas for positive emotion subjects (Aftanas *et al.*, 1998). Aftanas and Golocheikine (2002) found a drop in dimensional complexity in participants who were in a meditative state.

Cognitive States

Do brain states become more or less complex with the performance of a cognitive task? The answer, it seems, is that it depends. A number of studies have found increases in correlation dimension and other complexity measures for cognitive tasks such as arithmetic and attention (Micheloyannis *et al.*, 2002, Molle *et al.*, 1995). But at least one study shows a drop in dimensional complexity during execution of a working memory task (Sammer, 1996). The change is probably mechanism specific and will vary depending on the particular processes that are called upon in each task. However, EEG complexity seems to correlate positively with task difficulty, regardless of task specificity (Lamberts *et al.*, 2000; Muller *et al.*, 2003).

These results can be interpreted to mean that complex cognitive activity requires more complex underlying brain dynamics. When actively solving problems more brain areas may be recruited or more intricate processes in particular regions may occur. But one must be careful in interpreting these data. A complex cognitive process may also involve the decoupling or disconnection of certain mechanisms, reducing or producing a net decrease in complexity.

Sleeping States

A number of studies examining sleep EEG have concluded that deeper stages of sleep are demonstrative of lower complexity, as measured by lower dimensionality and lower Lyapunov values (Kobayashi *et al.*, 1999, Kobayashi *et al.*, 2001). Ferri *et al.*, (2002) found nonlinear structure during non-REM stage II and slow-wave sleep, stages III and IV. In contrast, infants show only intermittent nonlinearity, which occurs most often during quiet sleep (Ferri *et al.*, 2003). This nonlinearity is probably due to asymmetric coupling between separate brain regions that increases with deeper stages of sleep (Pereda *et al.*, 2003).

EEG readouts for slow wave sleep show high amplitude, low frequency wave patterns. It is tempting to conclude that this is the result of synchronization in large neural populations. However, when sleep dynamics are analyzed at a more global level, that involving a majority of the brain and not just the area around electrodes, there is less evidence in support of synchronization (Stam, 2005). Sleep, like the awake resting brain state, is not less active, but engaged in extensive information processing (Hobson & Pace-Schott, 2002).

Development

Changes in EEG activity during different periods in development have been observed. Meyer-Lindenberg (1996) compared the EEGs between a sample of healthy children and adults. Significant nonlinearities were detected in both groups, even in newborn children. The correlation dimension increased with age. Anokhin *et al.*, 1996 also found an increase in EEG dimension with age that was particularly pronounced in the frontal brain regions. They suggest that this is due to a greater number of autonomous synchronous brain networks.

Epilepsy

The most clear-cut results of studies investigating EEG have been obtained with epileptic patients. Stam (2003) summarizes the findings in this area. Before an epileptic seizure, neural activity is characterized by high dimensionality and low synchronization. During a seizure, it is low dimensional with a high amount of synchronization. In other words, there is what may be characterized as a substantial loss of complexity. Although there may be synchrony, seizure state EEG data indicate strongly nonlinear brain dynamics (Ferri *et al.*, 2001).

A chaotic attractor contains an infinite number of *Unstable Periodic Orbits* (UPOs). Part of normal chaotic behavior is when the system state moves into one of these orbits, then falls out of it, only to wander into another one later. The number of UPOs in a system can be used as a gauge of how complex it is. Govindan *et al.* (1999) analyzed the number of UPOs from EEG data obtained from normal subjects while awake, in different stages of sleep and during epilepsy. They found 12-14 UPOs during wakefulness while during epilepsy it was only about half this number.

Govindan and his colleagues additionally calculated the dimensionality for the waking, sleep and epileptic states mentioned above. They found an increase in the dimension value with an increase in mental activity. Sleep stage 4 had a dimension of 4, alert but restful had 6, while for people engaged in a mental task, the value was close to 10. The correlation dimension for epilepsy is lower and has been found to be near 2 (Pradhan & Dutt, 1993).

One question that can be addressed here is what the transition from non-seizure to seizure states is like. Stam (2005) mentions two aspects to this transition, alterations in local dynamics and changes in neural coupling between regions. The first aspect seems to be a bifurcation caused by a change in a critical control parameter. Velazquez *et al.* (2003) suggest that the ratio of excitatory to inhibitory synaptic action could be such a control parameter, with a drop in inhibition past some value triggering a seizure. With regards to the second aspect there are contradictory findings. Most seizures show increased coupling between brain areas, but in certain seizures, there is a decrease in coupling prior to seizure onset (Mormann *et al.*, 2003).

Alzheimer's Disease and Dementia

Alzheimer's disease patients show less dynamical complexity when their eyes are closed and they are in a resting state (Stam *et al.*, 1994, 1995). Lower global dimensional complexity in the EEG also correlates with lower scores on neuropsychological tests like the WAIS-R (Yagyu *et al.*, 1997). In addition, there appears to be less functional interaction between brain regions in Alzheimer's patients, especially for frontal and anterior temporal areas (Jeong *et al.*, 2001). It may be that in Alzheimer's disease and perhaps in dementia in general, there is a "disconnection syndrome" that takes place. Fewer brain areas may communicate with each other producing islands of activity that were formerly integrated and connected.

Dynamics and Olfaction

So far we have discussed general patterns of dynamical action that take place in the brain. But we have yet to outline a specific example of how dynamical functioning works for a given cognitive activity such as perception. We remedy that now by describing the most elaborate study of neurodynamics, that conducted by Walter J. Freeman and his colleagues examining chaotic neural function and olfaction (Freeman, 1991). In this research, EEG recordings are made from an electrode array covering the surface of the olfactory bulb. This activity is measured in rabbits presented with various odors. The output from each electrode represents the actions of thousands of neurons. They discovered that the baseline EEG activity when the rabbit exhaled was disordered. When a rabbit inhaled a familiar scent, however, the pattern was ordered.

Freeman found two distinct modes of neural activity. In *pulse mode*, olfactory bulb neurons responded to an odorant. At this stage they interacted in an unconstrained manner. This activity was magnified and produced a chaotic low amplitude attractor at the cortical level that was measured using an EEG. In *wave mode*, there was a sudden increase in the density of neural interaction. In other words, neurons begin to be coupled and influence each other's behavior more. This produced a high amplitude limit cycle attractor also measured at the cortical level with an EEG. However, in this case, the coupling happened at the global scale and traveled downward to affect the activity of individual neurons.

Olfactory bulb activity alternates between these modes. During inhalation, the system's chaotic pulse mode state makes it sensitive to incoming stimuli. Recognition of an odor occurs during wave mode limit cycle activity. Finally, the system returns to a chaotic attractor state during exhalation. These dynamics are a good illustration of how information processing can cascade both up and down the spatial scale of neural networks.

When the ordered activity during inhalation was analyzed, it revealed a spatial pattern of amplitude waves across the olfactory bulb surface. Whenever a rabbit sniffed a particular odor under similar conditions, this amplitude pattern turned out to be similar. This suggested that the amplitude pattern coded for the identity of a specific scent. These amplitude maps though are not completely invariant. They don't remain the same each time the rabbit smells the same odor. Rather, they change in response to the context, such as the animal's reinforcement history and internal state of arousal. For instance if a rabbit was first reinforced for responding to sawdust, a particular pattern of amplitude activity was produced. Then, when the rabbit was presented with banana another unique pattern resulted. Finally, when sawdust was presented a second time another third pattern occurred, different from the first exposure to sawdust.

This shows that the neural code for a stimulus is not static, but dynamic and constantly changing. The representation is affected by the global activity of other neurons, those that code for different odors and/or additional internal states. It means that the mental representation of a stimulus is not entirely defined by the structure or characteristics of the stimulus itself but is instead a result of the global neural activity of which it is a part. This poses problems for the traditional cognitive view that a mental representation must be made up of object features such as color, shape, or form and that these features are invariant with time.

What is the neural basis for this phenomenon? According to Freeman, the representation consists of a *nerve cell assembly*. This is a collection of linked neurons whose synapses are strengthened by simultaneous activity from input neurons during learning. The cell assembly codes for the identity of an odorant, but is subject to modulation from other input neurons that can subsequently change it. So any initial pattern of activity that forms to a stimulus will be constantly altered by inputs from other associated patterns such as those coding for more smells, how hungry the animal is, how much it has been reinforced for a response and so on.

Freeman (1991) has also recorded activity from the olfactory cortex. Information from the bulb is sent to this cortical region for further processing. This is convergent connectivity, since one cortical neuron receives inputs from thousands of bulbar neurons. As in the bulb, there is a coherent, but malleable cortical signal that represents the identity of odors. These signals are chaotic attractors, they are the pattern of activation that the system settles into when responding to an odorant. When a new meaningful odorant is perceived, it is added in as another attractor state. Since all this activity is connected, the addition of the new attractor causes modifications to all the existing attractors.

According to Freeman, olfaction and perception in general is an active dynamic process. He outlines the general perceptual process as an iterative series of steps. First, self-organizing activity in the limbic system sends a command to the motor systems instructing them to sample information from the environment. This causes the animal to sniff, look, or listen. At the same time, a reafference signal is transmitted to the senses, alerting them to prepare for sensory input. Neurons in the sensory systems then undergo synchronous activity in response to the stimuli. These signals are transmitted back to the limbic system and combined with similar signals, creating a gestalt. The entire process then repeats.

Notice that this perceptual process is one where the organism plays an active role, keenly seeking out information and generating expectations about what it will experience. It allows the animal to anticipate and plan for the future based on what it already knows. Once more, this is in stark contrast to the traditional cognitive view of perception as a passive process triggered and driven by the stimulus alone.

CHAPTER 7
The Fractal Mind

A *fractal* is a pattern that repeats the same design over many different scales. The details in a fractal look the same at different magnifications. Fractals are pervasive throughout the natural world. They can be found in the branching structures of rivers, trees, and blood vessels, in the contours of landscapes and mountains, in the flow patterns of clouds, smoke and mist, and in the shorelines of continents. Fractals can additionally be found in man-made phenomena such as income distribution and fluctuations in stock prices. Many different sorts of fractal geometric patterns can be generated using computer programs that are widely available on the Internet. Fractals and chaos often co-occur. A number of chaotic attractors have been found to have a fractal texture. The patterning of points on these attractors appears the same at different scales.

All fractals have four features in common (Williams, 1997). First, they have a statistical geometric regularity so they appear the same at different length scales. This property is called *scale invariance.* They need not be exactly the same, only approximately so. For example, the branching of blood vessels magnified ten times (10x) under a microscope will appear similar to, but not identical with, the pattern of branching at one hundred times (100x) magnification. It should be mentioned that fractal organization can be temporal as well as spatial. A process unfolding over time can appear the same when measured at one second, one tenth of a second, and so on. Turbulence in fluids and gases reveal similar patterns of variation across different time scales.

Second, a number quantifies the complexity of a fractal over a range of scales. This measure is referred to as the fractal's dimension and is described in greater detail in the next section. Third, fractals are generated by repetitions of an operation. The operation may be natural, such as the crumbling of rock pieces, or it could be deterministic, such as the iteration of a mathematical equation. Figure 7.1 shows several examples of computer-generated deterministic fractal patterns. Finally, fractals are always rough or jagged in appearance. They lack the presence of smooth lines or gradual changes.

Fractals and Dimensionality

The *similarity dimension* of a fractal refers to how much space it occupies. Euclidean dimensions are represented as integers: a line is one-dimensional, a square is two-dimensional, and a cube exists in three dimensions. Yet the more pervasive fractals found in the natural world have non-integer values and fall between these standard whole dimensions. The more wiggly a line is, the higher its fractional similarity dimension. A line that bends or deviates just a little bit from being straight will have a dimension just slightly greater than one. A line that wanders all over a two-dimensional page fills up more of

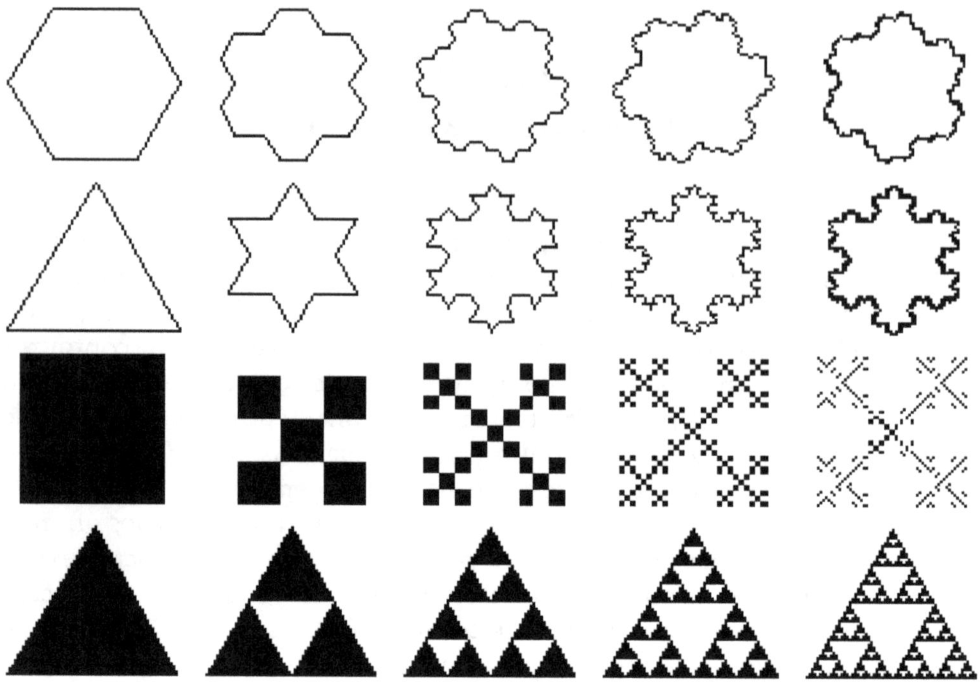

Figure 7.1 *Fractal patterns show similarities at different scales. From top to bottom these are the flowsnake, Koch snowflake, box, and Sierpinski gasket fractals. By Eric W. Weisstein. ©2005 Wolfram Research, Inc.*

the available space and its dimensionality will be closer to two. In this section we will describe how to calculate the similarity dimension of a wiggly line.

It is often not possible to measure the exact length of a line in nature. That is because the length will increase with increasingly more precise measurements. The finer or more exact the measurement, the more detail or small deviations we take into account and the longer the length. For this reason we can only estimate a line length with our estimate being dependent on the level of detail we wish to take into account.

One way to estimate a wiggly line is to use a measuring stick or ruler. We can place one end of the ruler where the line starts and then mark off where the other end of the ruler intersects the line. We record this length, then move the starting end of the ruler up to the intersection point and determine a new intersection, recording this second length. We repeat this process until we reach the end of the line and then add up all our measurements to obtain our estimated length.

By now you may have guessed that our estimated length varies with the length of our ruler. If we used a smaller ruler, we take more of the wiggle and bending of the line into account and our estimated length goes up. If we had used a longer ruler, we would miss more of these details and the estimated length would be less. If we let the Greek symbol epsilon ε stand for the length of our ruler and L_ε stand for estimated length, a plot of the two against one another

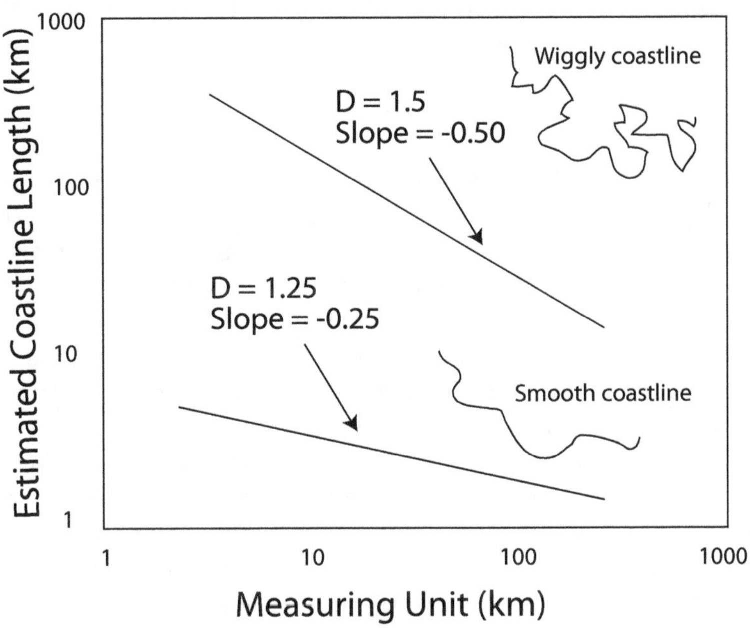

Figure 7.2 *The estimated length of a contour such as a coastline increases as the ruler by which we measure it gets shorter. This relationship is expressed by calculating the fractal dimension. A higher fractal dimension indicates the line is more "wiggly" and that more length is gained by measuring with a shorter ruler.*

in log space yields a straight line trending downwards (Figure 7.2). The similarity dimension *D* then becomes:

$$D = \log L_\varepsilon / \log(1 / \varepsilon)$$

The slope of the straight lines in Figure 7.2 is related to the dimensionality ($D = 1 - \text{slope}$). The steeper the slope, i.e., the more negative it is, the more the line meanders and the greater its dimensionality. The slope of the imaginary coastline depicted by the top line in Figure 7.2 has a slope of -0.5 so it has a fractional similarity dimension of 1.5 ($D = 1.5$). This could represent a coastline with lots of wiggle. The slope of the coastline depicted by the bottom line in Figure 7.2 is less, at -0.25, so it has a correspondingly lower similarity dimension ($D = 1.25$). This coast is therefore less wiggly. A perfectly two-dimensional figure has a slope equal to one. Lines with slopes greater than one will have fractional similarity dimensions greater than two.

Fractal Dimensionality and Magnitude Estimation

One of the major paradigms in psychophysical research is *magnitude estimation*. In this task, observers are presented with stimuli such as sounds or lights at different intensities and asked to rate their subjective intensity using a numerical scale. A log-log plot of physical intensity versus subjective intensity produces a straight line described by a power function. According to *Steven's power law*, perceived magnitude P, is equal to a constant K multiplied by the stimulus intensity s raised to the power n:

$$P = Ks^n$$

Based on the empirical data, we see there are three major classes of function, each depicted in Figure 7.3. For $n = 1$, there is a one-to-one correspondence between actual and estimated magnitude. Estimation of line length produces an exponent close to one. For these types of stimuli, human observers are very accurate at detecting changes in magnitude.

When $n < 1$, observers underestimate changes in stimulus intensity. A light for instance, may double in brightness, but observers will perceive it as less than doubling. This effect has been dubbed *response compression* because observers seem to compress the scale of values. They are insensitive to changes in stimulus magnitude. When $n > 1$ there is overestimation of intensity. These data are obtained in response to electric shock applied to the skin. Observers will report that a shock has more than doubled if its intensity is increased two-fold. This effect is called *response expansion* and demonstrates hypersensitivity

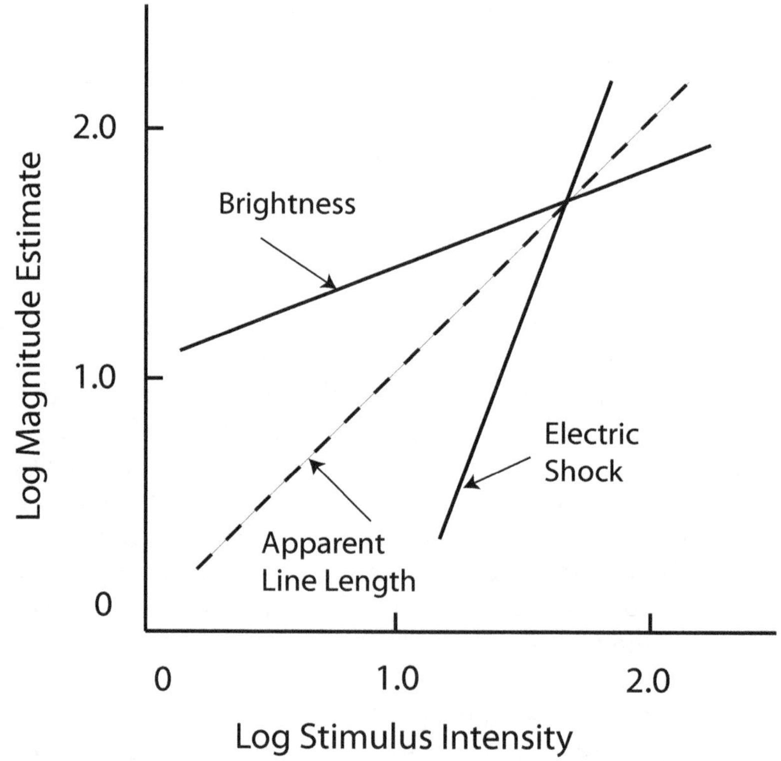

Figure 7.3 *According to Steven's power law, the perceived magnitude of a stimulus is not always a simple one-to-one mapping with actual stimulus magnitude. Observers underestimate changes in light intensity but overestimate changes in the strength of an electric shock. Line length estimation under normal conditions is fairly accurate (adapted from Stevens, 1962).*

Paradigm	Exponent
Brightness of 5 degree target (in continuous darkness)	0.33
Brightness of brief flash	0.5
Smell of heptane	0.6
Vibration of 250 Hz on finger	0.6
Loudness of 3000 Hz tone	0.67
Visual area	0.7
Tactile hardness	0.8
Vibrations of 60 Hz on finger	0.95
Temperature	1.0
Visual length	1.0
Duration of white noise	1.1
Pressure on palm	1.1
Vocal sound pressure	1.1
Thickness of blocks	1.3
Taste of sucrose	1.3
Taste of salt	1.4
Temperature (warmth on arm)	1.5
Tactual roughness	1.5
Handgrip	1.7
Electric stimulus in hearing	2.0
Electric stimulus to skin	3.5
Electric stimulus to teeth	7.0

Table 7.1 *Magnitude estimation exponents for different types of stimuli (after Utttal, 1973).*

to changes in actual magnitude. Table 7.1 shows exponents obtained in estimation experiments for various stimuli (Uttal, 1973).

The similarities between magnitude estimation and fractal dimensionality are obvious. The slope of the line in psychophysical estimation experiments can be seen as a measure of psychological complexity (Gentry, 1995). A higher exponent implies a more complex mental experience. Application of electric shock to the skin may activate more neurons or induce more complicated neural processing than simply looking at a light. In fact, there may be a direct analogy between complexity of neural activity and the complexity of a line. A very wiggly line can be thought of as tracing out a complex path through a neural network, yielding a more complicated perception.

Perception of Fractal Stimuli

A number of studies have investigated human perception of fractal stimuli. Cutting and Garvin (1987) generated fractal curves varying in fractional dimension and other features. Fractal dimension was found to be a reliable predictor for observers rating the stimuli in terms of their complexity. Knill, Field, and Kersten (1990) found that exponent values ranging between 2.8 and 3.6 produced peak sensitivity for discrimination of randomly generated noise textures. Westheimer (1991) found that participants could detect a 0.0085 shift in the fractal dimension of borders starting with $D = 1.15$ and a shift of 0.015 for borders where $D = 1.25$. These later results show remarkable human sensitivity to such differences.

In another study, Kumar, Zhou, and Glaser (1993) compared five standard algorithmic methods for estimating fractal dimension against human observers. They presented their stimuli to their observers in the form of jagged lines and one-dimensional luminance patterns. Only one algorithm, that implementing a maximum-likelihood method, produced better performance than human observers. Gilden, Schmuckler, and Clayton (1993) also tested algorithmic performance in the discrimination of fractals from other types of random contour. Their algorithms were designed to have the same sensitivities as human observers. They operated on the basis of distinguishing signal from noise and were not sensitive to the self-affinity in the patterns. The algorithms performed well, suggesting that humans may detect fractals on the basis of signal to noise also.

Fractals and Visual Aesthetics

A number of recent studies show that human observers find images with fractal statistics attractive and pleasing. Aks and Sprott (1996) found that aesthetic preferences correlated with both the fractal dimension and Lyapunov exponents of computer images generated from equations. Their results showed preferred patterns had a fractal dimension of 1.26 and a Lyapunov exponent of 0.37 bits per iteration. They also found that participants who reported themselves as creative had a preference for patterns with a slightly higher fractal dimension, while participants who self-reported themselves as scientific had a greater preference for patterns with higher Lyapunov exponents.

Spehar *et al.* (2003) found that human observers displayed a consistent preference for fractal images regardless of whether they were generated by a natural process, by mathematics, or by human hand. Peak preferences in their study were found for a fractal dimension of about 1.3 for natural images and simulated coastlines and about 1.5 for paintings by the artist Jackson Pollock.

The explanation for these preferences seems to stem from the similarity between such images and natural scenes. In particular both paintings and natural scenes have been found to share similar image statistics. These include power spectra distributed as $1/f^2$, sparse spatial structure, and similar edge co-occurrence statistics (Graham, Chandler, & Field, 2006). Graham and Field (2007) found that landscapes, portrait/still-lifes, and abstract paintings all shared basic statistical regularities similar to those of natural scenes.

Redies (2007) proposes a theory as to why these similarities would induce aesthetic qualities. He suggests that an artist creates a work of art so that it induces a specific resonant state in the visual system. This resonant state is based on the adaptation of the visual system to natural scenes. According to this view, looking at a painting activates our brain in much the same way as when we are looking at a natural scene. The sense of naturalness is part of what induces the aesthetic preference.

Richard Voss proposes a similar explanation (Voss & Clark, 1975). He believes that fractal noise mimics the patterns of the natural world, like water turbulence and the movement of tree branches in the wind. These patterns have meaning for us because we evolved in this world. The fractal structure and organization of the human brain thus resembles this aspect of the environment, making us sensitive and responsive to it. The neural process underlying our aesthetic response to fractal stimuli may also explain the therapeutic benefits of pink noise. We discuss in the section on noise.

The Fractal Brain

A considerable amount of data point to the fact that the brain has a fractal-like structure where neural structures at many scales are recursively embedded (Alexander, 1993). Alexander and Globus (1996) mention that attempts to model brain function could be augmented by including several scales of organization in the simulations. Traditional neural networks have only two scales, individual neurons or nodes and the entire network taken as a whole. Examples of these include almost all artificial neural networks created to date, such as three layer architectures that run using backpropagation methods. In these monolithic networks there is no way to partition the network into any other divisions.

A stratified network, though, consists of multiple scales of organization. Here we find networks within networks, where each smaller unit of organization is nested inside a larger one. Each of these networks can perform different tasks. For instance, feature detection models of pattern recognition might have features such as lines and curves represented at a small scale, these would be organized into networks representing individual letters at intermediate scale, which be nested within networks standing for entire words a larger scale.

The brain shows just this sort of organization. The visual system at the lower level consists of columns of cells responsive to lines at different orientations. These columns are organized into hypercolumns that contain cells responsive to inputs from either eye with other regions responding to color or non-color information. These hypercolumns in turn are part of blocks, which are part of areas, etc. all the way up to the characterization of the system as a whole. Kaas (1982) refers to this as *recursive modularity* because each level can be considered to have some degree of functional autonomy.

One reason for recursive modularity may be the *connectivity problem* (Laskoski, 1990). If all the neurons in the human brain were completely interconnected, the brain would require a circumference of 10 km! Nature's solution to this may be to fully connect all neurons at one level of spatial scale. Once this connectivity has reached its maximum, it gets incorporated as a functional unit into the next largest scale of organization. At this level, it is now wired up to other units. When the interconnections between these elements reach their maximum, they in turn are linked as networks into an even larger scale and so on. This process allows for a maximum amount of connectivity in a small volume.

What is activity in these systems like? According to Globus (1992) units at any scale can affect information processing through the entire system. A network at one level could turn on, turn off, participate in, or departicipate from the ongoing flow of computation. Stratified networks allow processing to cascade both down the network from the large to the small, but also up the network from the small to the large. In this sense, they differ from a strictly hierarchical network, where control is top-down only.

This type of top-down and bottom-up influence is seen in other natural phenomena. In laminar flow in liquids the macroscopic behavior of the fluid flows smoothly without interruption. This constrains the interactions the molecules have with one another, biasing them to move in a predominant direction. In turbulent flow, microscopic molecular interactions are magnified and affect the movement of water at larger scales.

Levels of Analysis and the Brain

How many levels are there in neural systems like the brain? In reality, the brain can be analyzed at multiple levels ranging across the atomic, molecular, cellular, network, and regional scales (Friedenberg & Silverman, 2006). Cyclical feedback probably exists between all adjacent levels but also between levels that are more spatially and temporally separate.

The influence of a lower level on a higher one has been dubbed *upward causation*. This can be seen in the case of epilepsy, where a seizure consisting of periodic oscillatory behavior between neurons spreads to the point where it can involve the entire brain in a grand mal seizure. The opposite process whereby a higher level regulates the behavior of lower one is called *downward causation*. A few case studies report that global patterns can also act on a local assembly of cells. This opens up the possibility of 'cognitive' control of epileptic seizures (Le Van Quyen *et al.*, 2001).

Varela (1999) distinguishes between three levels that are both spatially and temporally distinct. At the microscopic level we are dealing with cellular activity that occurs in the range of 10-100 msec. Here the dynamics are dominated by coupled limit-cycle oscillations. This activity is mostly periodic and linear. At the mesoscopic level there is network activity characterized by cell populations operating in the 100–300 msec time range. Neural synchrony and coupling occur here and activity is predominantly nonlinear. At the macroscopic level we now have communication between brain regions and the time frame is measured in seconds. This behavior is more random and stochastic in nature.

When discussing these levels of organization it is important to take note of several issues. The organization can be both vertical and horizontal. Vertical organization refers to neural circuits or patterns of activation that are embedded one inside the other. Here, "higher" regions, those that subsume greater spatial or temporal extent, regulate "lower" ones. In horizontal organization, there is information sharing or communication between circuits or patterns at the same level. Also, although these processes can involve different tissue and time scales, they must be considered holistically. Only the combined and integrated action of all the components can give rise to the coherent mental state.

The classical view of the visual system is that it is hierarchical, with ganglion and thalamic neurons early on coding for spots of light that are combined cortically into edges, moving edges, angles and other more complex features. In this conception, the farther back we go from the eye, the more complex and global the representations become. However, Hegde and Felleman (2007) cite a growing body of evidence showing that visual processing is not hierarchical and does not strictly follow the visual anatomy that does have a more hierarchical organization. They conclude that the hierarchical processing model of the visual system is overly simplistic.

Mind-Brain Interaction in Complex Systems

Chris Langton provides us with a sketch that aids in understanding complex systems (in Lewin, 1999). In this view, a complex system operates on at least two levels. At the local level, the agents or parts interact. This interaction gives rise to global structure and properties that in turn feed back down to the local interactions. There is thus a two-way street between levels. Local interactions have bottom-up effects on global structure while the global structure produces top-down effects on the local organization. Figure 7.4 shows this relationship.

Let us imagine that the local elements are neurons and their interactions consist of messages transmitted back and forth to one another. In certain accounts, mind then becomes the emergent global structure that is produced by this bottom-up activity. However, mind also has the top-down power to influence neuronal interactions, which have the power to influence mind and so on. What results is a dynamic synergistic relationship where mind is caused by the brain but also acts on the brain.

This brain-mind model has obvious connections to the schools of mind-body debate in philosophy. It conflicts with all dualist positions where the relation between body and mind is one way. Many scientists adopt a monist view, believing that mind is purely physical. This view, known as *physicalism* or materialism, equates the mind with the brain. The functions of mind are believed to be nothing more than the physical interactions that occur in the brain. A monist interactionist perspective would best approximate the model, since it views emergent global structure as a particular pattern of physical activity among neurons. This structure is a configuration or property of the neuronal activity itself and would have the power to feed back and alter its lower-level interactions

In philosophy of mind, *supervenience* is the term used to describe a type of relationship between the mental and the physical. By one definition, a mental property supervenes on a physical one if and only if two objects, such as brains, share all relevant physical properties. The relevant physical properties in this

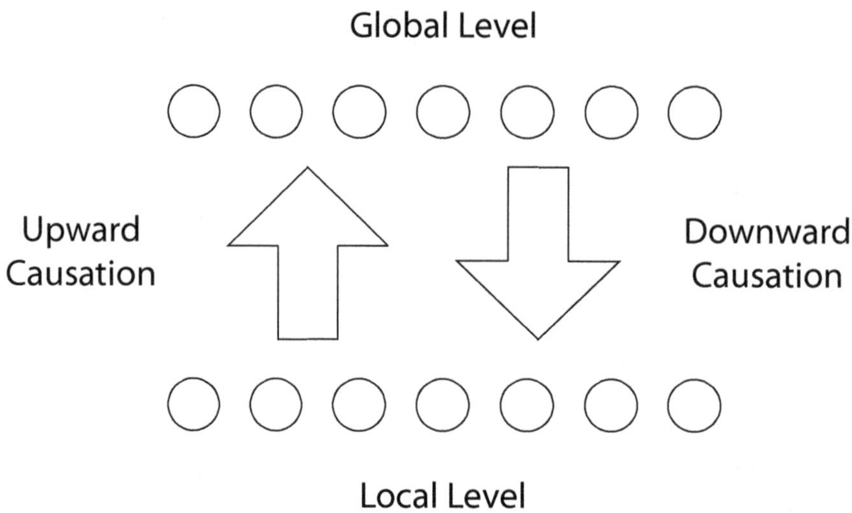

Figure 7.4 *In analyzing brain activity, we must take into account local and global activity. Both influence each other.*

case might be fully functional neurons instantiating a particular pattern of activity. If this was the case in two individuals, then it would follow that they must also share all relevant mental properties such as having a particular thought, being conscious, etc. Supervenience however, does not have to be symmetric. Although two people with indiscernible physical properties must have the same mental ones, it would be possible for two people to have the same thought with two different patterns of neural activation. In other words, mental properties may realized by more than one physical property. This view goes by the name of *multiple realization.*

So it is possible for a physical mind to supervene on itself. The "mind" may simply be a pattern of neural activation 'B' that results (emerges) from a preceding or simultaneous pattern of neural activity 'A'. B then affects (supervenes) on 'A', which subsequently alters 'B' and so on. This process may continue back and forth indefinitely. In fact, mind may be dependent on just this sort of cyclical feedback activity. It is important to note that the emergent global structure in this model, what we are calling mind, is entirely physical. What makes it global or emergent are its physical characteristics or properties. A global pattern of activation might differ from a local one in a variety of ways. It could, for example, be an amplitude-modulated pattern of activity rather than one based on oscillation (Freeman, 2000).

Fractal Geometry and Human Nature

Marks-Tarlow (1995) points out a number of correspondences between fractals and human nature. He states that people show self-similar behavior at different levels of observation. He gives the example of an aggressive person who acts this way by dominating a conversation, pushing to the front of a line, or speaking negatively of a colleague at work in order to get a promotion. In each case, the trait of aggression is demonstrated at increasingly more important levels of scale.

Fractals, especially those resulting from natural processes, display ordered unpredictability. They share this feature in common with chaos, which has order but cannot be predicted in the long run. People display ordered unpredictability as well. We retain an identity and continuity over time in that we can be expected to act in certain ways in particular situations. Yet despite this, we are also unpredictable. One cannot with any degree of accuracy predict exactly what human behavior may be alike even if we know the person and the circumstances they will be in.

Another feature of fractals that can be applied to psychology is the property of bounded infinity (Marks-Tarlow, 1995). As we discovered above in our discussion of coastlines, a fractal can be infinitely long, since the smaller the yardstick, the longer the estimated length. The same can be said with people. The closer one examines a person, the more there is to see. Because people are so complex and have had so many experiences with a changing world, investigating one aspect of a person reveals a further wealth of details and information. It may be impossible to every "get to the bottom" when describing a person.

Fractal dimensionality can also be related to psychology. Marks-Tarlow (1995) speculates that Euclidean dimensions characterize the basic or primary structures of the psyche. The fractional dimensionality represents how they fit together and is demonstrated in someone's coping style or potential, or the tendency of the system towards growth or disintegration.

He illustrates this using the example of regression. In regression, one reverts back to an earlier childlike way of dealing with a situation, such as crying or throwing a temper tantrum. It could be that the tendency to regress, which is repeating the past rather than experiencing the present and moving toward the future, can be represented using a fractional dimensionality number. Ordinary coping seems best represented with a whole integer value dimension, while behavior under stress may be fractional. In this sense, disorders such as obsessions, compulsions, and phobias, could be instances of the fractional dimensionality of a fragmented self. We discuss psychological disorders and their relation to dynamical concepts later.

CHAPTER 8
Statistical Mechanics and Noise

Statistical Mechanics and Magnetism

As we have just discussed in the fractal chapter, natural systems can be considered from multiple levels of analysis. From the simplest standpoint, we can analyze a system from at least two broad levels. There is the global level, where we can observe what happens by "stepping back" and looking at the system from a distance. Then there is the local level, where we can observe what is happening at some fine level of detail within the system. A major goal of reductionism is to be able to explain the global in terms of the local, to describe the whole based on an understanding of the parts.

This is also the goal of a field in physics called *statistical mechanics*. It is the study of the likelihood of a macroscopic system being in a family of microscopic states. Take for example the overall behavior of a gas in a container. This would constitute the macroscopic state. The gross behavior of the gas could be described independently of the actual movements of each individual gas atom. In other words, we could successfully describe the global behavior of the gas without having to resort to knowing what each and every atom was up to. One of the successes of statistical mechanics is that it can produce solutions that are the same as those that model atomic or molecular dynamics.

One global property that we can all observe is magnetism. At one point or another we might have played with magnets and seen how they can be used to pick up metal objects such as paperclips. What microscopic effects give rise to this phenomenon? In a magnetic material such as iron, individual iron atoms are arranged in a very regular way. They occupy the nodes of a three-dimensional lattice. Each iron atom has a polarity or spin: it can either be pointing up or down.

In the presence of a constant external magnetic field many but never all of the spins align themselves. That is, most of the atoms are pointing in the same direction and the material is said to be magnetized. If we then remove this external field, the spins will flop back into different directions and the material will have no net magnetization. These two situations are illustrated in Figure 8.1.

There are two other variables that influence spin direction other than the presence of the external magnetic field. The first is temperature. When a piece of iron is cool the iron atoms mostly align. But at higher temperatures, thermal noise knocks the atoms about, altering their spins and preventing magnetization. The second variable is nearest neighbor interaction. Each iron atom tends to copy its neighbors. The greater the number of immediate neighbors with a given spin, the greater the likelihood that an individual iron atom will adopt their orientation.

We can measure the total amount of magnetization in a piece of iron by simply summing over the spin states of all the atoms. If we assign "up" a value

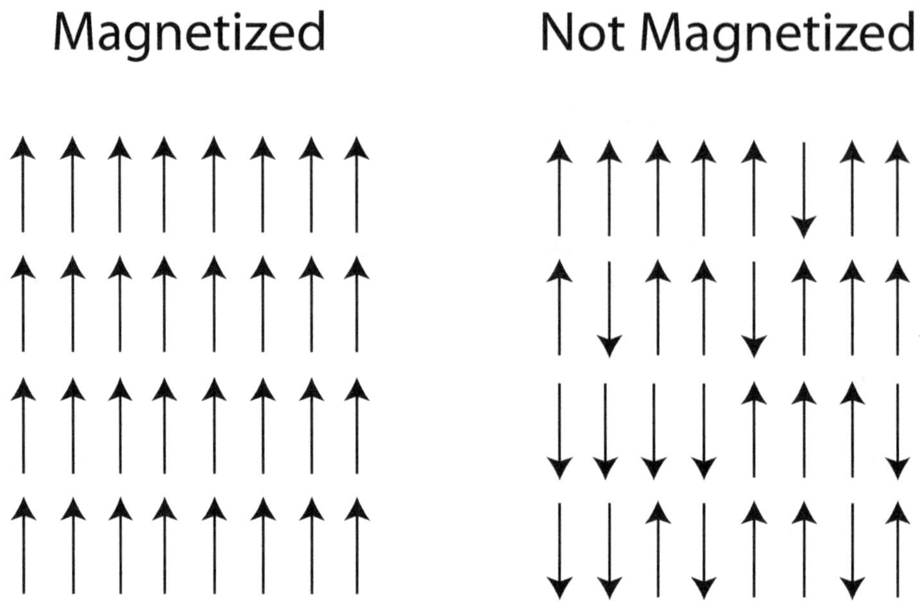

Figure 8.1 *Iron atoms have a spin, pointing either up or down. When many or all of these atoms point in the same direction the material is said to be magnetized. If the spins are at random orientations, there is no magnetization.*

of +1 and "down" a value of -1, then magnetization will be lowest at zero, since half of the atoms point up and the other half point down. The greater the deviation from zero, the more the atoms are aligning either in one direction or the other.

What happens if we take a magnet that is very hot and let it gradually cool down? As is the case with so many other systems we describe in this book, there is a critical point demarking a phase transition. When the magnet is very hot, the net magnetization is near zero. As it cools down, we reach a critical point, called the *Curie temperature*, where the iron atoms suddenly align themselves in the same orientation. The net magnetization rapidly diverges from zero and then levels off (see Figure 8.2).

In the high temperature regime above the Curie point we have disorder. There is little or no coordination between iron atoms. The heat overwhelms the material's ability to achieve a consistent orientation. In the low temperature regime below the Curie point we have an ordered state where the nearest neighbor rule dominates. All or most of the atoms are aligned. Only in the threshold region near the critical point is there a mixture of order and disorder producing a complex state characterized by the presence of fractal organization. Two- and three-dimensional plots of spin in this region show similarity at different spatial scales, clusters and patterns of aligned atoms that appear the same at different magnifications.

So does a statistical mechanics interpretation of magnetization constitute a victory for reductionism? The answer is no. A reductionist account would focus on the nature of the iron atoms. It would require a detailed description of what it taking place inside each atom and then use that knowledge to explain

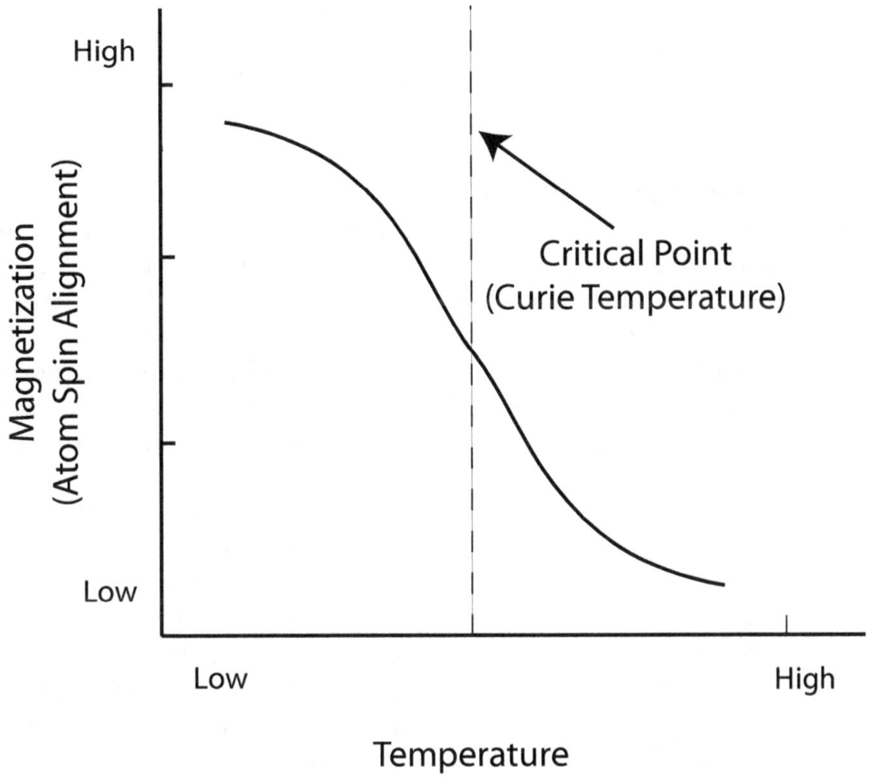

Figure 8.2 *Net magnetization in iron as a function of temperature. At the critical point or Curie temperature, there is a rapid change in spin alignment (adapted from Back et al., 1995).*

the global characteristics. This knowledge would include quantum mechanical interpretations and descriptions of electron behavior in each atom, for instance.

One statistical mechanical interpretation of magnetization is called the *Ising model.* It says nothing about what is taking place inside the atoms. Each atom acts merely as a placeholder with a given orientation and will switch to its opposite state with a transition probability, taking into account temperature and number of nearest neighbors. These two variables are 'outside' of the atoms. Mathematical simulations of the Ising model are able to reproduce exactly the phase transition plot of magnetization versus temperature. The Ising model therefore shows that it is the relations between the parts and not the parts themselves that can accurately and completely predict the global behavior of this system.

As mentioned above, magnetization is just one example of numerous systems that display similar features and behavior. Critical points and phase transitions pop up in systems all over nature, from physics and biology to psychology and economics. What are the implications of this? It means that if a system has certain microscopic features and relations, then we can predict how it will act macroscopically. It also means that in certain cases we can ignore the details of the system, i.e., the makeup of its parts. This notion is called *universality*. Simply put, universality states that simple rules can generate complicated patterns of behavior, and that the interactions among the different parts of a system, not the detailed properties of their parts, can account for large scale behavior. One of the goals of complexity science is to determine the different classes of rules that govern such behavior. These rules are known as *universality classes*.

A Statistical Mechanics of the Mind

We can use the concept of magnetization as a metaphor for mind. The mind can also be considered from global and local perspectives. At the global level we need to explain how coordinated brain activity occurs. This coordinated action consists of neural populations that synchronize their activity across vast regions. For example, when paying attention to a visual stimulus, there is activity in the frontal, parietal, and occipital lobes. These lobes in neural terms are located quite distant from each other, since they are located roughly at the front, top, and rear of the brain.

How does this occur? One way is to have a central authority that manages the process. This manager "knows" that the brain has to perform a particular task and turns the various areas on and off at just the right times. Although this idea could work, it runs into problems, since it requires a person with a mind inside of your mind who must then be explained. There is no evidence that any such coordinator exists. So the process must occur spontaneously. In other words, activity at the local level consisting of individual neurons, must somehow give rise to large-scale coordinated patterns of activity.

It is here where we can apply ideas from magnetization. First, we can imagine that iron atoms are neurons. Instead of a spin state, each neuron can have an activity state. This could mean that the neuron is either "on" or "off" or that it is firing at a particular frequency or not. The lattice specifying the locations of the atoms is equivalent to the location throughout the brain of the different neurons. Nearest neighbors are also at work. Since most neuron connections are local, a cell given the right type of connections would tend to copy its neighbors. If the connections are excitatory and the neighbors are active, their inputs would summate to increase the cell's chances of firing.

The final piece in this analogy is temperature. In magnetization, an increase in temperature reduces coordination. This is like adding noise into a signal. Neural "noise" could be inputs from some other brain process or perhaps an inherent tendency for neurons to revert back to different rates of firing. In either case, this noise would counteract the spread of coordinated action throughout the system by nearest neighbors.

A disordered brain state at high "temperatures" might then correspond to the situation where one is trying to perform a cognitive task, but is distracted. This scenario has been studied often experimentally. In the *dichotic listening task*, a participant is asked to focus on the content being played to one ear, while ignoring the content from the other ear. Performance is degraded under these conditions because the input from the unattended ear cannot be completely filtered out. Similarly, in *the Stroop effect*, it is more difficult to name the color of a word (say red) when the word spells out a different color (such as blue). The name is automatically processed and cannot be ignored. In these cases, information from the secondary task interferes with the primary task. It could act like thermal noise, interfering with the local spread of information through the brain. It might for instance, prevent the frontal or parietal lobes from communicating with the visual areas in the occipital and temporal lobes.

In contrast, there would be no interference in a brain at low "temperatures". In the absence of any distractions local neural action could propagate unimpeded. It can be easier to focus attention under these conditions and primary task performance improves. However, too much order can also interfere with proper brain function. People who are deprived of normal sensory inputs suffer cognitive impairments. Solitary confinement as we all know is a form of punishment. Noise can actually facilitate signal detection, a phenomena called *stochastic resonance.* This is a nonlinear cooperative effect where global random fluctuations are entrained by an independent and usually periodic fluctuation. The result is that the weak signal becomes amplified and is more perceptible (Gammaitoni *et al.*, 1998).

So it appears that optimal brain function takes place in a "middle ground" where there is neither too much nor too little noise. Like the proverbial Goldilocks, the brain seeks a compromise between extremes. This is of course the equivalent of the critical point near the Curie threshold temperature. In this regime, we see both order and disorder, coordinated and uncoordinated neural activity. It is the task of future researchers to determine the exact role that noise plays in different cognitive tasks. We have much more to say on the role of noise in the next section.

Noise

Characteristics of Noise

In a scientific context, *noise* is considered as variance or changes in the dependent variables that cannot be accounted for by manipulation of the independent variables. There can be many different sources for this variance. If it comes from individual differences then it can be factored out by employing a within-subjects design or by a statistical adjustment such as a repeated measures analysis of variance. This technique takes out error variance due to differences in responding between participants. Noise can also come from other variables that were not controlled for in the experiment. These can include such factors as time of day and room temperature. Even when these are held constant there is usually still some noise left in the data. It may not be possible to eliminate this residual noise since its origin is from microscopic quantum fluctuations (Baggott, 1992).

It can be useful to distinguish between external and internal noise (van Kampen, 1987). *External noise* originates outside a system but acts on it. For instance, it may be difficult to hear a radio broadcast in a room because of an ambulance siren wailing outside on the street. External noise does not affect the behavior of the studied system. In this case, the siren will have no effect on the loudness or content of the radio broadcast. External noise is connected or coupled to the studied system but can be switched off or bypassed. We could for example, close the windows or take the radio into the hallway. *Internal noise* is part of the system of interest and is more difficult to control. It would include for instance static or interference in the radio broadcast itself.

Why Nervous System Noise is Important

Rather than treat noise as a nuisance that needs to be gotten rid of, Ward (2002) suggests that we instead treat it as a legitimate phenomena. This is especially true if the noise is internal. In this case, the noise reflects some mechanism or process of the system we are studying. If we ignore it, we do so at our peril, since it may help to explain the very thing we are investigating.

It would seem that brain activity would have to be ordered since the thoughts and behavior it gives rise to certainly are. However, recent studies show that noise plays a key role in brain activity and is in fact necessary for normal functioning. For example, noise can help a neural system transition from one ordered state to another. It can help a neuron or neural circuit escape from a local minima. It may be the basis for how information is transmitted in the firing intervals of neurons, and it may increase a neuron's dynamic signaling range. The picture that is now starting to emerge is that too little noise is bad. Excessive order could spell disaster for an organism, reducing its flexibility and capability of dealing with a dynamic and changing world.

Sources of Nervous System Noise

There are multiple sources of noise in neural computation. To start, external noise is usually part of the stimuli that impinge on the perceptual systems. In this sense, noise is "imported" from the outside environment. Pelli (1981) argues that photon noise is the major source of noise in the visual system. If this were true, then it might not be possible to completely remove all external noise when studying visual system function. That is, we may never have a situation in which we can isolate and study internal visual noise by itself.

Smetters (2000) points out three primary sources of internal noise. The first comes from synaptic integration. Here, post-synaptic potentials that are generated from synapses on the dendrites travel across the soma or surface of the neuron cell body. These signals are integrated, typically at the axon hillock, as part of the neuron's "decision" to fire. There are several sources of noise during this stage of synaptic integration. Most brain neurons are spontaneously active and produce action potentials even in the absence of synaptic input from other cells. This baseline rate of firing can be used to transmit information, since excitatory inputs can speed up firing rate while inhibitory inputs can slow it down. The change in firing rate may serve as a code. However, baseline firing rates are typically considered to be random (Bernander *et al.*, 1991). Some neurons also generate "miniature" synaptic events, where transmitter is released without any action potential activity. In the central nervous system, these must compete with "legitimate" potentials induced by incoming messages.

This baseline and spontaneous activity may impede normal processing and reduce the clarity of any signal. It is possible that the receiving cell may need to integrate over a large number of synaptic inputs to average away the noise before initiating a response. Alternatively, the noise may be beneficial, pushing the neuron out of a local minima (Otmakhov *et al.*, 1993). Liljenstrom and Wu (1995) found that spontaneous activity actually improved performance by reducing recall time is associative memory tasks.

The second step where noise can happen is during action potential generation. As mentioned previously, the speed with which action potentials are created is quite variable. This makes a rate code, where average frequency is used to transmit information, unlikely. That is because it takes time to calculate an average, time the organism often does not have because it must react quickly to changing circumstances. An interval code, one based on the timing between spikes, is more likely the case. The timing of action potentials has found to be reliable under particular conditions (Reike *et al.*, 1996).

The third source of noise is during synaptic transmission. The arrival of an action potential at the terminal button does not always result in neurotransmitter release. It has been estimated that in some cases, this may occur only one time out of ten. Release probability varies based on type of synapse and in different synapses of the same type between the same cells (Dobrunz & Stevens, 1997). Rather than being a source of error in signal transmission, neurons may instead use this variability to increase their dynamic signaling range and flexibility. Neural plasticity may even operate on the basis of changes in transmitter release probability (Markram, 1997; Stevens & Wang, 1994).

One source of synaptic variability are ion channel proteins. These are the postsynaptic receptors that open up after a transmitter molecule has docked, allowing ions to flow into or out of the cell, thereby changing the dendrite's electrical potential. It was previously assumed that these channels could be in one of several states, such as open, closed, or inactivated, that they transitioned instantaneously between these states, and that the rate of switching was dependent only on the current state of the channel. More recent work shows that these assumptions are probably incorrect. The probability of a channel switching states is much more complex than previously realized and is probably dependent on numerous factors such as the amount of time the channel has spent in its current state (Liebovitch & Todorov, 2000).

Scannell and Young (1999) used functional brain imaging techniques to study the relationship between activity in single neurons and the population of neurons in which they participated. They found that population activity was strongly influenced by neurons with low frequency baseline firing rates. However, when neural representations were sparse and tuned to a number of different stimulus dimensions, the population activity was not influenced by these neurons and instead driven mostly by stimulus or task factors. The conclusion to draw here is that spontaneous baseline activity in small numbers of neurons (noise) can, under certain conditions, have a huge affect on global brain activity. Factors such as attentional focus may "set" the amount of such noise.

Major Types of Noise

We can describe noise by plotting it as a *time series*. A time series is simply a set of observations or measurements that are ordered in time. If we observed somebody's mood at regular intervals, say every hour throughout the day, we would have a time series. It is the fundamental way of representing data in dynamical analyses. Plotting a time series of noise measurements can help us to learn more about it.

Let us suppose that we have measured the static coming out of a radio tuned in-between stations. We then plot our values in the order they were recorded. The time series plot we might obtain is depicted at the top of Figure 8.3. We can see that there are many different frequency components here. Low frequencies show up as large deviations while higher frequencies appear as smaller deviations.

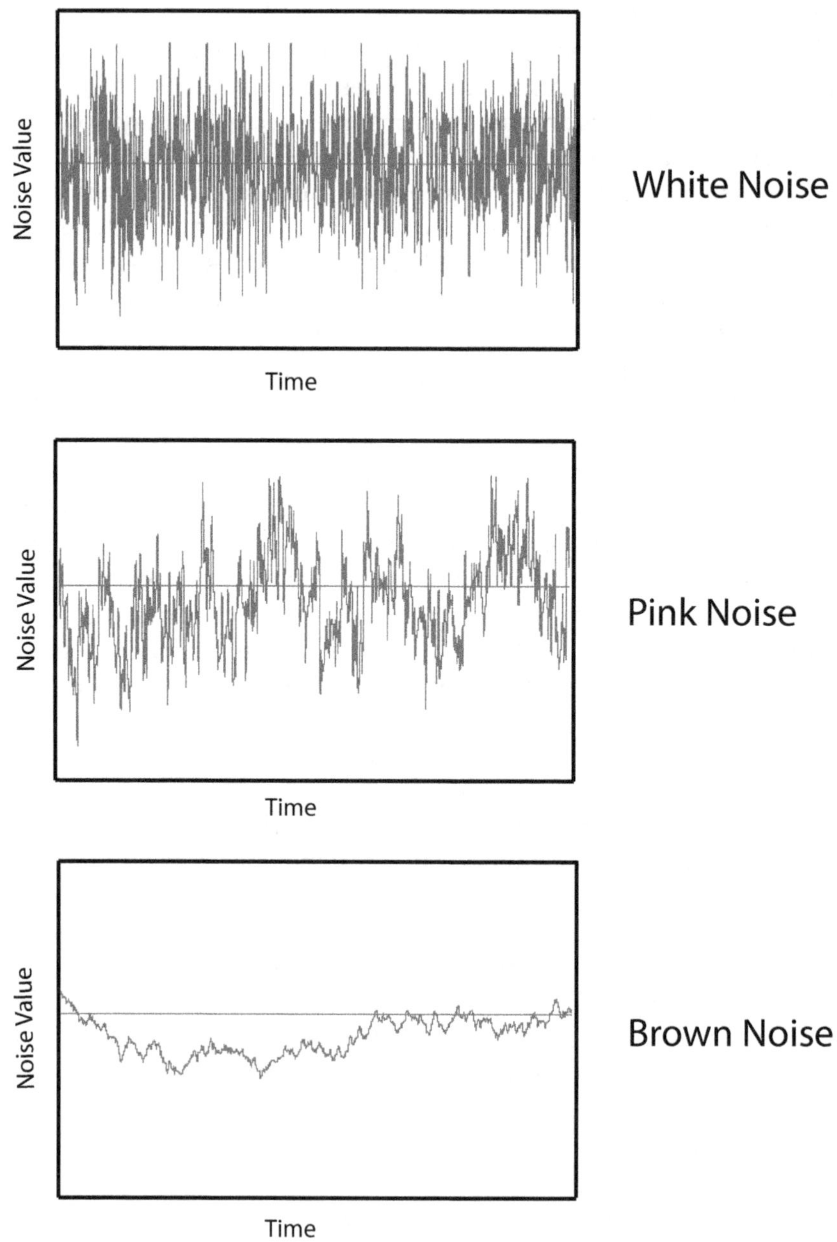

Figure 8.3 *Time series for white, pink and brown noise.*

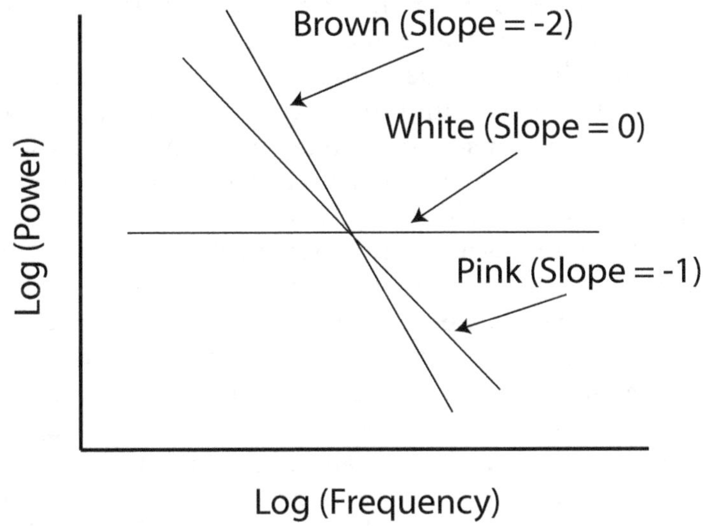

Figure 8.4 *Power spectra for white, pink and brown noise.*

We can get a better description of the noise's frequency components if we plot the power of each frequency in log-log space. This type of plot is called a *power spectrum.* Figure 8.4 shows the power spectrum for three different kinds of noise. Notice that the slope of the line for this noise is very close to zero. This means that this noise has roughly equal amounts of each frequency within a given range. This is the signature of *white noise.* White noise can be found in the natural world. Examples include the sound of wind passing through the leaves of a tree or water coursing through a stream.

The middle portion of Figure 8.3 shows a time series plot for measurements of another type of noise. Notice that in comparison to white noise there is a distinct low frequency trend in the data. This appears as larger wavelike fluctuations. However, there are some mid and high frequency trends present as well. The slope of the line in the power spectrum for this noise in Figure 8.4 is now very close to negative one (-1). This is the hallmark of *pink noise* that is dominated by low frequencies. Pink noise is also present in natural phenomena. It can be found in heartbeats, thunderstorms, and earthquakes. This power spectrum is also a signature for complex systems.

The last type of noise we will discuss here is *brown noise.* The time series is shown at the bottom of Figure 8.3. Low frequency fluctuations are even more dominant in brown noise than they are in pink noise. The slope of the line in the log-log plot is now negative two (-2). An example of brown noise in nature is the sound of waves on a beach. There are other types of noise with different characteristics, for instance "black noise" with a slope less than negative two, but we will not outline them here.

Pink Noise and Neural Processes

Pink noise occurs not just as a natural phenomenon. It has been found in human-generated sources as well. Power spectra with slopes near negative one have been found in Bach's Brandenburg Concerto no. 1 (Voss & Clark, 1975).

Other similar results have been obtained for classical music, rock music, and a news and talk radio station. Since people created these sources, it suggests that pink noise may be an inherent property of the brain and cognitive systems.

Novikov *et al.* (1997) found direct evidence for pink noise in human brain activity. They placed sensors in a circular pattern around each side of the head in several participants. A superconducting quantum interference device (SQUID) recorded magnetic field strength at each sensor location. The power spectra of the resulting time series showed slopes close to negative one.

Brain activity can also be measured using an EEG. In many of these studies, a stimulus is presented to a participant and the EEG is used to capture the activity that is generated in response to the stimulus. This technique is called an *event-related potential* (ERP) and can be used to infer what brain mechanisms are at work in the task. McDonald and Ward (1998) presented auditory tones to participants every two seconds. They analyzed the ERP data from immediately before the tones were presented and about one tenth of a second after. Power spectra from the time series displayed slopes indicative of pink noise. They conclude that the brain activity in response to the tones as well as the noise this activity was embedded in were both characteristic of pink noise.

Why should we see pink noise in the brain? You will recall that a negative slope in a log-log plot is the signature of complex systems that self-organize on the edge of chaos. We found this pattern in hub-based networks, in phenomena demonstrating self-organized criticality and in fractals. Here, the signature indicates the activity of different neural mechanisms operating at a variety of distinct temporal and/or spatial scales.

Low frequencies correspond to large-scale brain processes. These take longer because they involve more neurons or neurons that are spaced farther apart. Consequently the computations performed by these neural groups will take longer because messages need to be transmitted over a greater distance or between more elements. High frequencies alternatively correspond to smaller-scale brain operations. These proceed more quickly because they involve fewer neurons or neurons that are closer to one another. Computations in these neural groups are fast because messages only need to be sent short distances or between a small number of elements. Ward (2002) carves up neural timescales into three different categories whose characteristics are listed in Table 8.1.

Pink Noise and Cognitive Processes

Van Orden, Holden, and Turvey (2003) mention that background noise in cognitive experiments remains even after other variables like treatment and task effects are controlled for. They found the presence of pink noise in several experiments involving simple responses and word naming. The authors conclude this noise must be intrinsic to the operation of cognitive processes in living beings. They argue for the presence of what they call interaction-dominant dynamics where the mind and body are continually interacting with one another to produce pink noise. This noise is the natural consequence of complex systems (like people) that self-organize. In this section, we describe several other experiments that demonstrate the presence of pink noise in human cognition.

Timescale (msec)	Cycle Frequency (Hz)	Brain System	Cognitive System	Behavioral Task
1 to 2.5	1,000 to 400	1 to 10 neurons	Preconscious	Peripheral sensory processing
10 to 25	100 to 40	10s to 100s of neurons: within area circuit	Unconscious	Sternberg (1966) STM scanning task
100 to 250	10 to 4	1,000s to millions of neurons: between-area loops	Conscious	Crovitz (1970) strobe matching task

Table 8.1 *Neural timescales can be considered as falling into three categories (from Ward, 2002). Reproduced with permission of MIT Press.*

Gilden, Thornton, and Mallon (1995) had participants repeatedly estimate different time intervals running from 10 seconds to 0.3 seconds in duration. They created a time series of the errors and then plotted these as power spectra. The slopes of the lines for each duration estimated closely approximated negative one, indicating the presence of pink noise. Furthermore, the lower the frequency judged, the higher the intercept of the functions. The ten-second duration estimates had the highest intercept and the 0.3 duration had the lowest, what we would expect since there is more power at lower frequencies for longer durations.

The power spectra in this experiment however were not perfectly flat. As durations decreased, there was a rise at the high frequency end of the functions. These results suggest that for shorter durations there was a deviation away from pink noise, with an increase in high level frequency components. Gilden, Thorton, and Mallon (1995) accounted for this by creating a two-stage time estimation model. In the first stage, an internal clock made of pink noise is used to judge the duration. Then, in a second stage, a motor program with white noise is used to make a response. They varied the proportion of pink and white noise, adding more white noise in for lower estimations to reflect the increased role of the response component in those conditions. The results of the model were power spectra that fit the empirical data quite well.

Gilden (1997) analyzed reaction times from many classic experiments in cognitive psychology such as mental rotation, lexical decision, and both serial and parallel search. Pink noise best described the power spectra for all of these tasks. He then fit the same pink and white noise model described previously for time estimation to these reaction times. The results again provided a very good fit to the data. The cognitive component of these tasks seems to consist of activity with more lower frequencies than higher frequencies while the response component seems to be simply random error as indicated by white noise.

Clayton and Frey (1997) had participants perform three tasks varying in memory load. They created a time series from the reaction times in each task and then looked at the resulting power spectra. The slopes became more negative, i.e., they deviated away from zero more, with increased memory load. If the error variability were randomly distributed, then the slopes would have all been closer to zero, indicating the presence of white noise. The results seem to imply that memory was calling on more high frequency mental processes, which would produce a pink noise signature.

Ward and Richard (2001) performed an experiment similar to that of Clayton and Frey (1997). They again had three conditions, each one demanding more cognitive resources. Only this time the task differed in the complexity of choice, requiring participants to choose between one-, two-, or four items. The task was modeled after a memory-scanning model in which differing numbers of items must be scanned in short-term memory (Sternberg, 1966). In this task, the scanning operation takes about 10-25 msecs and so corresponds to the intermediary time scale from Table 8.1. Like Clayton and Frey (1997), the slopes in their power spectra decreased with an increase in the number of items that had to be scanned. Because this resulted in more processing at the intermediary scale with higher frequencies, it can account for the gradual lessening of the slopes.

Noise as Therapy

Stimulation by visual, auditory or transcutaneous noise patterns has been found to provide relief from chronic pain from trauma, inflammation, herpes, cancer, and backpain (Takakura *et al.*, 1987). The noise in this study came from a digitally filtered Gaussian white noise source that was used to generate white and brown noise. The pink noise came from long period frequency changes of classical music. The Gaussian white noise was not effective and in some cases produced fear reactions, perhaps because it was too abrupt. The participants in this study also habituated rapidly to brown noise.

The pink noise though, was found to be more effective than a comparison medical condition of constant electrical current. The effect was improved further when the participants listened to the auditory version concurrently. The patients preferred the pink noise stimulation to the electrical current and the therapeutic effect was greater if the patients reported that they enjoyed the music.

What is the cause of this effect? Anderson and Mandell (1996) propose several explanations. They suggest that pink noise may be produce an increased release of endogenous opiates, or perhaps block conduction in pain pathways. It may also be the case that the stimulation interacts with the cellular and subcellular mechanisms that produce pain. Evidence for this latter explanation comes from a study showing beneficial effects of magnetic noise stimulation. Muzalevskaya *et al.* (1993) found this type of stimulation caused normalization of cardiac function and EEG, immune system response, prolonged survival following radiation exposure, and reduction in cancer growth, both in humans and animals. The exact nature of this effect is yet to be explained, but magnetic stimulation penetrates tissues deeply and may interact with the dipoles in neuron cell membranes.

CHAPTER 9
Dynamics and Psychology

Dynamics and Psychology

It is clear that the "new" paradigm of dynamics is well embraced within the broad scientific community. This approach is being used in many disciplines, including physics, climatology, chemistry, geology and biology. But what about psychology? In the 1990s, there was a surge of interest in chaos in psychology as witnessed by the formation of societies that sponsored conferences as well as the publication of several books. However, chaos has not constituted a major paradigm shift in psychology. Currently, only a small fraction of psychologists actively do nonlinear dynamical systems research. The vast majority of investigators continue to use the classical scientific approach.

In this chapter we first address reasons why psychology has failed to adopt the dynamical systems approach. Then we summarize some of the research in different areas of psychology that have been done using the dynamical approach. The areas covered are development, social behavior, and industrial-organizational psychology. Because psychology is such a diverse discipline with so many subdivisions, we reserve the chapter after this one for discussing clinically relevant areas like personality, abnormality and therapy. Then, in the chapters beyond that, we focus exclusively on dynamics and cognition.

The Failure of Psychology to Adopt the Dynamical Approach

Psychologists have not adopted dynamics, either theoretically or methodologically. Why is this? Combs & Winkler (1995) provide several reasons. First, traditional psychology has focused on structures rather than processes. Many cognitive models emphasize anatomy over physiology. Nowhere is this more evident than in the formation of cognitive models of memory. Each type of memory in these models is labeled using a box and assigned a different location in the brain. Witness the separate structures for iconic, working, long-term, declarative, procedural, semantic and episodic memories.

Mental "items" such as concepts, ideas and facts in cognition are represented in various formats like features, symbols, words, propositions and images (Friedenberg & Silverman, 2006). Each of these representations is believed to be static and stored in memory, where it can later be retrieved and matched against stimulus input or used in a computation. In the dynamical systems approach these representations are not static and unchanging. They are activated in different ways depending on the context and task and are constantly subject to alterations with new knowledge.

The structure vs. process debate is as old as psychology itself and was manifest very early in its history. Wilhelm Wundt and Edward Titchener led the voluntarist and structuralist movements respectively. They emphasized a cataloging of the elements of mind. William James instead argued for studying the "stream of consciousness". He saw mind as a flow and thought that processes were more important than representations. Chaos falls squarely on the functionalist side of this debate and asks that we rethink our traditional view of representation. Spivey (2007) describes what a dynamical representation might be like in the chapter on neurodynamics.

A second major reason why psychology has not adopted chaos has to do with methodology. Psychology uses the traditional scientific method, where manipulations are done on different conditions in an experiment and then outcomes are measured. Statistical analyses are performed on the data to determine if there are any differences. If there are, it is concluded that the manipulated variable caused the change.

The dynamic systems approach does not rely entirely on traditional statistical techniques. The data are organized as a time series and then mathematical methods are applied to determine how the measures change over time. Variables are then found that form the dimensions of the space within which these changes occur. Oftentimes, a graphical depiction of this space is needed. Many other analyses can be done as well.

Unfortunately, these methods for determining the presence of chaos, such as fractal dimension estimates, Lyapunov exponents and phase portraits, if not performed under ideal conditions, can lead to less than reliable conclusions (Grassberger, 1986; Osborne & Provenzale, 1989; Theiler, 1991). To make matters worse, time series analyses often require a large number of data points, on the order of several thousand. This is far more than what is obtained in the typical psychology experiment (Ward, 2002). While psychologists are trained in the use of statistical methods such as correlation, regression, and analysis of variance, there are few formal training procedures available to learn the more advanced mathematics of dynamical modeling. To add insult to injury, even when the above conditions are satisfied, it can be very difficult to detect the presence of deterministic chaos in human data. In many cases, chaotic trends cannot be distinguished from random ones.

Psychologists and nonlinear dynamical modelers treat noise very differently. To the psychologist, noise is a nuisance variable. It is considered as error, what cannot be accounted for by the independent variable. Noise is treated as variance produced by other uncontrolled variables that must be kept constant in order to minimize their effects. In chaos theory, noise is extremely important and not something to be thrown away. Noise is considered to be part of the brain or mental mechanism being studied. Analysis of it can reveal interesting phenomena. There are specific techniques that one can use to study noise such as power spectra. These are discussed in the section on this topic.

Another pervasive debate in psychology is whether to study individuals or groups. *Idiographic analysis* examines the behavior of a single subject or few subjects over time. *Nomothetic analysis* examines two or more groups at a single point in time (Tryon, 1995). Although there are case studies of change in single participants in the clinical and developmental areas, the vast majority of research in psychology is nomothetic. In fact, most statistical tests are designed to look for such differences. Dynamical techniques are instead inherently idiographic and set up to measure individual change over time. Methods such as phase portraits are good at displaying complex temporal deviations and lend themselves well to the study of certain psychological phenomena like mood fluctuations where portraits are usually different for each individual (Winkler *et al.*, 1991; Winkler, Combs & Daley, 1994; Sabelli *et al.*, 1995).

Hypothesis Testing

Yet another difference between psychological and dynamical methods centers on hypothesis testing. In psychology and in the traditional scientific approach, one starts with a hypothesis and derives from it a prediction. This prediction is then tested with established statistical tests. In the dynamical approach there are no hypotheses. Instead, one attempts to understand a pattern of behavior. Hypotheses are then built around this pattern (Barton, 1994). One could for instance hypothesize what combination of variables in the state space would give rise to an attractor and what the shape and dynamics of that attractor will be.

Barton (1994) recommends three approaches in regards to nonlinear data analysis. First, if it is possible to model a dynamical system using nonlinear equations, then researchers by all means should attempt to do so. If a system can be described in a more simple linear fashion, then it should be. Nonlinear methods have shown to be useful in modeling neural systems (Freeman, 1990) while cellular automata and multi-agent models have proven fruitful in social psychology and economics. The choice of what method to adopt will be influenced by a number of factors, including what equations best describe the data, what methods have been used in that particular area of research previously, and the goals of the investigators.

Second, if it is possible to reconstruct a nonlinear attractor based on time series data, then this should be tried as well. As we have already discussed, researchers measuring EEG brain data in a variety of conditions have used this approach. There are a number of attendant problems associated with this technique, including estimating the true dimensionality of the data and the use of appropriate controls. Please the section on EEG for more on this.

Finally, Barton (1994) mentions that mathematical techniques focusing only on cyclical patterns are unlikely to be of much use. One needs to get at the dynamic processes that underlie and generate the observed oscillations. These provide a more satisfying and theoretical level of explanation and can sometimes even by analyzed using traditional statistical procedures (Wolff, 1987). In summary, Barton (1994) concludes that linear and nonlinear methods are compatible rather than contradictory and can be used to analyze different aspects of a system.

Development

Traditional and Dynamical Accounts of Development

Esther Thelen and Linda B. Smith in their book, *A Dynamical Systems Approach to the Development of Cognition and Action*, lay out a comprehensive and detailed theory of dynamical systems and development. They begin by refuting many of the major theoretical positions in development. One of these is the notion that development proceeds according to a plan that is somehow programmed into the organism. In this view, the growing child "reads off" maturational instructions that are encoded genetically or as information in the central nervous system. These instructions tell the organism how to perform certain actions such as reaching, grasping, or walking.

This view is untenable for a number of reasons (Thelen & Smith, 1994). It begs the question of where the instructions come from. If they are embedded in the genes, it is not clear how they are represented there and used by the organism. In fact, the genetic code vastly under-specifies even an organism's anatomical structure. It is now acknowledged that much of the biological complexity (structural and functional) inherent in animals comes from the unfolding of developmental processes (Holland, 1995). For example, cells differentiate and form into layers, organs and other structures based on chemical gradients and interactions with cells that have already been produced, not because they are following a set of step-by-step instructions.

It is also unlikely that there are pre-existing and specific instructions for how to perform actions that are embedded in the central nervous system. In the traditional view, these sorts of instructions take the form of a *central pattern generator* (CPG) which we discuss elsewhere in this book. A CPG is essentially a pattern of neural activation that when turned on can drive certain behaviors, for example by telling the muscles of a limb how to move when walking.

CPGs in the traditional view are fixed; they always produce the same pattern. How then can they explain the flexible and adaptive quality of behavior? Walking is hardly ever just the repetitive execution of muscles in a set order. It varies depending on such factors as the incline of the terrain, obstacles, and speed. A different neural pattern is required for each of these situations. Behaviors like walking instead seem to be regulated by the interaction between multiple body systems like the sensory nerves and the limbs as well as spinal cord and brain action (Cohen, 1992). Development of action as well as its regulation in the mature animal is controlled not by a single process like a CPG or motor code but instead by the interplay of many different processes.

Thelen and Smith (1994) summarize their views on the development of behavior in three points. First, any behavior such as walking should not be considered as unitary. It consists of many different subcomponents that must all interact in the right way to produce the behavior. Walking involves not just the ability to move the legs but other skills such as the intention to move and the capacity to support one's weight.

Second, each of these components are distinct and can develop at different rates. Although the ability to move the legs in a rhythmic stepping motion is present early in the first year of life, babies cannot bear their full body weight until much later. This means that all the components of a behavior must have matured before it can be expressed. In some cases, a single component such as bone density or muscle strength can underlie multiple behaviors and may limit their expression because it requires more time to grow.

Third, all behavior is context specific. Whether an infant shows walking behavior depends in large part on the environment and the way it has to be expressed. Walking movements may or may not be possible depending on the infant's emotional state, whether they are lying down or standing up, whether they are supported, and a variety of other factors.

The Time Course of Development

Development takes place over many time scales. Neural changes are measured in seconds or less. Motor behavior occurs over periods of seconds and minutes. The development of more complex abilities like walking takes even longer, over months or years. Because changes in any capacity reflect changes across numerous subcomponents, it is best to examine developmental change multi-dimensionally. When a given ability appears depends on the development of those systems upon which it is dependent.

As we have seen, most of the change that takes place in nature is nonlinear. Although some developmental changes may be linear, most are not. Stage theories of development such as those of Piaget postulate plateaus where very little change occurs followed by steep rises where a large reorganization takes place. These sudden changes may be thought of as phase shifts. Other theories postulate more gradual and incremental change. In reality, the maturational time course of different abilities is more complex than these possibilities and may consist of fluctuations, oscillations, and other trajectories (see Figure 9.1).

So what determines when a given ability will appear? That depends on the maturity of its components. The capacity to reach and grasp an object requires depth perception but also motor coordination. If neither of these capabilities has developed enough, the infant will be unable to perform the action. If we imagine that a complex Behavior D requires minimal levels of development in three component skills A, B, and C, then it cannot be expressed until all three have reached that level.

Thelen and Smith (1994) suggest that we think of change in terms of attractors that evolve and disintegrate. When the child is in a period of relative stability we can say their behavior is characterized by a system captured by an attractor. As change takes place, this attractor becomes unstable and eventually disappears only to be replaced with a different attractor. We should then see fluctuations and instability until a new attractor state is reached. During times of transition, the child's behavior is particularly susceptible to small changes in the organism, task, or environment.

We can illustrate this with the phases of behavior that to lead up to walking. When a child is crawling, they are in a relatively stable attractor state. Each of the component systems contributing to crawling has sufficiently developed. Change in one or more of these systems then allows the child to stand. During the learning to stand period, the child's behavior is more variable and suscep-

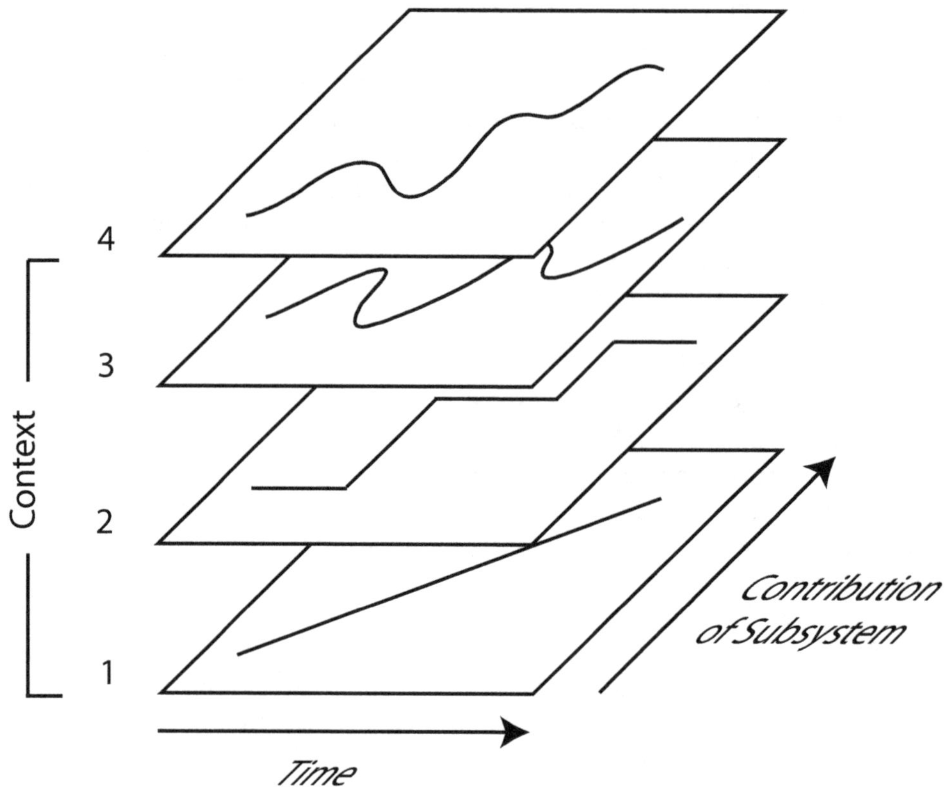

Figure 9.1 *A complex behavior may require a minimal level of development in component skills before it can be expressed (after Thelen & Smith, 1994).*

tible to environmental change. They may sway back and forth and fall down if bumped or distracted. Once the child is capable of standing, they may be considered as having reached a new attractor and are less likely to fall. However, this attractor, like the one before it, will soon dissolve and give way to the new attractor of upright stepping.

Social Behavior

Dynamics in Social Psychology

Social psychology is a field well suited to the application of dynamical methods. Since its inception, there has been a focus on dynamics, whether these be the dynamics of groups, interpersonal processes or attitude changes (Nowak & Vallacher, 1998). The introduction of the dynamical systems view in science has produced a spate of research projects from this perspective over the past two decades or so. Many topics, including social influence (Nowak, Szamrej & Latane, 1990) altruism and cooperation (Messick & Liebrand, 1995), and group dynamics (Losada & Markovich, 1990) have been studied using this approach.

Social processes are in constant flux as people communicate, act, are affected by and affect in turn those around them. Social psychology is in many cases concerned with the study of emergent phenomena that arise from multiple interacting levels. One can analyze social phenomena from the level of the individual, the family, the school, the corporation, the society as a whole, etc.

Societies are complex and thus need to be analyzed with complex methods. Modeling is particularly useful here. Individuals can be considered as agents interacting in a multi-agent system, so the application of multi-agent models can explain many emergent phenomena. Cellular automata can also be fruitfully applied. Grid elements and interaction rules in these simulations can represent people and they way they interrelate. In addition, connectionist models of social behavior have been formulated where nodes stand for individuals and connections represent social relations. Nowak and Vallacher (1998) mention that person perception, stereotyping, societal categorization, causal attribution, attitudes and beliefs, and social influence have all been studied as connectionist models (Kunda & Thagard, 1996; Read & Miller, 1998; Smith, 1996).

Social phenomena can be both ordered and disordered. The behavior of many social relations for example is periodic (Newtson, 1994). Most human activities show daily and also weekly periodic structure. One interesting finding shows that people who are poorly connected with others demonstrate greater periodicity in their social contacts (Nezlek, 1993; Nezlek & Wheeler, 1984). Social behavior can also be characterized using more complex attractor states.

Dyadic Processes

A *dyadic relationship* is simply that between two individuals. A number of such relationships can exist, from coworkers to friends, to romantic partners to parent-child relations. All of these relationships require some form of coordination where the couple coordinate and synchronize their behaviors toward one another. This coordination can take many forms, from turn taking in conversation, to the relative timing of household duties such as childcare and shopping.

Nowak and Vallacher (1998) outline the basic forms of behavioral coordination in couples. It can be positive in which case the behavior or cognitive/emotional state of one person induces a similar state in the other person. It can also be negative, where such states induce opposite effects in the other partner. This latter case can occur at the end of a relationship when one person might become happy at the other person's sadness. Coordination can also shift abruptly between being in phase or out of phase, demonstrating hysteresis transitions.

Behavior coordination between dyads can take place at many different time scales (Nowak & Vallacher, 1998). The examples they give include in-phase coordination like eating dinner together or watching a movie. They can also include out of phase behaviors such as taking turns using the bathroom. There are of course more complicated forms of coordinated action. A couple may follow a pre-set plan as to who will go shopping and clean the house prior to the arrival of visitors. A division of labor is also possible, with a husband who always agrees to do the laundry and a wife who always makes dinner.

Hysteresis in Relationships

How close two people in a relationship are depends crucially on their history. If a couple has been dating for some time, they are less likely to split and it may take a large degree of dissatisfaction to get them to do so (Rusbult, 1980, 1983; Rusbult & Martz, 1995). The opposite situation also holds. It may take a lot of dating and familiarity before a couple decides to marry. In each case, we have a variable such as mutual satisfaction that must take on either low or high values before a relationship dissolves or forms, respectively. This pattern of behavior is hysteresis.

Tesser (1980; Tesser & Achee, 1994) have created a model to study hysteresis in close relationships. This model is depicted in Figure 4.13 and was discussed earlier. If you will recall there are two control parameters driving change in their model. These are the social pressure against the couple being together and the couple's emotional involvement with one another. The main dependent variable was the strength of dating-mating behavior. In the model, when there is low social pressure against dating then dating behavior increases linearly with increased emotional involvement. However, when there is high pressure against dating, no dating behavior occurs until a critical high level of emotional involvement.

Dynamics of Romantic Relationships

Rinaldi and Gragnani (1998) have modeled dyadic processes, those that occur between two individuals. In a number of studies they have applied nonlinear dynamical systems techniques toward the understanding of romantic relationships. Their models describe the course of a love relationship between a couple and can mimic the temporal changes that occur, such as initial starting points, trajectories, stability and fluctuation. They have several different classes of model, based on the personality characteristics of the two individuals.

In one of their earlier nonlinear models, they assume that both the man and woman in the romantic relationship are securely attached, meaning they feel confident about and trusting of their partner. They also assume that both members of the pair are non-synergic. Synergism is the extent that an individual's reactions are enhanced by love. The equations take three primary factors into account: the pleasure of being loved (return), the reaction to partner's appeal (instinct) and the forgetting process (oblivion). These factors were assumed to capture the essential dynamical aspects of any romantic relationship.

The dynamics of the model are shown in Figure 9.2. The dimensions of this space are x_1 and x_2, a measure of the love of individual 1 and 2 for their partner. This variable x can be positive, zero, or negative. Positive values reflect positive feelings that can range from friendship to passion. Negative values indicate negative feelings varying from antagonism to disdain. A zero stands for complete indifference.

When the couple is secure and non-synergetic, there are no limit cycles. Depending on the initial starting points, the relationships end in one of two fixed point attractors. Figure 9.2 shows the trajectories of each individual for a robust couple who converge onto point $E+$, representing mutually positive feelings toward one another. It also depicts a fragile couple who converge toward point $E-$, representing mutually antagonistic feelings. The dashed line in Figure 9.2 shows the boundary for the two basins of attraction.

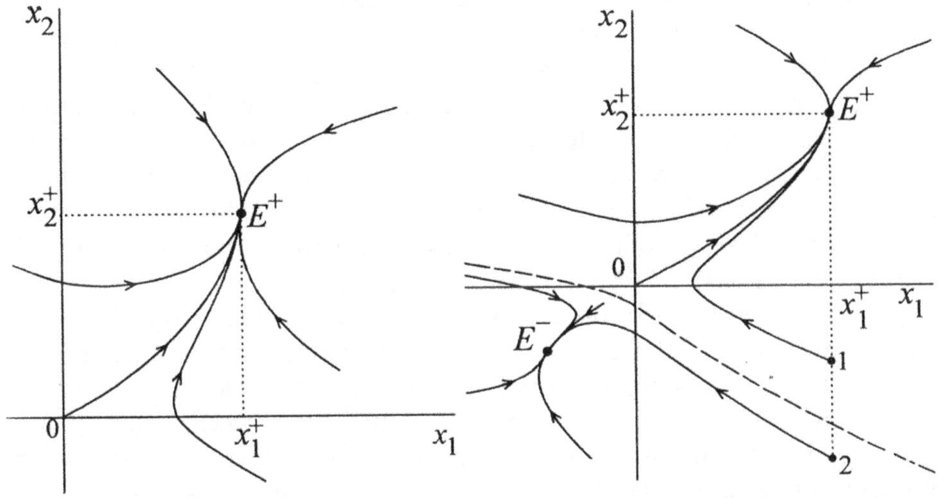

Figure 9.2 *A dynamical model of romantic relationships (Rinaldi & Gragnani, 1998). The left figure shows the evolution of feelings in a robust couple. The right figure depicts the dynamics for a fragile couple. The dashed line indicates the boundary of two basins of attraction. Permission to reproduce figure granted by World Scientific Publications.*

A couple is robust if they trend toward the positive outcome even after a relatively large displacement. As long as the disturbance does not decrease their love below the boundary between the two basins, they will gravitate back to the high level of mutual love at point *E*+. However, if the disturbance is great enough to bring them into the other basin of attraction, they will move and then stabilize at point *E*-. A fragile couple is one that is unable to recover from this kind of disturbance.

Rinaldi and Gragnani (1998) ran the model for other different combinations of personality types. A non-secure, non-synergetic couple and a secure synergetic couple also show simple dynamics without cyclical patterns of behavior. However, when one member of a couple is secure and synergetic and the other is non-secure and non-synergetic the landscape of possible trajectories becomes more complex. Depending on initial starting conditions, a relationship can now be either steady, moving toward one of several fixed attractors, or cyclical, moving as a limit cycle.

These models have been able to capture much of the dynamical nature of romantic relationships. Perturbation and return to a stable positive or negative state occur often in real world couples. It also explains why relationships can be so volatile, that being the presence of synergism or insecurity within the couple. However, the model fails to take into consideration many other factors that impact on relationships, namely other personality characteristics such as extroversion, environmental factors such as income level, and long term changes such as aging. The authors point out that men and woman could be modeled using different state equations and that the dimensionality of the model could be expanded to account for the dynamics of larger groups of individuals. Several other examples of nonlinear dynamics in social relationships exist.

Cellular Automata Models of Social Influence

Nowak *et al.* (1990, 1993) have used a CA model to help understand the spread of social influence. For definitions and other examples of CA the reader is referred to the model section of this book. In the Nowak model, each individual is characterized by three variables: their opinion on a topic, persuasive strength or ability to convert others to their opinion, and their location in a geometric space. In the basic version, each individual occupies a box location in a two-dimensional grid. The color of each box corresponds to a person's opinion, and the height of each box to the strength of their opinion.

People interact with those around them. Any person's opinion is influenced first and foremost by his or her own opinion, but then secondarily by the opinions of their neighbors, weighted by distance so that closer neighbors have more influence. A simulation run consists of picking an individual at random and calculating influence for each opinion in the group. If the resulting strength of an opinion is greater than the strength of an individual's own opinion, their opinion changes to that of the group. This procedure is then iterated for everyone in the group and continues until there are no more changes.

The results are characterized by polarization and clustering. Polarization occurs when a majority of the group adopts a single position. Clustering occurs when there are small groupings of those holding a minority opinion. Both of these effects are mirrored in real life (Noelle-Neumann, 1984; Festinger *et al.*, 1950).

Nowak *et al.* (1993, 1996) discuss several situations that give rise to clustering. In the leaders and followers situation a circle of followers surrounds a leader. The leader and the followers mutually reinforce each other's influence and so are shielded from the surrounding majority. In the stronghold arrangement weaker individuals ring a group of leaders. This is also a very stable arrangement. If one of the leaders changes their mind, they are quickly reconverted back to the local group's opinion. In a wall situation, leaders are aligned along the edge of a cluster and protect the weaker interior members from outside influence.

There are three control parameters in this CA model that when manipulated cause significant qualitative changes in the dynamics (Nowak & Vallacher, 1998). These are bias, noise, and self-influence. Bias refers to how desirable an opinion is. If one opinion is vastly more desirable than another it would presumably spread rapidly and everyone in the group would be converted. When bias for two or more opinions are approximately equal, then the other two parameters exert their influence.

Noise is the equivalent of some outside influence such as a magazine article or news clip that could sway opinion. Small amounts of noise may be enough to change the opinion of weak individuals but if they are near a leader, their opinion will revert back and the overall configuration of opinions remains unchanged. If noise is powerful enough though, it could sway even strong-minded individuals and lead to the eventually dissolution of minority clusters. Self-influence refers to the relative strength of a person's belief. A "stubborn bastard" or person with a very high belief value could resist the effects of noise or an adjacent leader with different views.

Industrial and Organizational Psychology

Organizational Change

Industrial and organizational (I/O) psychology involves the application of psychological theory, method, and intervention to the workplace. The goal in this discipline is to make organizations more productive while at the same time ensuring employee happiness and satisfaction. Organizations can be of any sort, such as corporations, companies, schools, and governments. There are a variety of topics addressed, including motivation, leadership, employee selection, training and development and organizational behavior. The field of *organizational development* can be considered a sub-discipline of I/O psychology. Its focus is on how organizations change over time.

The traditional approach to examining change in organizations postulates two states. There is a period of stability with little change or innovation. Lewin (1947) referred to these constant periods as quasi-stationary equilibrium. They are interrupted by more unstable chaotic periods where change and restructuring occurs. A more fruitful paradigm, and one that sits better with empirical studies, is the dynamical systems view of organizational change. Here, the concepts of control and order parameters, bifurcation, attractors, and chaos are applied to the study of organization change (Guastello, Dooley, & Goldstein, 1995). Change here can take many forms and is not seen as a simple alternation between order and disorder.

The dynamical approach to organizations sees them as self-organizing. In this view, organizations are seen as neither resisting change nor is change necessarily forced upon them from outside influences. Instead, organizations become susceptible to change when they are under far-from-equilibrium conditions. For example, a domestic company that manufactures athletic shoes may be pushed into a new organizational state when foreign tariffs are removed, allowing an influx of cheap competing products produced in China. The domestic company can then move to a number of different possible states. It could for example reduce operating costs by laying off or reassigning personnel, or it might merge with another company to become more competitive.

The Dynamics and Structure of Organizational Change

Meyer, Gaba, and Colwell (2005) argue that contemporary organizational science is still fixated on the notions of equilibrium and linearity. They encourage researchers in the field to analyze change within the principles of complexity and network theory. In previous studies, when examining hospitals and technology corporations they found that change never occurred in a linear manner. They instead documented discontinuous spike-like jolts, turbulent step functions, and oscillatory boom-and-bust cycles (Meyer, 1982; Meyer *et al.*, 1990; Gaba & Meyer, 2005).

Sudden and dramatic changes in policy or economic conditions can push organizations away from any possible equilibrium and can serve as natural opportunities for them to adopt unorthodox moves and learn new ways of coping. They found that organizations would often react at multiple levels in such situations, adopting changes at the organization, industry, and inter-organizational levels. For example, formerly competing hospitals collaborated and formed symbiotic networks when a malpractice insurance carrier cancelled a large number of California physician's group coverage. In addition, they found that change could start at small scales (individual persons) and work its way upward to large scales (the field as a whole), but that this direction could also be reversed, with change happening globally and percolating downward through the system.

Meyer, Gaba, and Colwell (2005) also examined the coevolution between venture capital and technology firms. They found that investments made in each population moved in cyclical patterns, where each affected the other but was also linked to movements in the capital markets. There was a positive feedback relationship between private and corporate venture capital that fueled an investment boom. However, they note that positive feedback mechanisms, if they continue for too long, can cause an economic crash and in fact can accelerate a decline in investment. It is probably the case that in markets, both positive feedback and negative feedback mechanisms exist, the former driving dynamic change, the latter promoting stability.

Colwell (2003) performed a network analysis of the nanotech investing community. He found that the links among nodes in this network were distributed according to a power law with hubs. Nodes included organizations such as nanotech start up firms, government agencies, universities, and professional associations. This organization was not imposed by a central controller but emerged from the internal characteristics of the network itself. He concludes that the nanotech investing community is a complex adaptive system organized as a scale-free network with small-world properties. See the section on network science for more on these topics.

Organizational Decision Making

Guastello (1992) examined staffing decisions under conditions where there were few qualified job applicants. He found that winning strategies involved hiring a diverse workforce and being prepared for fluctuations in the number of employees, their performance, and higher termination rates. Organizations whose staffing policies remained constant went out of business. Successful policies in comparison changed their selection and training procedures about every four years.

In this same study, Guastello (1992) found that the workforce census of a small construction contractor oscillated according to the following equation:

$$N_{t+1} = bN_t e^{-aN}$$

N was the number of employees, b was a bifurcation parameter that was a function of the hiring rate, and a was a crowding parameter. The value t of course stands for different time steps. A more stable census was obtained from long range staffing policies that took future staffing needs into account as opposed to a policy where new members were hired only as openings became available.

Work Performance

Guastello (1987) describes a catastrophe cusp model of job performance. The dimensions of ability and task variety form the two dimensions of the control surface. When task variety is low, increases in training for example, produce gradual and continuous increases in ability level. This can be seen as the back portion of Figure 9.3. However, when task variety is high, performance level will increase gradually with training until an intermediate level after which there is a sudden jump to a much higher performance level. The cusp is the region where this shift occurs. In these types of models, ability is an asymmetry parameter, so named because changes along it define movement from low to high levels of performance. Task variety is the bifurcation parameter.

The stability of job performance over time is more variable when rewards are not tied to performance and when there is more task variability. This is indicated in Figure 9.3 as an orbit shown by the larger circle on the control surface. This situation can be found under conditions of extrinsic motivation as

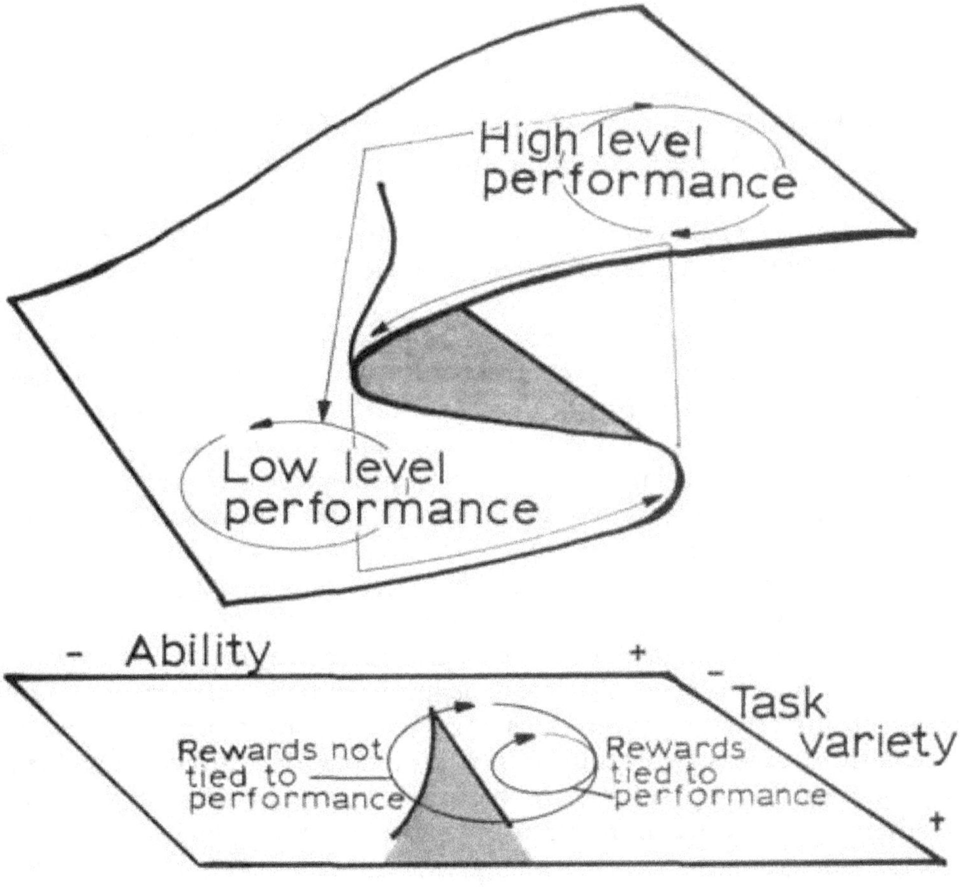

Figure 9.3 *A catastrophe cusp model of job performance. Guastello, S. J. A Butterfly Catastrophe Model of Motivation in Organizations: Academic Performance. The American Psychological Association. Washington, D. C. 1987. Adapted with permission.*

in the case of being paid hourly versus piecework rewards. Job performance is less variable when rewards are tied to performance and when there is less task variability. This corresponds to the smaller orbit and circle on the control surface in Figure 9.3 and occurs for intrinsic motivation, as when one is interested in the task.

The cusp model described above is just one subset of a more general butterfly catastrophe model for motivation and performance dynamics in organizations (Guastello, 1981, 1987). In this model, work behavior is described as a five-dimensional geometry with three attractor states separated by two repellors. The attractor states make up regions of the performance variable. They are innovation, competence, and high performance as the first, adequate attendance and performance as the second, and inadequate performance, high absenteeism, and imminent termination as the third. There are four control parameters. These are ability, intrinsic and extrinsic motivation, and a management parameter. The management parameter is a continuum running from being rigid and authoritarian on one end to flexible and tolerant of individual differences at the other end.

Leadership Emergence

Guastello (2007) makes the important point that there is no single type of leader. Instead, a social structure emerges made up of primary leaders, secondary leaders, and nonleaders. He defines primary leaders as those who apply a variety of social skills to a situation while secondary leaders are those who play more specific roles. The behavior of a leader is context-specific and depends on the group's goal, whether that is production, creative problem solving, or coordination-intensive activity.

The distribution of leaders and nonleaders is determined by a swallowtail catastrophe with three control parameters (Guastello, 2007; Guastello & Bond (2007). The asymmetry parameter differentiates the leaders from the nonleaders while the bias parameter distinguishes between the two leadership types. Bifurcation determines how a person will evolve into either a primary or secondary leader. All three of these parameters work in conjunction to decide how leadership emerges in a group.

Asymmetry consists of several behaviors of which the most important to problem solving and creative groups is control of the task. Task control also makes up the bias parameter for coordination-intensive groups. For production-oriented groups, the most significant asymmetry behaviors are promotion of optimism and acting to reduce tension. The bifurcation parameter manifests differently for each type of group. It is creative input for creative groups, controlling aspects of the work situation for production groups, and is mostly limited to whether the group can communicate openly or in a more constrained fashion for coordination groups.

Guastello (2002) concludes by saying that leadership emergence occurs under conditions of uncertainty when a group or organization is under strong pressures to produce a new result. In these situations, there is often a good deal of unstructured interaction between participants. At a certain critical level, the group self-organizes and becomes ordered by producing leaders or perhaps by producing some other changes like stratifying into a hierarchical organization. This change can occur suddenly and is best characterized as a phase shift.

CHAPTER 10
PERSONALITY, PATHOLOGY, AND THERAPY

Lewis and Junyk (1997) outline how personality self-organizes over time. Their ideas are based on the application of dynamical systems principles. They first propose the concept of a *personality attractor*, defined as the familiar, predictable states people exhibit in their social and emotional behavior. In their scheme, a personality attractor replaces the previous psychological concept of a trait. A *trait* is a characteristic pattern of behavior or a disposition to feel and act a certain way. Extraversion is an example of a trait. People who are extraverted are sociable, fun loving, and affectionate. They will be more likely to have a large number of friends, go to amusement parks and kiss and hug in public.

Traits are persistent and enduring. They explain how a person thinks, feels, and acts throughout his or her lifetime and are believed to vary little with the situation. If a person was calm and level headed, they could be expected to act this way in most situations, including those that would normally induce anxiety, such as taking an examination. Unlike traits, however, a personality attractor is not independent of the situation in which it occurs. They are best thought of as temporary resting points for ongoing personality self-organization.

A personality attractor can be global, such as anxiety, or it can be content-specific, such as test-anxiety. In state space these can be represented as wells. The global attractor is the larger well, with the more specific, context-based attractor existing as a smaller well inside of it. The system tends to gravity toward these states. In contrast, there are also transient or temporary states that people don't experience for very long. These are called *personality repellors*. A stingy person, for example, might one day feel generous and buy an expensive gift for a friend, but this state wouldn't last long. A personality repellor can also be an internal state that is avoided because it causes the individual pain or suffering. This notion of a defense is discussed in greater detail elsewhere.

The personality landscape can additionally include unpredictable transitions between stable attractor states. These are regions where unstable, chaotic behavior occurs. The system in these regions is transiting from one mood or state of mind to another. A child that is hungry may alternate between sadness/crying and anger/screaming as well as other states before finally "giving up" and accepting the situation. Transitions can also occur when new personality patterns develop. In these cases, the system becomes sensitive to fluctuations, deviates substantially from previously established stable states, and ultimately settles down into a novel pattern.

A Model of Personality Self-Organization

Lewis (1995) outlines a general model of personality self-organization. In this model cognitions and emotions mutually affect one another through a feedback process and ultimately become coupled together. The cognitive elements include thoughts and beliefs, while the emotional elements can be states such as happiness, sadness, and anxiety. The process is governed by appraisal. *Appraisal* is the cognitive monitoring, interpretation and evaluation, both of what is going on in our external environment, but also of what transpires in our internal mental world. Appraisal is a constant, ongoing process.

Appraisal automatically triggers emotions when the current situation is relevant to a person and their goals. These emotions then guide perceptual and cognitive action by directing attention to relevant features of the environment and motivating a response. The results of the action in turn feedback and influence both the appraisal process and the emotions. This cyclical process is shown in Figure 10.1.

Let's illustrate the model with an example. John is an undergraduate student attending a small liberal arts college. In the past, he has done well on his exams, receiving straight "As". He gets a grade of "D" on his first calculus exam. This causes him to appraise the situation. The appraisal consists of think-

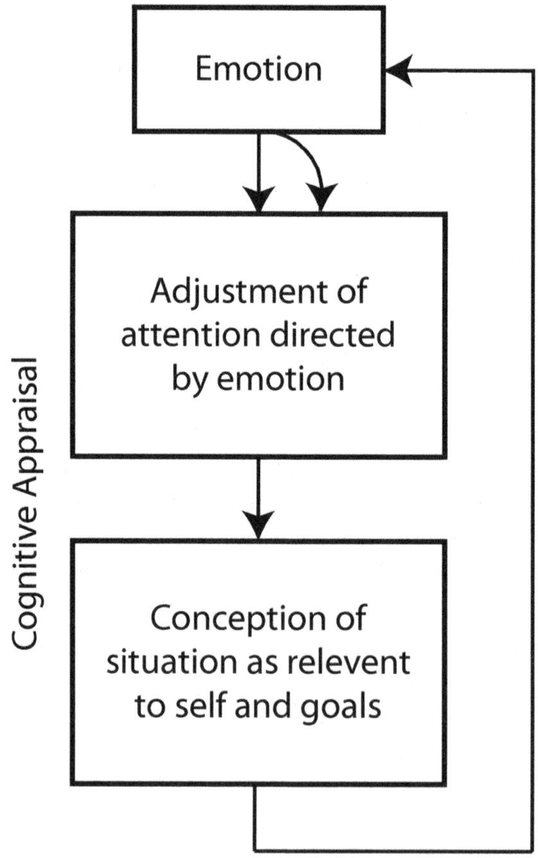

Figure 10.1 *A self-organizing model of personality based on appraisal (by Lewis, 1995. Copyright 1995 by S. Karger AG, Basel).*

ing about why he did so poorly. He concludes that it was because he did not study sufficiently. The appraisal also triggers feelings of shame and inadequacy because his self-concept is that of a good student and his goals are to obtain a 4.0 G.P.A. These emotions cause John to pay more attention in class and to spend more time studying for the course.

On his next calculus exam John gets a grade of "B". This causes new appraisal, emotional, and behavioral responses. He is relieved that he did better, but still feels that he can do better. He further alters his study habits, hires a tutor and attends review sessions. John's self-concept as a result of these experiences is now permanently changed. He considers himself to be a good student, but with trouble understanding calculus. His expectations are also changed in that he anticipates getting "As" in each of his classes with the exception of calculus and perhaps some other math courses.

In this model nothing remains static. John's appraisal, emotional reactions, and behaviors are altered with each cycle or iteration. The components of the model are dynamic and continually updated based on experience. Also, although we are considering cognitions, emotions, and actions separately, they are actually linked together. John's self-concept of being a good student will automatically and inevitably activate feelings of shame if he does poorly. These elements become coupled, providing a coherent way of responding to situations. Finally, note that change in this scenario is being driven by disequilibrium. It is the mismatch between expectations and reality that motivates behavior and changes in personality.

Psychopathology

Psychosis and Dynamics

Schmid (1998) proposes a new approach to psychiatry based on nonlinear dynamics. His system, called Process-Oriented PSYchiatry (POPSY) focuses on a time series evaluation of a disturbance and its treatment. It would allow for mathematical models of psychological disorders as well as the potential for short-term individual prognosis and therapy.

POPSY has many similarities to chaos theory. When studying chaos, we attempt to identify chaos by performing an analysis in state space. We then quantify chaos through identification of attractors. Then there is predicting chaos in the form of forecasting and finally controlling chaos in the sense of being able to manipulate a system's parameters to regulate attractor stability.

When approaching psychopathology we can also identify these same four features. We must identify the disorder in terms of a state space of relevant variables. We must then perform a diagnosis in terms of the presence of an abnormal attractor or attractors. A prognosis maps out a time course of the disturbance. Treatment or therapy involves control over appropriate parameters or factors in the individual's life to alter the dynamics of the disorder.

Schmid (1998) draws an analogy between psychosis and chaos. He lists the defining features of chaos and shows that the equivalent of each of these exists in psychosis. First, there is the outbreak of the disease or its remission if it has already occurred. This outbreak is equivalent to an "attractor jump" from a healthy pattern of behavior to a pathological one. Second, as is the case with chaos, it may be easy to diagnose a change in behavior but very difficult to predict its future outcome. Short-term predictability ought to be possible.

Third, we see sensitivity to initial conditions. A small change in a starting situation can produce large "downstream" effects. Psychotic individuals are extremely sensitive. They can be easily frightened or disturbed in response to stimuli such a remark someone might make about them. We also see recursion or feedback in psychotics in that a change in their internal or external environment produces an abnormal response. This manifests itself as vigilance to the outside environment and cyclical transitions between psychotic and normal episodes.

Instability is also demonstrated in psychotics in the form of sudden and unexpected reactive thoughts and behaviors such as free-associated thinking. Psychotics have difficulty making decisions and alternate between different thoughts and emotions. In other words, they can be characterized as ambivalent, confused, and disorganized. Disorganized thoughts and speech are in fact the hallmark of schizophrenia.

Finally, we see fractal structure in psychosis where there is self-similarity over different time scales. The course of a mental disorder can vary dynamically on a least three distinct time levels. At the social level it can be measured by medical or work records over the course of years to months. At the psychopathological or psychodynamic level of the individual it is measured over shorter periods lasting from months to seconds. At the bio-pathological level variation occurs over even shorter time spans of hours to milliseconds. According to Schmid, there is similarity in the dynamics between all three of these levels.

Dynamical Disease

Schmid (1998) next introduces the idea of a *dynamical disease*. This is a disease, either physical or psychological, where the normal pattern of functioning in the organism changes. This is replaced with a new abnormal functional organization. A dynamical disease is characterized by:

- Constant parameters developing into large amplitude oscillations;
- New periodicities arising in previously stable periodic processes;
- Rhythmic processes being replaced with constant or periodic dynamics;
- Chaotic or aperiodic behavior transitioning into constant, periodic, or other aperiodic dynamics.

Examples of dynamical diseases in the medical category can include gastrointestinal disorders like irritable bowel syndrome (IBS) as well as circadian hormonal, cardiac, and respiratory diseases. In the psychological category they include bipolar disorder, depression, and schizophrenia. In each of these cases, there is a healthy pattern of behavior characterized by recurrent activity such as heartbeat, breathing, hormone release, or psychological states.

It is important to note that these dysfunctions are not caused by the continued presence of a pathogen or environmental factor such as a virus, bacteria, or childhood sexual molestation. Rather, the new state occurs because of a shift from healthy to abnormal functioning in the organism. Parasites might cause IBS, but the symptoms continue after they are removed. Similarly, a natural disaster such as a tsunami might trigger post-traumatic stress disorder, but flashbacks, nightmares and anxiety continue long after the event is over.

Schmid (1998) also proposes that psychological dysfunction could be the result of a linear information-processing style. If a person's attention is too narrowly focused, an overly parallel stimulus environment could leave them overwhelmed. Such a person could not form gestalts effectively or make appropriate inferences. Schmid suggests that this type of linear cognitive system could not filter out irrelevant information or focus on essentials. The result could be stress, emotional disruption, hallucinations, delusions and other symptoms.

Psychological Disorders

Most traditional theories of psychological disorders acknowledge that they emerge over time. The behaviorist view is that maladaptive behaviors are conditioned. A person may develop a phobia because an aversive stimulus, like a slap across the face, is associated with a previously neutral stimulus such as the sight of the person administering the punishment. In the psychoanalytic paradigm, disorders arise as a result of fixations during development to different erogenous zones. A child that is overindulged or deprived of pleasure for a particular zone, such as the mouth, may develop maladaptive traits, such as excessive dependence or independence. In the cognitive view incorrect attributions, such as blaming yourself for failures like failing to obtain a raise, can result in depression.

Each of these theories explains disorders as a temporal process, whether it is the associations between two stimuli, the experience of pleasure, or attribution of blame in response to some incident. Implicit in these theories is the idea that dynamical processes create disorders. These processes typically involve a stimulus or environmental situation that is interpreted by an individual, producing a cognitive and/or emotional reaction. This in turn can produce a behavior that feeds back to influence their response, either directly through a change in thoughts or feelings, or indirectly through the alteration of their environment. Psychological disorders and personality traits, as we discuss elsewhere, are formed by a cyclical iteration over time between these elements. This means they can be modeled and understood within the dynamical systems framework.

A psychological disorder can even be defined in terms of change. Noam (1992) argues that the pain associated with psychopathology is due to a self-system that is encapsulated. This encapsulation or insulation causes it to resist transactions with the environment that might cause a reorganization toward a more healthy state. Implicit in this theory is the view that a healthy person must be open to change. They must be able to alter their internal cognitive and emotional structures in response to the ever-changing environment. As we have stated earlier, the mind is best viewed as an open system, one that includes reciprocal transactions with the environment.

The role of the therapist within the dynamical view is to "de-encapsulate" or "de-insulate" the individual. Once this resistance is broken down, they will be more receptive to environmental influences. Then, the natural chaos and variability that exists in the world will cause them to reorganize and create more complex and hopefully adaptive responses. However, it should be noted that exposure to especially chaotic environments can also have a negative effect. Unpredictability and lack of the daily routines and rhythms in life can also cause psychological disorder. Techniques such as waking up people up at random times during the night have been mentioned as one way of helping to extract information during torture.

Traditional versus Dynamical Views of Abnormality

A disorder in the traditional perspective is viewed as a fixed structure (i.e., a defense "mechanism" in psychoanalysis). The goal of therapy is to destroy this structure and replace it with a new one that is more adaptive. The dynamical perspective instead sees abnormality as a process rather than a structure. The goal here is to increase attunement to chaos and variability in the environment. Resonance to the environment should result in the creation of new responses that better enable a person to deal with change. People with chaotic schemas or emotional models are those best able to adapt to change.

In the traditional view, the therapist is the source of change. They provide insight into hidden mental processes or deliberately structure the conditioning features of the environment. In the dynamical view, the environment itself is the source of change. The central nervous system uses chaos to generate novel patterns of response (Freeman, 1991). In this understanding, environmental chaos is the input that drives the brain to produce better ways of dealing with novelty or problem situations.

Traditional views focus on the past. They see early childhood as the crucial time during which personality and abnormality are formed. Witness, for instance, the fixation-regression model in psychodynamics that posits the existence of critical periods or stages of personality formation. The dynamical view instead states that reorganization can happen at any point during the lifespan. There is evidence that this is in fact the case (Colarusso & Nemiroff, 1987).

Another way the traditional and dynamical views part ways concerns their understanding of equilibrium. The traditional account equates equilibrium and stability with mental health. Disequilibrium and instability are the source of disorder. The dynamical perspective turns this notion on its head. It sees disequilibrium and adaptability as necessary for mental health. Internal chaos is not a pathological state, but a state of maximum readiness for an emerging self-organized system.

Defenses

A *psychological defense* is a thought or behavior intended to prevent or curtail painful emotional states. The idea first originated in psychodynamic theory but has been extended to other areas in psychology, including social-personality (Carver & Scheier, 1994) and emotion theory (Folkman & Lazarus, 1985). Sigmund Freud named a number of defense mechanisms. In repression, one banishes anxiety-arousing thoughts, feelings, and memories from conscious awareness. In projection, people disguise their own threatening impulses by attributing them to others. Defenses are a sign of abnormality. Although they represent an adaptive solution to a problem, they often create more problems than they solve. They can impair personal functioning as well as social relationships.

Defenses can be considered special cases of the personality attractors discussed previously (Lewis & Junyk, 1997). They can arise the same way personality traits do, as a coupling between thoughts and feelings driven by cognitive appraisal to ongoing events. Because they arose as a function of avoiding intense negative emotional states, they probably lie near them on the personality landscape. In the self-organizing personality model, a defense starts with a painful appraisal of a situation, such as a fear of failure, which if experienced repeatedly, produces a corresponding attractor. Repeated exposure also results in an anticipation of the negative appraisal. The emotion attached to this anticipation is anxiety, as when one fears having to give a speech in front of many people. This produces a *core apprehension,* a large attractor that encompasses the original attractor (Figure 10.2).

In this way, the attractor basins around painful events are widened and can include many other attractors and in some cases, perhaps the entire landscape. This can explain generalized anxiety disorder, where anxiety is not attached to any one specific stimulus. People with this disorder are said to have "free floating" anxiety, because they seem to worry about everything. In comparison, with a phobia anxiety is focused very narrowly on a single stimulus such as a snake or spider.

To avoid the anxiety associated with an event imagine a repellor superimposed on top of the core apprehension. The repellor redirects the system state away from the core apprehension and toward the adjacent defensive attractor, which resides nearby. The result is that attention is pulled away from thinking about the event and from experiencing the anxiety associated with it. Defense mechanisms are thus at least initially adaptive. Without them, the system would enter an anxiety-induced chaotic state and become unstable. In the long run, though, they are maladaptive because they redirect attention away from important events and restrict the area within the personality space that can be traversed.

Specific Disorders

Anxiety Disorders

Distressing, persistent anxiety or maladaptive behaviors that reduce anxiety characterize *anxiety disorders.* Anxiety is the behavioral manifestation of excessive signal propagation through the nervous system. From a physiological perspective, there is too much activation and too little inhibition between neurons. Some anti-anxiety medications act to facilitate the release of inhibitory neurotransmitters such as GABA, slowing the rate of information transmission through neural networks. Epilepsy can be considered as an anxiety disorder because it too is characterized by excessive and uncontrolled neural activation.

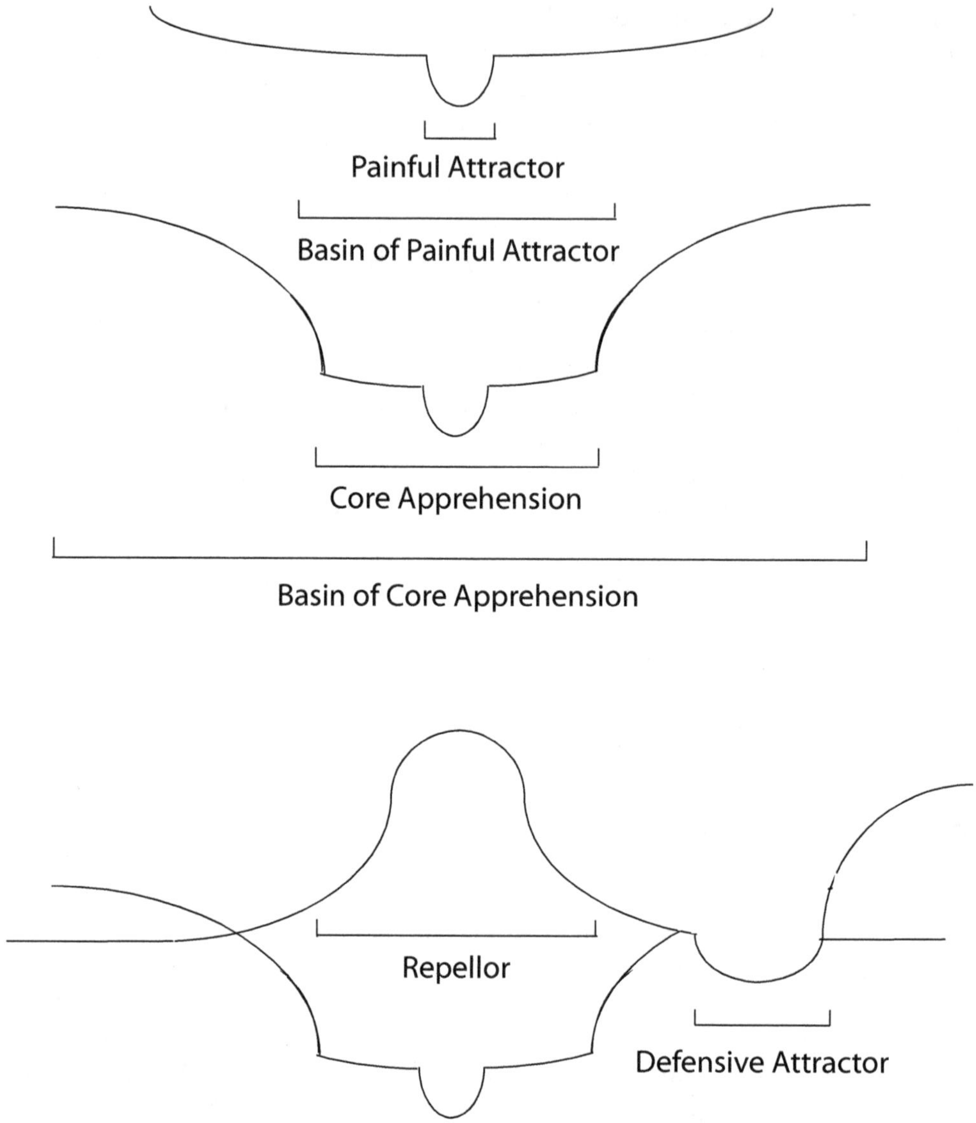

Figure 10.2 *An attractor model of psychological defenses. A description of the process is given in the text. Lewis, M. D. & Junyk, N. The Self-Organization of Psychological Defenses. In The Psychological Meaning of Chaos. Masterpasqua, F. & Perna, P. A. (Eds.) The American Psychological Association. Washington, D. C. 1997. Adapted with permission.*

Anxiety has been likened to chaos (Butz, 1990). Neural activity is more chaotic and disordered during anxious states and more ordered while calm. Butz (1997) points out the interesting parallel between sensitivity to initial conditions and anxiety. Anxious individuals often over-react to minor disturbances. A small input, such as an off the cuff comment someone might make, can often elicit large responses such as worrying, nightmares, and arousal. In a dynamic system, this would mean a tendency for two points starting at the same location to diverge rapidly, i.e., the system could have a high Lyapunov exponent.

There are a several different types of anxiety disorder. Two in particular, *obsessive-compulsive disorder* (OCD) and *post-traumatic stress disorder* (PTSD) lend themselves well to a dynamical interpretation. In OCD, a person continually obsesses about some thought, one usually involving contamination. They temporarily relieve the anxiety associated with the obsession by engaging in a compulsion, such as washing hands. In PTSD, there is a continual re-experiencing of a traumatic event in the form of flashbacks and memories with associated anxiety. Both the obsessions in OCD and the recurring trauma in PTSD can be considered as powerful attractors with large basins that dominate personality dynamics. The attractor wells are so deep and wide that it becomes easy for the system state to wander into them and consequently very difficult to escape.

Mood Disorders

Mood disorders are psychological disorders characterized by emotional extremes. There are two major types. In *major depressive disorder*, a person experiences feelings of worthlessness and diminished interest or pleasure in most activities but typically returns to a normal mood after some time. In *bipolar disorder*, patients alternate between a depressive and manic state. The depressive state is indistinguishable from the symptoms of major depression. The manic state is characterized by euphoria, hyperactivity, and wild optimism.

Hannah (1990) examined daily mood fluctuations in normal individuals. He used a lagging method to reconstruct strange attractors, estimate fractal dimensions, and attain a Lyapunov exponent. The moods he studied were of positive and negative affect. He found that although mood could fluctuate, there was considerable order in the fluctuations. His findings indicate that mood variability is best characterized by low dimensional chaos.

In depressed individuals, the state of depression is more severe and longer lasting than the temporary feeling of being "down" experienced by non-disordered people. Butz (1997) suggests depressive states in those who suffer from mood disorders are a low dimensional attractor and contrasts these with anxious states that are more open and have "looser" boundaries. Other researchers have also discussed depression in terms of these sorts of chaotic dynamics (Heiby, 1991; 1995a, 1995b).

Sabelli and Carlson-Sabelli (1990a, 1990b) produced three-dimensional models of mood fluctuation. They conclude that conflict behavior is a "trifurcation" that produces three outcomes: "fight", "flee", or "submit" that correspond to the emotions of anger, fear, or depression. Bipolar disorder has the obvious implication of being a limit cycle attractor, where the system alternates back and forth between the two extreme moods. It can also be interpreted according to bifurcations (Sabelli, Carlson-Sabelli, & Javaid, 1990).

Abraham, Abraham and Shaw (1990) have created perhaps the most detailed model of mood fluctuation. They define a dynamical system characterized by a vectorfield in two dimensions: self-image and mood. The field is structured so as to pull extreme moods that have been displaced by some event back to normal (the midpoint of the two-dimensional image-mood space). Different starting points in this field produce different trajectories in the space. They next introduce a dependency parameter. Different values of this parameter produce different patterns of mood movement in a bifurcation sequence. When dependency is low, movement in the two-dimensional space is characterized by a radial point attractor. As dependency increases, there is a transition to focal point, single cyclic attractors, two cyclic attractors and finally to chaotic attractors (Figure 10.3).

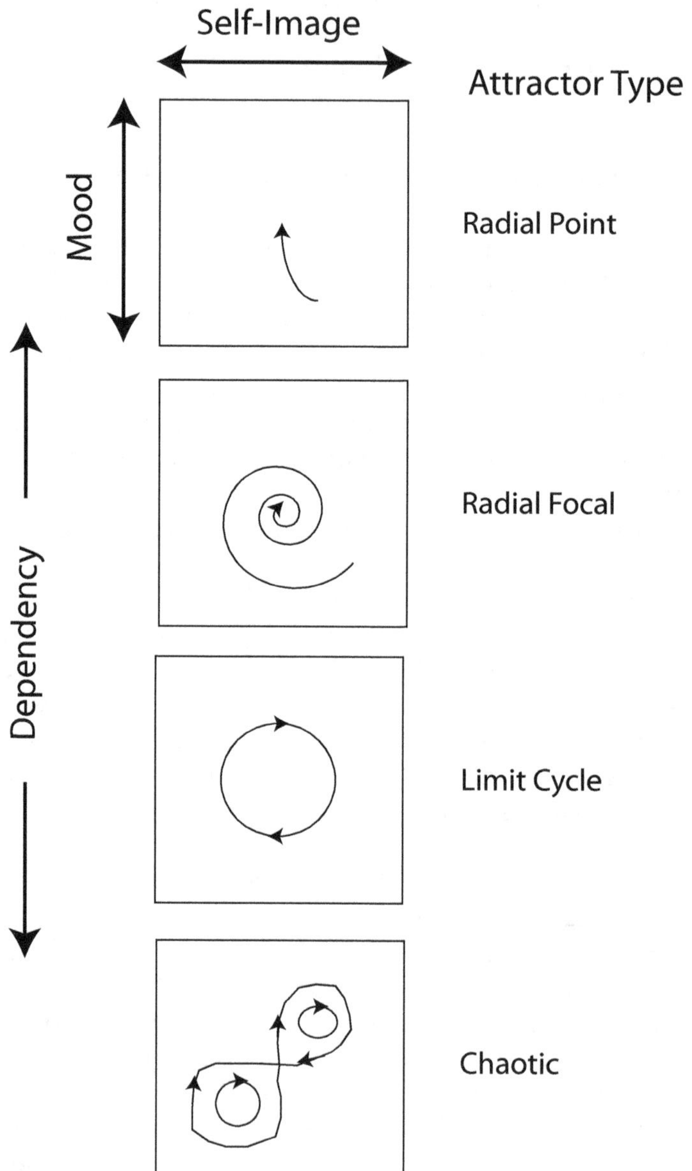

Figure 10.3 *A dynamical model of mood fluctuation (adapted from Abraham, Abraham & Shaw, 1990).*

Dissociative Identity Disorder

Dissociative Identity Disorder (DID) also goes by the name of multiple personality disorder. It is characterized by the presence of two or more distinct identities or personality states that control behavior. Current theories for its origin state that the identities form in response to childhood trauma. A person may form a strong identity in order to stand up to an abusive parent. This personality may then alternate with another more passive identity. Formation of the alternate identity compartmentalizes negative feelings and memories. It allows the trauma to be contained outside normal awareness.

Derrickson-Kossmann and Drinkard (1997) provide an explanation for DID in terms of chaos and complexity. They view the process of alternate identity formation as a bifurcation in the self-organizing development of personality. Each resulting identity can be thought of as an attractor to which the system gravitates depending on circumstances. This compartmentalization is harmful because it limits the flexibility and adaptability of the person to circumstances. In other words, the personality system is limited in the way it can react and deal with situations.

Because each personality is compartmentalized in the self, therapy consists of breaking down these internal barriers. This allows greater flow of information and energy between the selves and the environment. The capacity for healthy adaptation increases with greater flow. Cognitive therapy introduces information into the system that facilitates the dissolution of these internal barriers.

Another therapeutic technique uses *abreactions*, the emotional release of painful experiences. These also help to break down barriers and promote flow. Abreactions may be bifurcation points where established system organization may change unpredictably. They serve to move the system toward randomness and away from a more ordered and frozen way of being.

Schizophrenia

Schizophrenia is a group of severe disorders characterized by disorganized and delusional thinking, disturbed perceptions, and inappropriate emotions and actions. Many schizophrenics have disordered thoughts. In some cases, they feel they are being persecuted or spied upon. Many also suffer from hallucinations that can be visual, but are more often auditory and come in the form of voices. In many cases these patients also demonstrate the absence of appropriate emotional responses. They will have expressionless faces and speak in a monotone voice, without any prosody or affect.

Xu, Tong, and Liu (1999) analyzed EEG time series data from normal and schizophrenic patients. Both groups were tested with their eyes opened and closed. The results showed more information transmission between parts of the cerebral cortex in normal individuals when their eyes were open. Conversely, this pattern was reversed for the schizophrenics. They showed more cortical activity with their eyes closed. One of the symptoms of schizophrenia is called word salad, a random stream of conversation in which ideas are only loosely associated. This uncontrolled train of thought seems to manifest itself as higher cortical activity in these patients with eyes closed. Looking at an object or visual scene seems to constrain or limit this stream of consciousness.

From a physiological standpoint, schizophrenics have excessive activity in the parts of the brain that use the neurotransmitter dopamine (DA). Medications used to treat them are DA antagonists, and act to reduce this activity. King, Barchas and Huberman (1983, 1984) have formulated a dynamical model of brain function in schizophrenics. The model takes as variables the firing rate of DA neurons in the substantia nigra as well the concentration of released and stored DA. Utilizing feedback loops between the substantia nigra and the caudate, they obtained complex bifurcational behavior among static, cyclic, and chaotic attractors. The reader is advised to consult the source material for more details of their model.

Learning Disabilities, Self-Organization, and Multi-Component Cognition

Learning disabilities (LD) make up a diverse and extensive variety of disorders. To date there is no universally agreed-upon classification system or unified theory to bind them. Zera and Lucian (2001) propose that the self-organization perspective can serve to integrate and explain much of this field. One current way of classifying LD is to focus on specific disorders such as difficulties in language, reading, or mathematics and to focus on specific brain areas that may underlie these deficiencies. This approach may actually confuse the issue as in most LD cases, the functional problems are much more complex.

A self-organizing perspective instead emphasizes a global approach, where the interaction between parts is as important or more important than a focus on individual deficits or brain areas. LD is at heart a neurological problem. Its symptoms are the manifestation of brain dysfunction. We know that the brain is a complex self-organizing system. Therefore, it makes sense to interpret LD through a self-organizing perspective.

To illustrate this, take the example of reading. Rather than examine reading as a specific deficit, it is better to consider it as a symptom of a more diffuse and complicated problem. Reading calls on a number of different skills, including attention, memory, and problem solving. Its operation is the result of multiple interacting brain sites. Thus, research in reading should be extended to include other language-based and nonverbal-based LD categories, as well as more specialized reading subcategories like dyslexia, dysnomia, and dyscalculia (Zera & Lucian, 2001).

There are additional reasons to consider LD from a self-organizing view. The brain, as we have already described, reorganizes itself in response to damage and to changed environmental circumstances. This plasticity ensures that there will be individual differences in ability and dysfunction. There is also the phenomenon of crowding. In *crowding*, one brain hemisphere is called upon to execute tasks normally performed by the opposite hemisphere that has suffered damage (Loring *et al.*, 1999). The intact hemisphere is now overburdened or "crowded" and so less able to perform efficiently. Brain plasticity and reorganization effects like this demonstrate the complex interplay between brain areas that can give rise to LD.

Every major type of LD (as assessed in the traditional narrow sense) is plagued with classification problems. Reading LD can be based on several component skills such as phonology or orthographic processes. Investigators differ as to whether there are two, three, or four subtypes (Kolb & Whishaw, 1996). There are a variety of different symptoms across individuals, who also differ in their other cognitive capacities. Language capacity impacts on reading and on almost every other cognitive function so deficits here can contribute to many sorts of LD.

The same situation holds for mathematics-related LD. Mathematical cognition is impacted upon by numerous skills, including visual-spatial processing, attention deficits, grapho-motor dysfunction, and memory deficiencies (Rourke & Conway, 1997). Each of these may contribute in unique ways to math ability. For instance, visuospatial processing is necessary for accurate number alignment in decimal problems and when moving left to right or vice versa in algebraic problems. Reasoning and decision-making skills are called upon to solve applied math problems. These skills are necessary for an overall understanding of the problem, to filter out inessential information, and to generate goals and subgoals.

Executive dysfunction is another problem with widespread influence on cognitive ability. Individuals with executive dysfunction typically suffer frontal lobe damage. Frontal brain systems have multiple reciprocal connections with other areas and so regulate attention and working memory. It is these systems that seem to underlie attention deficit hyperactive disorder or ADHD (Barkley, 1997). They also play a role in metacognitive skills, which broadly refer to the ability to monitor, organize, and alter one's cognitive processes.

Zera and Lucian (2001) conclude that we should abandon trying to classify LD. The current categorization schemes are too inaccurate, too broad and detract from the large differences between individuals. The symptoms of one person will never be the same as those of another because in a self-organizing system, the effects of a deficit will spread throughout the cognitive system in a random and nonlinear fashion. This means we should treat LD in a specialized, case-specific manner.

Dynamical Therapy

How then is a dynamical disorder best treated? Schmid (1998) believes that therapy should be a form of "chaos control". Just as we would shift our arms and hands when juggling a set of balls in the air, we ought to shift our treatments dynamically to restore a proper balance. This consists of weak but exact treatments, such as perhaps a small dose of drug at just the right time. Timing is crucial here. If a drug is given at the wrong time, say before bedtime rather than in the morning, it may have little or no effect. Treatments must be synchronized with the organism's state to work properly.

So the role of a therapist is to transform the patient's dynamics away from a "syndrome" attractor back to a "healthy" attractor. This can be achieved in a variety of ways in addition to the use of medication or traditional talk therapies. It can be accomplished with audiovisual stimulation such as light or music therapy. It may even happen without formal therapy at all. Witness the positive changes that can sometimes come about in people who undergo major life-changing events such as moving, getting married or having a child.

Paar (1992) incredibly suggests that confusion can be used a therapeutic tool. Imagine an individual who is in an unhealthy but stable attractor personality state. Creating a state of confusion would introduce a new problem for the person. This would cause them to move toward a bifurcation point from which the person could either return to their old behavior patterns or more hopefully, transition into a new pattern of behavior that would be different and healthier. Moving to this new state would be easier because of the inherent instability near the bifurcation point.

This implies that a dramatic life change such as getting married or moving to a new country could lead to the resolution of certain psychological problems. The change in circumstance would challenge the individual and force them to adopt new ways of dealing with life. It does not guarantee this, of course, since the added stress of the confusion and novelty might cause them to react negatively. So the use of this kind of treatment would depend in large part upon the individual and the severity of the induced change.

Group Therapy

Order and Disorder in Group Therapy

Brabender (1997) outlines the ways order and disorder can occur in group therapy. Order can be seen when members discuss a single topic and when their associations follow one another in a coherent and logical way. Conversations that stay on topic are more predictable in the sense that one can at least generally anticipate what someone might say. These conversations also imply that there is some coupling among the members, since they are all listening to each other and exchanging related information. A common shared idea or theme in a group is the equivalent of an attractor in a multidimensional conversational state space.

Chaos or disorder manifests in a group when the topic of conversation changes frequently. What one person says on any occasion may only be peripherally related to what was just said previously. Prediction of content under these conditions is more difficult and is a case where there is only a loose informational coupling between members. It is more the idiosyncratic and personal content of one's own mind rather than the current topic that determines what gets stated next. Other instances of chaos are the member's confusion about the purpose of the group. Note the similarity between group coupling here and the neural coupling discussed earlier. In both cases, linkage of the different group elements and sharing of information create coherence and order.

In group therapy sessions or in regular discussion groups, there can be alternating periods of order and disorder. The conversation may pass through ordered attractor states and more disordered chaotic states, winding its way through a semantic landscape. It is also possible for a group to be both orderly and disorderly at the same time. Although there may be confusion and miscommunication, members can still simultaneously converge on answers to problems and relate to common themes. Some amount of disorder may be necessary to generate ideas and solutions that can then be tied back to the original problem or topic.

Complex Adaptive Systems and Group Therapy

Brabender (1997) relates a number of features of complex adaptive systems to group therapy behavior. The first obvious similarity is that groups are composed of elements that are in constant interaction, the elements in this case being the individual people that make up the group. Control of the group emerges as a result of these interactions. As we saw in self-organizing systems, ordered and purposeful behavior can arise without a leader or central coordinator. In group therapy, the leader is the therapist. He or she does not dominate the conversation but instead serves as a guide or facilitator.

Another feature of group interaction is the formation of subgroups. During the course of conversation, there may be a split amongst the members, one group supporting the therapist or some particular viewpoint, the other opposed. These alliances can shift, with some members defecting or changing their minds. The formation and dissolution of subgroups also occurs at the neural and cognitive level within a single individual. In this case, it is populations of neurons representing percepts, ideas, and memories that may coalesce and dissolve over time. The formation of a subgroup or the introduction of new ideas can be likened to a bifurcation or phase shift.

Prediction or anticipation of the future is another hallmark of complex systems. Group therapy members construct hypotheses about the consequences of certain actions and then test them out. For example, a person might think that it is inappropriate to make negative comments to other members, but then see that this is acceptable by observing somebody else do it.

As mentioned previously, complex adaptive systems are in a more or less constant state of disequilibrium. From an informational point of view, disequilibrium generates novelty. It promotes the exploration of new topics that in turn open up additional new topics, and so on. In therapy there is often no single best answer. What works for one person might not work for someone else. This diversity and ambiguity means that a single topic or question can be explored indefinitely.

CHAPTER 11
COGNITION

Cognition and Dynamics

What is the nature of a cognitive system according to the dynamical systems perspective? Heath (2000) lists several characteristics. First it must be highly flexible, with the ability to generate a large variety of thoughts and behaviors. Second, since it is a complex system, it should be difficult to accurately predict its future behavior. Third, it should be able to create novel information processing mechanisms that result in emergent phenomena. Finally, its behavior must be seen to evolve over time given the appropriate resources. This list in essence is saying that human cognitive abilities are complex, adaptive, emergent, unpredictable and capable of generating variability. These are all features that we have examined in some detail already.

In this chapter we discuss theoretical issues related to cognitive dynamics. In the cognitive sciences the dynamical perspective forces us to re-evaluate many long-held assumptions. It forces us to rethink the notion of symbolic representation, concepts, learning, and concept formation. It provides us with a view of mind that is continuous rather than categorical (Spivey, 2007). We then contrast five traditional conceptions in cognitive science with their dynamical counterpoints. In conclusion, we discuss global models of cyclical processes in cognition and the relation between information-processing and the edge of chaos.

Representation

The traditional view in cognitive science is that mental representations are static and discrete. They can be represented by symbols, such as letters or words. These representations are said to be intentional, they are "about" the thing or things they refer to in the outside world, which are called their *referents*. In this classical view, if you thought about an elephant, then somewhere in the mind there would be a static enduring symbol that would stand for the elephant. If it were a pictorial representation, it might even resemble the elephant in some way. Thinking about the elephant would then correspond to the transformation or application of some process to the elephant symbol.

This perspective is based on the way computers process information and has been a very fruitful approach in the study of cognition over the past few decades. Recent theories however, are now beginning to challenge this view. Peschl (1997) in particular opposes this view. He does not believe that a certain state in a representation system like a symbol can refer to a certain state in the environment, such as an elephant. He provides three reasons for this.

The first reason is epistemological. According to Peschl, a representation is always a subjective construction created by the observer. It is produced in a unique way by the sensory and nervous system of an individual organism. Because the nervous system is highly dynamic, the representation is in constant flux, depending on the organism's state of mind. For instance, the representation we have may differ depending on how tired we are, what other information we have learned and other factors. This means that no two representations will be alike, either between individuals or within an individual at different times.

The second reason he provides is empirical. Experiments and simulations fail to show a stable referential relationship between a representation and a referent. If all networks were feed forward and sent the result of their computations forward, representations based on sensory input might be preserved. However, feed forward networks are the exception rather than the rule in most nervous systems. The vast majority of neural mechanisms are recurrent. Information in recurrent networks feeds information back to previously processed information in an iterative fashion. The result is a distortion of the original sensory input. In addition, he notes that the environmental signal is already distorted by the transduction of sensory receptors before it even gets to be processed more fully by "downstream" mechanisms.

A final reason is that there is a better alternative account of representations. This is the constructivist framework. In this view, representations are constructed by a physical nervous system and are always system relative. Constructionists believe that the primary purpose of a representation is to generate a behavior, not to depict the environment. This behavior is generated in the service of the organism's functional adaptation to its environment. A representation in this perspective maps stimulus inputs onto motor outputs.

The constructivist view is not alone. Evolutionary approaches such as the field of artificial life (A-life) already approach representation from a functional perspective. The genetic content of organisms in A-Life simulations can be thought of as a form of representation, but what they code for is a behavior that either helps or fails to assist the animal in surviving or reproducing. This point of view is also seen robotics. In the subsumption architecture, representations exist as production rules (if-then statements) or reflexes that map a stimulus onto a behavior (Brooks, 1991). These rules exist in a hierarchy, where lower level rules can call on higher level ones. Robots with this architecture are surprisingly adept at navigating about complex environments and performing other tasks. Knowledge in both of these views does not exist in isolated representations, but in the entire structure and architecture of the organism. The knowledge is embodied.

What is the substratum or stuff of which representations are made? Peschl (1997) makes several suggestions. They could be neurons or a single pattern of activation in a neuron population. Alternatively, they could be the synaptic weights between neurons. This second option corresponds to representations in artificial neural networks where the matrix of weights depicts the strength of connections between nodes. Representations can also be seen as sequences of neural activation patterns. This is the view favored by the dynamical system school. The sequences are patterns of activity over time in a neural system and correspond to trajectories and attractors in a neural state space. Finally, Peschl mentions that representations could also be genetic material. Genes are an indirect or second-order form of representation, since they aren't directly mediating perceptual or motor action but instead code for the structure and functioning of the body and brain that later would embody such representations.

Symbolic Dynamics

Part of the debate between the traditional cognitive science view and the new nonlinear dynamical systems view stems from the nature of thought. On the one hand, we can think of thoughts as discrete and bounded, having a definite beginning and end. If this were the case, we could use symbols to designate mental entities like "red", "justice" or "happy". Symbols are a great way to represent things because they can be plugged into and operated on by formal systems like language and mathematics. This has been the classical approach in cognition as we mentioned earlier. But what if the brain didn't represent things symbolically? Then we have to rethink the whole notion of mental representation and computation. It is in fact just this difference that divides the connectionist parallel distributed processing approach from the classical view. The dynamical systems approach favors the network view of mind in that representation is seen as distributed activity among numerous nodes in a network.

There may be a way to reconcile these two views. In *symbolic dynamics*, a bounded region of state space is assigned a symbol (Bollt, Stanford, Lai & Zyczkowski, 2000; Robinson, 1998). When the system is in this region, the symbol is activated. When the system state is outside this region, it is unactivated and dormant. Attractor basins are a good example of such regions. We could draw a circle around a basin and let the area inside this circle stand for a given letter or word. When the trajectory of the system wanders inside the bounded area, the symbol "turns on" and can be part of a computational process. The two basins in the vectorfield of Figure 4.8 shown earlier can illustrate this. One basin could correspond to the letter "A" and the other to the letter "T". Reading the word "AT" might then correspond to a trajectory that follows a path from the first basin to the second.

One problem with this account is that it is difficult to determine exactly where to place the boundary line. If an attraction basin is sufficiently steep and deep, the line would best map onto the outermost "lip" of the depression. But if the basin was gently sloping and shallow, we could place the line farther out, in which case the symbol would be triggered more easily, perhaps corresponding to a lowered threshold. Alternatively for this type of depression, we could place it closer in toward the center. In this case, activation of the symbol becomes more difficult and is the equivalent of insensitivity, i.e., a high threshold. It is likely that the region would fluctuate, both between and within individuals based on various factors. Because of these uncertainties, Spivey (2007) suggests we think of these as "fuzzy" rather than discrete symbols.

Concepts

Thelen and Smith (1994) outline the classical cognitive view of concepts and contrast it with the dynamical systems perspective. In the classical objectivist view, concepts represent reality; they form an internal model of an external reality. With development or learning the internal concept becomes refined over time to more accurately approximate the "true" notion of what the concept is. This true concept is an absolute definition derived from the physical structure of the world and from logic. For example, the concept "whale" could be defined as an instance of the concept "mammal" with certain distinguishing features that it shares in common with its superordinate category and additional features that would differentiate it from other categories.

Concepts in the objectivist view are formed using symbols. The symbols are arranged using syntactical rules. The use of existing concepts or the formation of new ones must therefore follow predefined linguistic or propositional laws. The focus in this view is how concepts represent reality. In the objectivist notion, all of our concepts are similar since they approximate an externally validated definition.

In opposition to this, the dynamical perspective focuses not on the structure of concepts and their relationship to reality but on how they are used in different contexts. A concept here is not fixed and immutable, it does not approximate some idealized logical conception of what the concept should be and the organization or structure of the internal concept need not be based on symbolic representation or the use of features.

In the dynamical view a concept is a pattern of neural activity that arises as part of the organism's previous and current experience. It is constantly changing based on context. For example, a person's concept of a whale would change after each new experience that individual has with a whale and would be altered every time that person reads about whales, goes whale watching, etc. Furthermore, each person's concept of a whale would be different, since individuals have a unique and irreproducible history of interaction with the world.

According to Lakoff (1987a, 1987b) concepts are *embodied*: they emerge from each person's perceptual and motoric interaction with the world. This makes them inherently subjective. It makes no sense to ask whether a concept is right or wrong because there is no external standard by which they are judged. A concept has "validity" only in the sense that it used efficaciously by an individual, either in their own actions or in their interactions with others. To illustrate, two people might think of the concept whale in very different ways, but this doesn't prevent them from performing whale-related actions or in talking about whales to each other. Language, rather than structuring concepts internally, instead allows us to communicate about them in a social context.

A concept, if it is dynamical, can take on multiple non-contradictory meanings. We may, for instance, represent a whale as a memory of a boat trip in California where we saw one, as a recollection of a scene from the movie "Moby Dick", and as a collection of facts acquired by having read a book on marine biology. Each of these can correspond to a different pattern of neural activation represented as an attractor in a state space. A person could alternate between these conceptions, which would be a transition of the system from one basin of attraction to another.

Supervised and Unsupervised Learning

A dynamical theory of concept learning posits that they form all by themselves, i.e., they self-organize. The acquisition of concepts should not require the presence of an external teacher who provides a standard of what the concept should be. The teacher idea forms the basis of a popular connectionist model known as back-propagation (Gurney, 1997). Here, the difference between input-generated activity and a pre-programmed standard feeds back into the network and is used to adjust the weights between nodes. After a number of iterations, the network converges on the appropriate response.

In cognitive science, learning can be generally classified as supervised or unsupervised. Supervised learning, as just mentioned, requires a "teacher" to provide the appropriate response. In unsupervised learning, the answer is not known ahead of time and the organism or device must determine the appropriate concepts and behaviors on its own. A dynamical self-organized account of learning must of course be unsupervised.

There are existing models in the field of machine learning that specify how an unsupervised process takes place. These models are known as *reinforcement learning* (Sutton & Barto, 1998). They take place in the form of a trial-and-error process, where the consequence of an action feeds back to affect future actions. In this scheme, an agent (human or otherwise) produces an action. The result of this action produces either a reward or a value. The agent acts in such a way as to maximize its rewards. Occasionally, the agent will engage in a behavior that increases value, even at the expense of minimizing reward, when for example, a student stays in on a weekend night to study for an exam rather than going out to drink.

Reinforcement learning can explain how we acquire concepts simply by pursing rewards and values. All that needs to be "programmed into" the agent is the desire to do things that feel good or have value. These sorts of tendencies are probably innate and include low-level drives such as the desire to eat and engage in sexual activity but probably also include higher-level drives such as the need for social interaction and attention. Thus, much learning can take place in the absence of a teacher who instructs or models actions.

Concept Formation

A model by Reeke and Edelman (1984) shows how a concept of the letter "A" may be acquired. In their model, an agent experiences different instances of the letter A. They extract local information in the form of features, but also global information based on the actions of tracing or writing out the letters. The featural and motor impressions then are associated, so that over time, the presence of different types of letter As evokes activity in both systems. Eventually the learner generates a coherent pattern of activity that corresponds to the presence of many different kinds of As.

This model was successfully able to learn letter categories without any instruction or teacher. It was able to extract the similarities among different versions of letters all on its own. Furthermore, the model works regardless of the specific examples of As shown or in how the agent might write them down. The type and order of features or motor actions are irrelevant to the learning process. One person could trace an A by drawing in the left diagonal first. Another person could trace out the right diagonal first. In either case, the resulting patterns, because they are being associated with the stimulus and with each other, would still result in the appropriate concept.

Thelen and Smith (1994) summarize the results of this model. The letter categories self-organize without external instruction. They also occur without the presence of a pre-existing template in the head of the learner. All that is required are certain motivational tendencies such as the desire to look at certain types of stimuli in the environment. What is learned is not a definition of a letter, but an individualized and unique pattern of neural activity based on the agent's history with the stimulus. Also, the concepts are not stored in any given location in the mind but consist of a correlated global pattern of activity, in this case from the perceptual and motor regions.

If we conceive of a concept as an attractor, then how does the attractor get formed? According to Thelen and Smith (1994) it occurs due to repeated conjoint stimulation in different neural populations that are active during experience. If a person looks at an object, there would be simultaneous activity throughout the different areas of their visual system. The form, color, and motion areas could all be active. In addition, there would be patterned activity in auditory regions if there were sound present, in tactile regions if touch was involved, in motor regions if they were manipulating or moving the object in some way, etc. Each time the object is experienced, there would be similar patterns in these areas. The repeated activation would cause them to become associated, since according to Hebbian learning, synapses that are active simultaneously are strengthened. This mechanism can explain how both percepts and concepts get formed.

From a state space perspective, each object experience would trace out a trajectory. Similar experiences would trace out similar trajectories, in effect "carving out" paths. Each experience would reinforce the one that came before it, since its path would be overlaid or adjacent to its predecessor. In this way a landscape is formed so that the next time the concept is experienced, the system will move along these paths, effectively being drawn into the attractor that corresponds to the concept.

The result is that a multitude of different experiences, each related to the concept in some way, can all produce the same mental response. This is exactly how concepts are used. When we look at a robin or a blue jay or a sparrow, we realize in each case that they are instances of the concept bird. The different paths the system can take to the bird attractor correspond to the mental instantiation of each of the individual birds.

Dynamical vs. Classical Cognitive Science

The Continuity of Mind

Michael Spivey, in his 2007 book titled "The Continuity of Mind", makes a strong case for the adoption of the nonlinear dynamical systems approach. Cognitive science, in his view, is making a mistake in assuming that thought is categorical and discrete. Cognitive researchers, by adopting the computer metaphor of mind, have come to view concepts and representations as symbolic and localized, mental functioning as modular, and processing as necessarily passing through a set of serial stages.

He proposes instead that all mental activity is fuzzy, graded and probabilistic. Thoughts should be characterized by trajectories in a state space. The dimensions of this space are the activity levels in large populations of neurons. Thought will sometimes gravitate toward attractor basins in this space that correspond to a given percept or concept, but usually pass through quite quickly. The emphasis in Spivey's view is on the movement of the trajectories, not on the locations in state space they dwell on only briefly. In this section, we outline Spivey's continuity approach to mind in greater detail, contrasting five traditional conceptions of mind from classical cognitive science with the dynamical concepts that, according to Spivey, should replace them.

Modularity vs. Distribularity

One of the assumptions in the classical cognitive science view is that the brain is modular. A *module* is hardwired, domain specific, fast, automatic, stimulus-driven, and not subject to control by a central authority (Fodor, 1983). Another feature of modules is that they are encapsulated with respect to information. This means that other mental processes can only access a module's output, not influence or access its inner workings. There has been much criticism of the modular approach in recent years. Studies of patients with aphasias, or language deficits, show that damage to many different areas of the left hemisphere can produce aphasias with identical symptomology. This refutes the longstanding Wernicke-Geschwind model of language processing, which states that certain left hemisphere regions perform very specific functions, each passing their outputs off to the next area in a serial fashion.

The assumption of modularity is necessary if the reductive approach in cognitive science is to succeed easily. If the brain consists of many special purpose mechanisms, then we can take out or dissociate some and study the others in isolation. Removing or controlling one module in this kind of system allows us to see the role it plays in its absence, much like a missing piece in a jigsaw puzzle where all the pieces fit into one together in only one way. Because most of the processing occurs inside a module, the interactions between them are straightforward, minimizing the difficulty of constructing a "causal map" or process model of the system's functions.

If, however, the brain is non-modular, then the reductive approach runs into problems. We could for instance, postulate that the internal operations of the mind are open to influence and feedback from other such mechanisms. This certainly is true if we reduce these mechanisms to the individual neuron level, since neuron firing rates are modulated by excitatory and inhibitory inputs from neighboring neurons. Now, we have a much more difficult time diagramming the system. Each module in some sense has lost its "boundary": its functions are local but also global, being tied to the operations of other things around it.

So is the brain modular or not? It is perhaps most accurate to say that it is partially modular. There are portions of neural tissue that are anatomically and morphologically distinct from one another and that exhibit some degree of functional independence. By the same token we also see areas that are more highly interdependent. Spivey uses the term "soft modularity" to describe this notion. The term *distribularity* has also been suggested as a new phrase to encompass this idea (Kingsbury & Finlay, 2001).

Component-Dominant vs. Interaction-Dominant Dynamics

If the brain were modular in the Fodorian sense, it would display *component-dominant dynamics*. In this scenario a module or functional unit waits until it receives enough input and has performed enough localized computations to reach a stable state. Only then does it produce an output that would be passed on to another module for processing. In a sense, the unit is being "selfish", keeping information to itself for as long as possible. This type of dynamic is seen in serial systems where computations are performed one at a time.

In contrast, the brain could demonstrate *interaction-dominant dynamics*. Here, functional units continually update their neighbors. Unit A, for example, would activate Unit B even before it had finished its computation. These systems may never reach a stable state since the units are continually activating each other. The units can be thought of as "altruistic" because they each "help" or share their information with one another. This dynamic is more characteristic of distributed systems that utilize parallel distributed processing techniques.

Spivey (2007) outlines several lines of evidence supporting the interaction-dominant view. Traditional views of the visual system have separated streams where different stimulus characteristics like color, form, and motion are processed in a mostly independent fashion. In the tilt illusion, the perceived orientation of bars inside a circular center region is influenced by the orientation of bars in a surrounding ring-shaped region. The inner bars are perceived as tilted in a direction away from those in the surround. The effect is enhanced when the color and luminance of the surround region matches that in the center. This implies that the channels for orientation, color and luminance are not separate, but interconnected, at least in the primary visual cortex where the basis for the illusion is said to occur.

Transcortical interactions also occur between modalities. That is, interactions take place not just between features within a given modality such as vision, but also between two modalities, such as vision and touch. The visual system interacts with the touch system during tactile discrimination of orientation. We know this because the visual cortex becomes active during this task (Sathian, Zangaladze, Hoffman, & Grafton, 1997). Disruption of the visual cortex by transcranial magnetic stimulation also disrupts the task (Zangaladze, Epstein, Grafton & Sathian, 1999).

Internalism vs. Externalism

Another way of describing the classical modular view of the brain is to say that it is a closed system. Closed systems are insulated from what is occurring around them. In other words, they are spatially and functionally bounded. It seems fairly certain that the brain is in fact an open system at almost every level of analysis. At a small scale, neurons are influenced by the larger neural circuits in which they are embedded. These neural circuits are influenced by more global and widespread brain activity. The brain in turn is influenced by the body of which it is a part and the body is bound up with the world around it.

The view of brain as an open system, of being more than just what is inside the skull, is not a new idea. In the ecological approach, visual perception includes a brain-body-environment interaction (Gibson, 1986). Take for instance size judgments. The horizon line cuts objects off at the eye height of the observer. So if a utility pole is bisected by the horizon and the viewer is six feet tall, the pole must be twelve feet tall. In the ecological approach, this estimation occurs directly and rapidly. The information that is needed to determine height is actually in the environment and in its relation to the observer. All the observer needs to do is learn about these regularities and then access them. Little or no computation is necessary.

In contrast the traditional view in perception, influenced by Hermann von Helmholtz, is that perception is inferential and computational. One takes information from the environment, performs a computation on that data and infers or determines a result. This computation takes place entirely "inside the head". The environment only comes into play initially, in terms of providing a starting stimulus. In the above example, size can be estimated by Emmert's Law, which states that object size is equal to retinal size multiplied by perceived distance adjusted by a constant. The image provides the values for the equation, which is then "solved" by the visual system.

This debate about whether the mind is just brain or something more has coalesced into two differing philosophies. *Internalism* holds that mind can be entirely explained by a description of the brain and of brain states (Segal, 2000). In opposition to this is *externalism*, advocating, like the ecological psychologists, that mind is the result of interactions between the brain, the body and the immediate physical environment (Clark & Chalmers, 1998).

Situated vs. Embodied Cognition

This philosophical divide is echoed not just in psychology but also in the fields of robotics and artificial intelligence. Brooks (2002) differentiates between robots that are situated and embodied. According to him, a *situated* creature is one that is "embedded in the world, and which does not deal with abstract descriptions, but through its sensors with the here and now" (Brooks, 2002: 51-52). The information a situated creature receives directly influences its behavior. An *embodied* creature, on the other hand, is "one that has a physical body and experiences the world, at least in part, directly through the influence of the world on that body."

Brooks (2002) then gives two examples. He says an airline reservation system is situated but not embodied. It receives queries about flights by users that constitute real time perceptual inputs and then produces outputs that are received and further processed by the users. But since the reservation system is a computer program, it lacks a true body. That is, it has no real sensors to directly perceive the world and no effectors like legs or arms to locomote, manipulate or interact physically with the environment.

An assembly line robot that spray-paints parts in an automobile manufacturing plant is embodied, but not situated. The robot has a body, consisting of a base and articulated robotic arm. It may have sensors and effectors to perceive and act on the environment to execute its task. But it doesn't interact dynamically or adaptively with the environment. Instead, it simply executes the same routine over and over again, following its internal programming.

Feed-forward vs. Recurrent Pathways

One last point to be made concerns the direction of interactions between the different levels of brain, body, and world. The causality here is bi-directional. It flows between these systems both global-to-local and local-to-global. In other words, the world acts on the body through afferent perceptual inputs, while the body acts on the world through efferent motor actions. It is the continuous loop or cycle between these that constitute mind, not any of the one-way paths. It is thus best to think of mind as a recurrent pathway, one that loops back on itself, instead of a feedforward pathway that goes in only one direction.

Examples of this bi-directionality are evident within the brain. For many years it was believed that the visual system was a feedforward system. The image projected onto the retina was sent by ganglion cells to the lateral geniculate nucleus (LGN). From there the signal was sent to cells in primary visual cortex (area VI), from VI it went to temporal and occipital regions. It turns out that these way stations in the visual pathway send projections back as well as forward. For example, there are numerous pathways that run from VI back to the LGN. The image sent from the eye is indeed passed upwards, but then also triggers downward cascading signals as part of the normal process of visual perception.

Cyclical Processes in Cognition

The Perceptual-Cognitive Loop

Goertzel (1994) provides a broad dynamical framework to understand attention and recognition processes. One of the primary roles of attention is combine features into wholes. Features extracted from the senses need to get grouped together to form objects and object parts before they can be recognized. Some of this grouping is preattentive and takes place automatically based on the gestalt laws. For example, parts that are close group by proximity, those are similar in shape or color group by similarity, etc (Goldstein, 2002).

But in some cases, overt attention is required to group features together. This happens when we are looking for the conjunction of two or more features in a particular location when those features also exist at other locations in the visual field as distractors. For example, trying to find your red Honda in a parking lot filled with other Hondas of the same shape as well as other red cars. According to feature integration theory, this task requires focused attention to "glue" the features together at a specific object location (Treisman, 1986). Once this happens, the percept can be recognized by comparing it against a stored memory representation.

In Goertzel's account this "coherentization" involves a feedback loop between brain areas coding for lower-level featural information and those areas dealing with higher-level conceptual information. He calls this the perceptual-cognitive loop. The dynamics of the loop can be summarized in a few steps. To start, the perceptual apparatus sends the cognitive end information on disparate features, with some tentative connections between them. The cognitive end modifies these features and links them with other entities stored in memory, trying to form a coherent whole. The cognitive end then sends its hypothesis of what it thinks the organization should be back to the perceptual mechanism, which now determines whether the conjectured whole is consonant with the feature information. It in turn sends a second report back to the cognitive end, with suggestions. If there is a good match, the process may end, but if there is disagreement, information will flow back and forth between them until some sort of decision is reached.

The Thought-Perception-Action Triad

Bateson and Bateson (1990) conceive of another similar loop, which we can refer to as the thought-perception-action triad. In this model, thought, perception, and action are linked together in a triangular configuration. A percept generated from the environment is fed to a cognitive center that then generates from it a thought. This thought decides upon an action that is then converted into a behavior by an action process. The results of this action alter the environment, producing a new percept, which starts the process all over again.

This model bears some similarity to the ecological approach to perception, which views an organism as causally and reciprocally embedded in an environment (Gibson, 1986). In the ecological approach, the linkage between perceiver and environment is even more direct. Perception is believed to take place rapidly, automatically, and effortlessly without any or much internal computation. In the ecological approach, the cognitive component of the triad could be removed and most perceptual abilities would remain intact. Freeman (1991) suggests a similar approach where iterative interaction between the motor and perceptual systems produces odor recognition.

Goertzel (2004) expresses the triad algebraically, in which one component of the triad is equal to the combination or interaction of the other two. In this notation, perception of an action leads to a new or revised thought, a thought about a perception leads to an action, and an action based on a thought leads to a perception. The flow of mental activity in this model is not restricted to thought-perception-action or even perception-thought-action. It can travel in any direction between the three.

Both the perceptual-cognitive loop and the thought-perception-action triad break with older cognitive science models where the flow of information is in one direction only. These feed-forward models typically saw the cognitive process as flowing inward from stimulus to perception and cognition and then outward through motor systems to action. These traditional models also had end points where processing terminated. The cyclical models in contrast don't ever stop. They see the cognitive act as perpetual and ongoing. The analogy between a process that tends toward equilibrium and one that is in constant disequilibrium is obvious.

Other ways in which these dynamical models break with their forbearers was discussed earlier in the section where we compared the dynamical vs. the classical cognitive science approaches. The older models are modular, component-dominant, closed, and situated. The dynamical models are distributed, interaction-dominant, open, and embedded.

Edge of Chaos and Cognitive Function

If the human cognitive system is complex and adaptive, then what would be its optimal mode of operation? Should it operate in a highly structured and orderly fashion as the classical computer-based cognitive science perspective suggests? Or is actual brain behavior closer to chaos?

A cognitive system should be able to adapt to a changing environment. This means it should be an open system that self-organizes by generating novel responses. The continual and adaptive generation of varied responses to deal with a dynamic environment means it must be at least somewhat disordered. Too much order and it would be unable to create such variability. The solution seems to be a system poised on the edge of chaos. This type of system is flexible, but not too flexible, ordered, but not too ordered.

Heath (2000) mentions that a cognitive system that is far from equilibrium can return from a state that is too simple. The disequilibrium allows it to "pull back" from simple attractor states to more complex patterns of action. The opposite may also be true in that a chaotic system can be pushed to a more ordered state. Ordered cognitive states might correspond to the application of a *control of chaos* procedure. These are procedures that stabilize chaotic systems. It seems that human cognitive function has the capability to go in either direction, either more or less ordered.

Langton (1992) suggests that the most productive information processing happens in complex systems poised at the edge of chaos. It is in this regime where stable cognitive structures such as memories or thoughts are best able to adapt to changing environmental pressures. He noted that in cellular automata (CA) simulations, the most highly structured and interesting output depends crucially on a fluidity parameter ϕ, which represents ease of information flow. Different qualitative patterns of CA activity are observed for different values of ϕ. When it is zero, the system is frozen and there is only a fixed point state. Subsequent increases in this value up to its maximum produce more and more complex and diverse CA states, eventually reaching chaotic behavior. Based on these simulations, Langton makes four important conclusions:

1. The most complex ordered behavior happens at the edge of chaos, at moderate values of ϕ. This region represents a phase transition from regular to chaotic activity.
2. Information processing is slow in this critical transition region, but rapid on either side of it.
3. Information decay in this system is greatest at the transition. This may explain why memory systems must return to ordered states to facilitate information storage.
4. Cooperative interactions between activity states in the system are maximal at the transition region. The neural coupling and coordination across brain regions that may underlie much cognitive activity is facilitated here.

In the Game of Life CA, optimal information processing was found to occur for $\phi = 0.273$ (Langton, 1990). The complement of this value, $1 - 0.273 = 0.727$, can then be thought of as a stability parameter (Heath, 2000). This implies that cognitive systems seem to operate optimally when they are about 27% fluid and 73% stable. Surprisingly, the value of the stability parameter obtained in Heath's (1994) associative memory model was 0.72. In this context, it represents resistance to information flow in an associative network model. This model is described in greater depth later in the section on memory.

CHAPTER 12
COGNITIVE PROCESSES

Cognitive Processes and Dynamics

Most textbooks in cognitive psychology are organized according to the various cognitive processes. These are usually categorized as perception, attention, memory, language, decision-making and problem solving. Following this organization, we discuss the research on dynamics in each of these areas in this order. A select number of studies are shown to provide the reader with the gist of what has been done, since there is not enough space to produce a comprehensive review.

Perception

Ambiguous Figures

An *ambiguous figure* is one that can be perceived in more than one way. One of the classic examples of an ambiguous figure is the Necker cube (Figure 12.1). The cube can be seen with the lower face closest to the observer in which case it is perceived as pointing down and to the left. Alternatively, it can be perceived with the upper face closest making the cube point up and to the right. Ambiguous figures are great stimuli for testing the dynamical concept of mind. Each possible percept can be thought of as an attractor or low point in a landscape. Varying a parameter of the stimulus or introducing other types of bias corresponds to pushing the state of the visual system closer to one or more of these attractors.

Kelso *et al.* (1994) recorded alterations between the two possible percepts and durations for each percept for observers viewing a single Necker cube. They recorded the data as a time series. Under these conditions there is no observable pattern. Switches don't occur at regular intervals. There was also no tendency for long durations to follow short ones, etc. This is what one would expect if the visual system were unstable. This instability is good, because it allows the system to transition very quickly from one state to another. Switching speed is an essential requirement for a visual system that must rapidly process multiple objects in a complex dynamical world. The instability of the visual system in fact seems to mimic the rapidly changing environment of which it is embedded.

Kelso *et al.* (1994) also varied the orientation of the Necker cube to bias the perception one way or the other. They then plotted the inter-switch intervals as histograms. For cubes strongly biased toward one percept, the distribution of switch times were on the long side. Participants tended to perceive in the biased way for some time before an alternation occurred. The histograms for these conditions had long tails. For more ambiguous cubes the switch times were shorter reflecting greater instability. The results suggest that the visual system remains poised at the brink of instability with competing forces of attraction and repulsion pulling and pushing the system so that it can rapidly switch from one stable state to another.

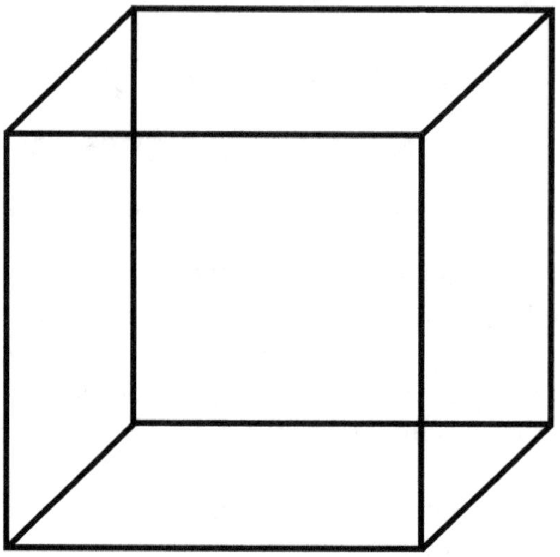

Figure 12.1 *The Necker cube is a classic example of an ambiguous figure.*

Glass Patterns

If one throws down onto a surface a random dot pattern and then shifts the location of all the dots in a consistent way, the result no longer appears random but instead highly structured. For instance, if the dots are rotated by a given degree the pattern now appears as a vortex with circular structure present (see Figure 12.2). Other transformations produce the formation of saddle-shapes, outward radiation from a central point and spirals. These are known as *Glass patterns* (Glass & Perez, 1973). The perceived structure arises because the visual system notices the correlation between local points: it links corresponding dots together to form wholes. The mechanism underlying this capacity may be the same as that underlying perceptual grouping.

Kelso *et al.* (1994) presented Glass patterns on a trial-by-trial basis. Across trials, the patterns could either change from disorder to order or in another condition, from order to disorder. Participants had to judge whether each pattern was ordered or not. They found that a greater number of frames was needed to elicit a perceived change from disorder to order than for the reverse direction. In other words, it was more difficult to build a percept or see structure emerge from these patterns than it was to break the existence of an existing perceived structure down.

The results demonstrate *hysteresis*, which is a perceptual persistence despite changes to the pattern that favor an alternative. Hysteresis occurred on 72% of all the trials in the study and constituted the dominant effect. This persistence provides "strong evidence for nonlinearity and multi-stability in visual perception" (Kelso, 1995: 203). The visual system in this case is poised between two possible percepts, but biased toward a given state based on prior experience or other factors.

Hysteresis is found elsewhere in perception. In the method of ascending and descending limits, the intensity of a stimulus such as a tone is gradually varied. If the tone starts off soft and gets louder, it needs to be made more intense

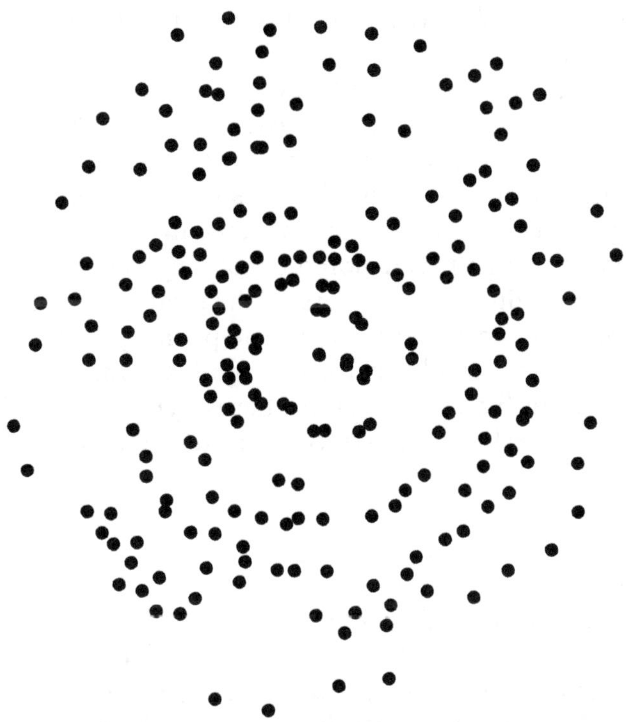

Figure 12.2 *A rotational Glass pattern.*

than the actual threshold to produce detection. If the tone starts off loud and is made softer it needs to be made softer than threshold before it is perceived as disappearing. To overcome this bias the actual threshold is calculated as the average of the crossover points across equal numbers of runs of each type.

Apparent Motion

In the set-up for *apparent motion*, a light or stimulus appears in one location, is shut off and then reappears at a different location. The relative timing and distance separating the two events affects the likelihood of perceiving the shape move. If the distance is short and the inter-stimulus interval is between 60 to 300 ms, there is a strong percept of motion (Graham, 1965). The visual system interprets the display as a single shape moving rather than simultaneous flashing or two separate events in different locations. It is this capacity that allows us to see motion in moving pictures, where each frame consists of the same scene with just slight changes in position between them.

In the motion quartet paradigm circular-shaped lights can appear and disappear in different configurations (Kruse & Stadler, 1990). The positions of the lights are at the corners of an imaginary rectangle. Changing the aspect ratio of this rectangle biases perception of apparent motion. If the horizontal distances are short and the vertical separation is long, horizontal motion is perceived: the dots alternate back and forth from left to right. If the converse is true, with shorter vertical separations, up and down apparent motion is perceived. This is an example of hysteresis because of the abrupt transitions between the two percepts. Aspect ratio is thus the control parameter that influences the multistability of the percept.

Hock, Kelso, and Schöner (1993) varied the speed of transition through the critical point or aspect ratio that produced a switch in the perceived direction of apparent motion, either from vertical to horizontal or vice versa. When they moved through this critical value quickly, hysteresis was preserved and participants persisted in perceiving motion in the same direction they had previously. When they moved through this region more slowly, the participant's perceived direction of motion switched more often. The gradual movement allowed for more spontaneous fluctuations to occur. The primacy of the stimulus, i.e., its greater ambiguity in these conditions, outweighed the historical bias. These data are again consistent with a visual system that is dynamic and able to transition between different states based on external stimulus characteristics and internal expectation.

Attention

Attention and the Structure of Time

Jones and Boltz (1989) formulate a temporal model of attending based on the order and predictability of events. Highly coherent events like speech, tonal music, and body gestures tend to be ordered. They have characteristic rhythmic patterns that unfold over perceptible time spans. In their model, high coherence events afford what they call *future-oriented attending*. The structure of the event allows an individual to predict for example, when the next note in a melody will occur and what that note will be.

In contrast, low coherence events have less order and consequently are more difficult to predict. Examples include lists of unrelated words and concurrent cocktail-party chatter. These events afford *analytical attending*, a focus on locally adjacent elements to try and organize them. This form of attention allows for grouping or counting items or other changes in the event. Estimating the duration of these events will be influenced by local details. Consequently, events with more items will be judged as longer.

This model implies a resonance or coupling between an attender and an environmental event. The observer in effect internally generates his or her own event that synchronizes with the external stimulus. This internal event can take the form of a neural rhythm that is phase-locked to the different time spans within the event. For example, an internal beat may synchronize with the start of every note in a melody or at the start of every several notes forming a measure. Jones and Boltz (1989) call this interactive attending *attunement* and mention that it is easier to attune to high coherence events with a hierarchical organization.

There are several key features of attunement. An attender is initially assumed to entrain with regular time intervals in the event. These can be the accented beats in a rhythm or a speaker's articulation rate. These intervals can be considered to form the attentional framework or skeleton by which the outside event and the observer's internal event are attached. Attunement shifts can then occur when the attender directs attention away from the referent period to some other level. This involves a shift in attention to higher or lower temporal durations. For example, if a listener wishes to catch the "gist" of what somebody is saying, they may spread their attention more globally across longer time periods. On the other hand, they can also narrow their attentional focus if they wish to concentrate on a speaker's vowel pronunciation.

Future-oriented attending makes two predictions regarding perceived duration. If a note or other temporal event ends sooner than expected, it will be judged as having a relatively short duration. If a note lasts longer than expected, it will be judged as having a longer duration. This model was tested in an experiment where participants were asked to compare the total durations of two melodies. Expectancies were generated and then violated systematically. The results supported this contrast model and were better able to account for the data than three alternative models.

The contrast model can be accounted for by an internal biological clock that breaks an event up into an interval time base. In order to account for attentional shifts between hierarchical or nested levels, the clock would have to contain "clocks within clocks". In other words, there would need to be multiple clocks, each running at different, but relatable frequencies to one another. A "master" clock could then control shifts or synchronization between them. This explanation sits well with a dynamical account of the brain where different neuronal groups coding for different frequencies interact.

It is important to stress a few points regarding a neural clock mechanism for attention. Any such clock would need to synchronize with the temporal event. It would be driven primarily by entrainment with this event and not, at least during the attending period, by any internal settings or tendencies toward innate time keeping. Also, it would apply to ordered events only, where a prediction can be generated. Unstructured events with little periodicity that induce analytical attending may call on different mechanisms. Finally, further elaboration is needed to formulate the role "top down" or cognitive factors play in this process. It is not clear how, for instance, attention shifts when cognitive goals change.

Memory

Stability and Plasticity in Memory Processes

One classic finding of cognitive psychology concerns the capacity of short-term or working memory. It has been found to be on average seven items, plus or minus two. This has since become known as the "magic number" (Miller, 1956). But why should working memory capacity be limited to this number? Nicolis and Tsuda (1985) provide a dynamical explanation. In their view, human cognition is represented by activity within a chaotic strange attractor. Because any two close trajectories in this type of attractor will diverge, information is lost locally. There is thus no advantage in storing large amounts of information in this type of system. They determine the optimum encoding length for information to be 4-8 bits, but note that it will vary depending on the specific type of nonlinear dynamical processing involved.

So it appears that encoding of information requires a more ordered state than one would find in chaotic brain activity or a strange attractor. The memory system, in order to store data, seems to require dropping down to a more structured, low dimensional pattern of behavior. In this state, it can repeat the neural patterns representing a piece of information over and over again, strengthening synaptic connections and "imprinting" the data into a circuit so that it can be retrieved for use later. If the memory system were too chaotic, any trajectories through it would quickly diverge. The result is that no simple attractors could occur, preventing the learning of novel items. Both fixed point and limit cycle attractors have been associated with the storage and retrieval of information in nonlinear dynamical memory systems (Chappell & Humphreys, 1994; Hopfield, 1982).

Hayashi and Ishizuka (1999) found that synchronization of hippocampal neurons was facilitated in response to afferent input. This resulted in both phase-locked synchrony and chaotic-type responses depending on the stimulus parameters. The functional role of the hippocampus is to transfer information from short-term to long-term memory. The oscillatory behavior of hippocampal neurons appears to be a key part of this consolidation process. Chaotic neural behavior may be the "default" hippocampal setting. When information needs to be stored, it then shifts to an oscillatory operating mode. Alternatively, it could be that encoding involves both synchronous oscillatory behavior and more chaotic behavior that alternate in some fashion.

This relationship between order and disorder in memory can be characterized as that between stability and plasticity. A memory system needs to be stable in order to store information. But too much stability and it cannot generate the variety of patterns necessary to represent and store novel information in the first place. Heath (1993) has formulated an associative memory model that takes stability and plasticity into account. In his model, memories are stored based on their novelty. Novelty is determined as the amount of similarity between a current input and memory contents. If the correlation is high, nothing is done because the data already exist. If the correlation is low, the input is stored.

In Heath's model, there are three temporal parameters. $\alpha(t)$ is memory stability and reflects the likelihood of storage. $\beta(t)$ is the learning rate and reflects the likelihood of new information getting into the system. A third parameter $\gamma(t)$, represents novelty sensitivity, the extent to which any change in the current memory depends on previously stored information. One version of this model provided a good fit to results obtained from human memory experiments on item recognition (Heath, 1989, 1994). The value of α that produced the best fit to the data was approximately 0.7, in line with the stability factor for CA simulations mentioned earlier.

Heath (2000) speculates that stability and plasticity may change developmentally. During initial learning, when encoding and storage are important, plasticity may be greater. Later in the learning process, when stable memory representations are necessary, stability may be greater. Children, for example, seem to demonstrate greater plasticity in their cognitive processing style than do adults. It may also be the case that individuals may have some degree of cognitive control over these factors. Concentration of attention seems to promote neural stability and the formation of memories. The lack of it, as when daydreaming or "spacing out" seems to favor a state of plasticity.

Language

The Lexicon

In the traditional cognitive science view, a word exists as a discrete and context-free symbol: the symbol stays the same regardless of how it is used. Words in this conception are static and thought of as passive objects that are submitted to linguistic processing. They are considered the operands that operators act upon (Elman, 1995). When words are processed in artificial networks however, this traditional notion is challenged. We see that under these conditions, there are no discrete symbols. Word representations become distributed as the set of weights or pattern of activation among nodes in the network. These "representations" are also highly context-dependent. The position of a word in a sentence, for example, determines its meaning and functional role.

Information about words in the classical cognitive view is stored in a lexicon. A *lexicon* is essentially a dictionary that contains word entries along with their various features. For instance, a word's meaning (semantics), function (syntax) and pronunciation (phonetics) might all be listed in a lexicon. The lexicon in the traditional view is something like an array or list. Once a word is activated, its corresponding values are "looked up" for use, much the same way a computer might look up a value in a table. A network conception of the lexicon is quite different. This is the issue that we explore next.

Elman (1990) trained a simple recurrent artificial neural network to predict the next word that would appear in a sentence. This network was a standard three layer one, with input, hidden, and output nodes. Twenty-nine nouns and verbs were used to form sentences. A total of 10,000 sentences were presented to the network, each word presented one after the other as input. The network's activity or output was compared against the actual next word that would occur in the sentence and the weights were adjusted using the back-propagation learning method. When the training period was finished, the network was tested by comparing its predictions to the words that would follow.

For each word, a mean vector of the pattern of activation among the hidden units was calculated. The average was calculated across all instances or contexts in which the word occurred. These mean vectors were considered the network's way of coding for the words. They were submitted to a hierarchical clustering analysis that ordered the vectors according to their similarity. The organization of this tree diagram shows that the network had "discovered" the crucial features of the 29 word set. It was able to separate them based on verbs vs. nouns. Within each of those categories it also made finer level distinctions. For the nouns it further categorized them into animate objects that were either animals or human and inanimate objects that were either foods or breakable items.

These results show that words are not the objects of processing, but inputs that drive processing. They are operators that alter the state of the network.

The different states that the network can be in correspond to different positions in state space (Elman, 1995). Furthermore, each state takes into account the context in which the word currently appears. The same word used in two different sentences will produce two different, although related, patterns of activation. This is because in real life, words are rarely learned in isolation. They are acquired in relation to other words and linguistic data.

A lexicon can thus be thought of as the structural relationship between activation patterns corresponding to a word in context. This can be represented as a state space, where similar words will be closer to each other and dissimilar words farther apart. In the cluster analysis of Elman's (1990) data, the activation pattern for the word "sandwich" was close to that for "cookie" since both are inanimate food nouns. The patterns for "sandwich" and "move" are distant because one is a noun and the other is a verb. Presentation of the word "sandwich" would cause the network to move into the noun region of the state space and from there to the inanimates portion of the space and ultimately, in a veridical response, to the food subsection of the inanimates region. Presentation of the word "move" would cause the network to move into the verb section of the state space and eventually to the particular region of that space coding for that type of verb.

Syntax

Syntax corresponds to the rules a language follows. The classical way of representing such rules is to draw a *phrase structure tree.* This depicts the hierarchical relationships between words in a sentence. For example, an overall sentence can be broken down into a noun phrase and a verb phrase. The words in the noun phrase could be decomposed further into a determiner, an adjective, and a noun. The verb phrase might then consist of a verb and another noun phrase. This decomposition would continue all the way down to labels for each of the individual words. How in the brain is a phrase structure tree such as this represented?

Elman (1991) performed another network training study, this one specifically designed to see if grammatical rules could emerge as distributed representations. The sentences used this time had more complex grammatical structure. There were sentences in which nouns and verbs had to agree in terms of number, being singular or plural, where verbs differed in their argument structure, and where nouns could be modified by relative clauses. The procedure was similar to the one described previously in that the network was fed the sentences one word at a time.

Once again, the network, after sufficient training, was able to predict verb number and other aspects of grammar. It could do this even for complex sentences like: "The boys who the dog chases run away". The network had clearly encoded these syntactical rules in an abstract way, since it could successfully generalize its performance to novel sentences.

As before, the hidden unit activation patterns were considered the network's representation of the grammatical rules. A principal components analysis (PCA) was performed on the hidden unit activation vectors. The PCA analysis provides the dimensions along which these vectors vary. It also allows one to visualize these vectors in a coordinate system, i.e., it provides a simplified state space for which one can see change in the vectors over time.

One principal component corresponded to the number of the main clause subject noun. Variation along this dimension represents whether the subject is singular or plural. A trajectory through this space showed how the network acquired the rule. Other components corresponded to sentences with verbs that have different argument expectations and for sentences that varied in their degree of embedding. In each case a trajectory through the space showed the network's acquisition of the rule.

This work by Elman shows that traditional linguistic constructs can exist as distributed representations rather than as localized symbols. These representations are better thought of as activation patterns among nodes in a neural network. The patterns can be represented in a state space whose dimensions are defined by word features or syntactical rules. Additionally, we can examine change in these activation patterns over time as trajectories in the state space. The traditional structures such as lexical tables and phrase structure trees are useful for understanding language. But they have biased our perception of how this information is represented and processed in real brains for far too long. The dynamical account is a much closer approximation to actual human language processing.

Conversational Behavior

Individuals engaged in a conversation under normal circumstances will take turns, demonstrating anti-phase coordination. But communication involves a motor component as well. People will frequently synchronize their gestures, such as head nodding, to the other's voice. Jaffe and Feldstein (1970) found that conversational turn taking is very often not symmetrical. In pairwise conversations between people in a small group, each pair had stable turn lengths for each speaker, typically with one person spending more time speaking. This result implies that each speaker has a characteristic frequency and that they adjust to the frequency of their partner.

Newtson (1994) reports another phenomenon called *shadowing*, where speakers adopt the posture of their conversational partners. For instance, if one person crosses their legs, within 10-20 seconds, their partner will also usually do so. Synchronous postural coordination during positive verbal exchange is also demonstrated by anti-phase activity of speech and movement (Condon & Ogston, 1967). A person might utter terms of acknowledgement such as "yeah", or "that's right" during a lull in their partner's speech. These forms of synchrony are believed to facilitate smooth and efficient informational exchange (Dittman & Llewellyn, 1969).

Stereotypical behavior patterns are also found during negative communication. If one partner is having difficulty comprehending what another is saying, they typically adopt a "frozen" posture, remaining motionless and bracing the body by for example, planting the hands on their hips or their chin in one hand.

Decision Making

Investigators in the field of decision-making attempt to model how humans decide between two or more choices. These choices can be important, such as whether or not to get married, or they can be more mundane, like choosing which of several movies one wants to watch. Making a decision takes time and is a process. We know this intuitively based on our introspections, which suggest we consider pros and cons, worry about outcomes, and vacillate between options before acting.

Many traditional models of decision-making are static and fail to take the time course of decision-making into consideration. For instance, in *prospect theory* each alternative can be judged on several criteria that have an associated score reflecting their objective importance (Kahneman & Tversky, 1979). These criteria are then assigned weights reflecting their subjective importance to the decision maker. The decision then becomes a simple matter of multiplying each criterion score times its weight and summing to calculate an overall score for each alternative. The one with the highest overall score is chosen

Another classical approach to decision-making involves *utility theory* (Neumann & Morgenstern, 1947). In this approach a decision-maker is assumed to be a rational agent who selects choices that maximize expected utility. Utility is a subjective value the option has and corresponds roughly to happiness. The agent assigns utility values to different choices and then follows specified rules to determine which one will return the greatest value. One of the problems with these theories is that people in real-life situations often make decisions which aren't rational, i.e., that fail to maximize utility.

Neither of these approaches specifies how the decision-making process unfolds over time. The decision is simply a question of calculating a score for the alternatives and picking the one that has the highest. They don't describe the consideration, hesitation, and deliberation that underlie our intuitions. These older accounts gloss over the fact that human preferences can change. They also don't consider how human decisions are influenced by context. Most importantly, the amount of time taken to reach a decision is not part of the calculations (Townsend & Busemeyer, 1995).

Decision Field Theory

Decision field theory (DFT) overcomes some of these obstacles (Busemeyer & Townsend, 1993). In this account, a decision maker is presented with a choice between say two risky alternatives A and B. At each moment in time, the person thinks about the payoffs for each alternative. This produces an emotional reaction to each, called a valence. These valences then determine preference states reflecting their likelihood of being acted upon. Attention also factors into this model. The decision maker may think about the gains associated with a particular alternative but then later switch and think about its associated losses. Attention can also switch between the advantages and losses for different alternatives.

Figure 12.3 depicts a typical decision process. The x-axis reflects time. The y-axis shows preference strength. The trajectories in this space show the preference states for alternatives A and B at each moment in time. A threshold

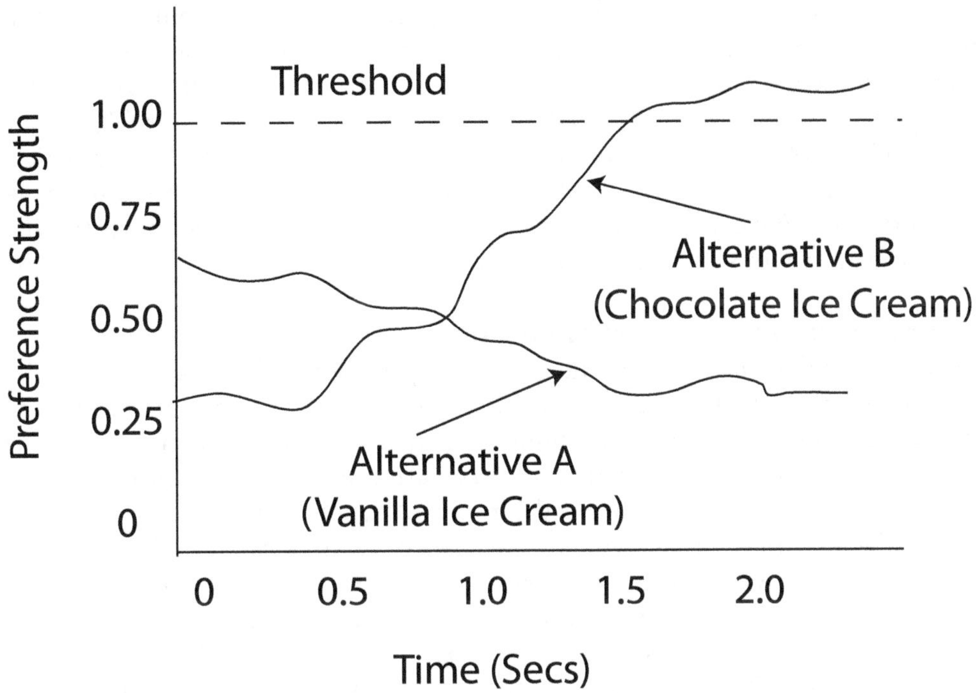

Figure 12.3 *A model showing how a choice between vanilla and chocolate ice cream might unfold according to decision field theory (after Busemeyer & Townsend, 1993).*

determines when the process stops. In this case the high threshold value is set at 1.0. The first alternative to reach this threshold is accepted. In Figure 12.3 this is option B after 1.5 seconds, which is the deliberation time for the decision process. In effect, this model is like a horse race, with each option as a horse. Preference states reflect the speed of the horses and can speed them up or slow them down, either pushing them closer to the finish line, or retarding their progress towards it.

The participant can set their threshold to different values. This crucially affects the outcome. If the threshold is reduced, then a lower preference value will be enough to reach it. In Figure 12.3, a lower threshold could be set to 0.5. Since option A reaches this at 0.75 seconds, it now becomes the chosen alternative. The threshold setting can explain why decisions change under time pressure (Diederich, 2003). When making a speeded decision, there may be very little time to consider and accuracy could suffer. Emotions may play a greater role here and this could reflect the decision making style of impulsive individuals. In contrast, when lots of time is available, more information about the alternatives can be considered and subjects may be more accurate. Under these conditions, cognitions may be more influential than emotions. This could reflect the decision style of people who are more careful and deliberate. These conditions can also explain speed-accuracy trade-offs.

There is lots of support for DFT. Different mathematical models of DFT can be formulated in which parameters are set to different values. These can then be run and the data compared to experiments with human participants.

In many cases, the model has been able to predict empirical deliberation times under various conditions (Busemeyer & Townsend, 1992). It has even been able to explain findings from neuroscience such as the firing rates of neurons during perceptual decision-making in primates (Schall, 2003; Shadlen, 2001). In summary, DFT is a good alternative to traditional static decision-making theories. It can account for the dynamical and temporal changes that take place during decision processes. It can also explain speed-accuracy trade-offs, personality differences and alterations in preferences over time.

A Dynamical Model of Problem Solving

Torre (1995) describes a dynamical model of problem solving as part of his triadic theory of mental processes. In this theory, there are three primary mental processes. They are the cognitive, which involves analytical and rational thought, the affective/perceptive, which involves creative, artistic, and emotional thinking, and the pragmatic, which is utilitarian and experience-based.

Each of these processes calls on three problem solving skills. The cognitive process involves an understanding of the problem, information gathering, and diagnosis. The affective/perceptive carries out tasks involving prognosis or anticipation of the future course of the problem, generation of solutions, and decision-making. The pragmatic tasks are planning, implementation, and evaluation.

Torre's model helps us to understand the dynamical interaction among these nine processes during problem solving. They can be conceptualized as nine different regions in an interactive state space. Problem solving then becomes a trajectory or trajectories that visit these regions. He stresses that they can be visited in any order, but the path is influenced by the specific goal of the overall problem situation. Any given problem will require multiple, perhaps cyclic, visits back to the same tasks. For example, an understanding of the problem might require information gathering and decision-making might require planning. The model is holistic in the sense each of the nine sub-tasks are to some extent mutually dependent on one another.

Torre (1984, 1987) tested the triadic model as part of a course to assist first-generation college students at Northeast Illinois University. He discovered that the students were not up to the cognitive demands placed on them, partly because of their weak pragmatic skills. With increases in stress, they showed a bifurcation in their mental functioning, where they vacillated back and forth between trying to succeed and giving up. With continued stress, Torre notes they demonstrated period-halving bifurcating behavior in the form of reduced attention span and ignoring of potentially helpful environmental cues. Additional bifurcations showed more alternating behavior between opposites such as introversion/extroversion and partying/studying. Notice that in this context, stress acts as the control parameter.

The students next demonstrated fears and insecurities as part of their affective/perceptual process. Although one may think of their poor cognitive adaptation as the root cause of all this, Torre states that there is no primary cause. The model is holistic and each process can affect the others. In this way, the deterioration can spread from one component to another. Cognitive breakdown can lead to emotional and pragmatic problems, but emotional problems can also lead to a breakdown of cognitive and pragmatic skills.

As part of the intervention procedure, Torre (1984) employed a variety of techniques that draw on principles from chaos theory and nonlinear dynamics. In the feedback process, students are made aware of the discrepancies between their intended and actual outcomes. The outcomes are the new attractors that develop from problem solving efforts and the discrepancies are the bifurcations between the intended and actual outcomes. In the feedforward technique, students are informed of the actions they need to take, these actions being the trajectories that will lead them to the new attractors. Torre refers to these two techniques together as self-organization because they involve learning the control parameters and pathways that will allow them to navigate the state space toward improvement.

This method met with success. The students showed a marked increase in problem solving ability and were in some cases able to develop ingenious solutions to problems. Torre recommends that these methods be adopted and applied within the framework of educational psychology, although more research needs to be done.

CHAPTER 13
Problem Solving and Evolution

Evolutionary Processes and Problem Solving

In the last chapter we finished by talking about problem solving. People it turns out aren't the only systems that solve problems. Here, we take an evolutionary perspective and show that the process of biological evolution can also be viewed as a form of problem solving. In this case, the problem is the fit between a species and its environment. Because environments are always changing, organisms must change with them if they are to survive.

Evolution can be conceptualized as a search for optimum fitness on a landscape. We present this framework and talk about the process of optimization. We then introduce basic ideas in genetics and describe Kauffman's (1993) model of genetic adaptation on correlated landscapes. These theoretical ideas are then applied to aspects of human problem solving.

Problem Spaces and Search

Many types of problems can be represented in the form of a *problem space*. This space represents the different possible solutions to the problem. The solutions are arranged so that similar ones are adjacent to one another. In two-dimensional depictions, the width and depth of this space represent ways in which the solutions are related. The height of the space can then stand for how good each solution is with higher points corresponding to better solutions.

The geography of the terrain in this map depending on the type of problem can quite varied and consist of plains, mountains and valleys. The highest peak represents the best solution to the problem and is called a *global optima*. There can also exist a number of smaller hills or peaks that are less than perfect solutions called *local optima*. The valleys of course are lower down and are bad solutions. Figure 13.1 shows hypothetical problem spaces.

When trying to solve a problem we must start somewhere. This starting point corresponds to the first attempted solution. The question then arises as to how we can get from this point to a good or best solution. We can do this in a number of different ways. The particular algorithm we choose determines how we move across this landscape. Mathematical or computer-programming procedures that attempt to find the high points in the landscape are known as *optimization* techniques. The path they trace out is a *search* through the possible solutions.

One particularly inelegant way to find a solution is to use brute force. In this method, one goes systematically through all the solutions, notes how good they are, and then at the end simply picks the best one. This works for problems where there are relatively few solutions. However, it is impractical for problems with a large number of solutions. The time needed to look through these spaces is too great.

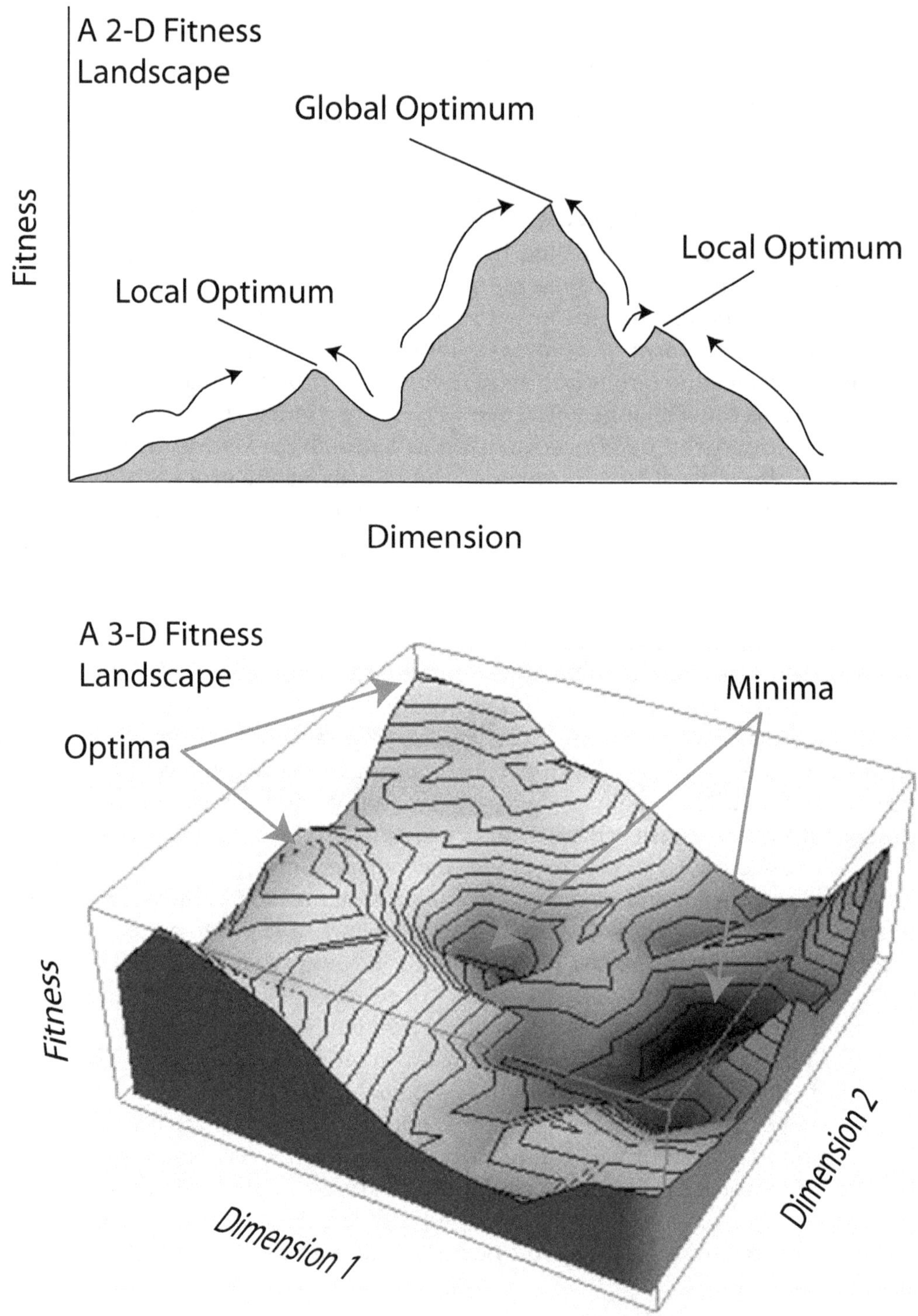

Figure 13.1 *Problem spaces conceptualized as fitness landscapes. Dimension refer to solutions solutions that vary systematically. The height of the surface at a given point represents how good a given solution is. Problem solving becomes a process of "hill climbing" to find the highest peak.*

Another simple way to find a solution is to always "walk uphill". Whatever one's starting point, always take a step in a direction that results in a better solution. This technique has a major flaw. If the landscape has more than one peak, it does not guarantee that you will end up on the highest one. Depending on the landscape and the starting point, the end point may be a local optima. This leaves one "trapped" on top of a small peak with no way of finding their way down.

Simulated Annealing

How then do we escape local optima to find a higher point? One way is to introduce random movement. Instead of always going up, we could occasionally go down. This "leap" off the local peak might be enough to get us into a new region of the space where we could start climbing again, this time to a higher peak.

This technique is known as *simulated annealing*. There are two kinds of moves one can make in this procedure. The first is the gradual uphill move of always picking the higher and locally better solution. But there is also a given probability of making a move in another direction, to a solution that might be worse than where we were before but which could lead to an even higher point. The probability of making these random leaps is a parameter analogous to temperature in chemistry.

Natural physical systems undergo this kind of random shaking up. Molecules of many substances tend to settle in low energy states. A low energy state is a configuration that is stable. But if they are repeatedly heated and then cooled, they move about their state space and ultimately settle into even lower and more stable energy states. In these kinds of spaces, the best "solution" is the lowest rather than the highest point, but otherwise the landscape analogy is the same.

Sword makers are able to produce very hard blades utilizing this process. The key component is annealing, which is a gradually cooling. In this case we are talking about a real physical annealing of a material rather than the simulated annealing of an algorithm. Alternate heating and annealing of a metal makes it extremely hard and durable because the molecular structure ultimately settles on very low and stable energy minima.

Genetics

In the following sections we talk about fitness landscapes and how evolutionary processes can act on them to produce optimum outcomes. These models are based on biological evolution and genetics. In order to make sense of them, it is necessary to review the basic principles of genetics. Those familiar with this topic may skip ahead.

A *chromosome* is a threadlike molecule of DNA (deoxyribonucleic acid) that carries genetic information. As *diploid* organisms, each of us contains two copies of each chromosome, one from our father, the other from our mother. Human beings have a total of 46 chromosomes, 23 from each parent. A single chromosome has many genes or segments that code for a particular trait such as eye color. Each gene is in turn made up of *alleles*, which can exist in one of two versions. In the case of eye color the gene coding for this characteristic might have an allele for brown eyes and another for green eyes.

Diploid organisms reproduce sexually. In the mother there are cells that will eventually become eggs and in the father there are cells that develop into sperm. In these cells in both parents, maternal and paternal chromosomes line up and split, forming recombined chromosomes that contain a mix of traits from both parents. This process is called *recombination*. Half of these (23) are randomly chosen in each parent and go into the egg and sperm. These two halves unite during conception to form a new full complement of 46 chromosomes. The recombination and halving creates offspring with genetic diversity. The children will be somewhat, but not exactly like their ancestors. It is this genetic diversity that selection acts upon.

In comparison, a *haploid* organism such as a bacterium has one copy of each chromosome. A typical bacterium will reproduce asexually by dividing, passing a duplicate copy of its genetic instructions to an offspring. Genetic diversity is created in haploid organisms by *mutation*. A gene in one of these chromosomes might change by chance. This altered gene is then passed to the next generation. If this gene has greater fitness value, it will allow the bacteria that have it to survive and reproduce more effectively.

The number of genes that an organism has can vary widely. The *E. Coli* bacteria has around 3,000 genes. Certain plant species have on the order of 20,000 genes. Humans posses roughly 100,000. If an organism has1,000 genes each with two alleles, it would contain $2^{1,000}$ possible genotypes. This constitutes a genotype space that can be quite vast for even relatively small numbers of genes. *Genotype* refers to the actual genetic content of an organism, the "instructions" that code for traits. These instructions aren't always identical to the actual physical makeup of the organism, known as the *phenotype*. Recessive genes, for example, won't get expressed unless there are two recessive alleles, while dominant genes get expressed if there is at least one dominant allele.

Random Fitness Landscapes

Kauffman (1993, 1995) provides us with the framework within which evolutionary processes operate. He starts with the previously introduced concept of a *fitness landscape*, the ordered space of possibilities that organisms evolve through. If we consider the genetic makeup of an organism, we can describe this space as all the genotypes that it can assume. In this hypothetical space, similar genotypes are adjacent to one another. Each genotype is assigned a fitness value. The higher the fitness value, the higher the height of the landscape at that point. Evolution selects those genotypes that are the most fit in every generation. A population of organisms can thus be thought of "walking" through this sea of possibilities trying to find the highest point, i.e., trying to become as fit as it possibly can.

We can order the genome or space of genotypes by placing similar genotypes next to each other. Let us illustrate using an organism that has four genes, each with two alleles. We represent each allele as a 0 or a 1. In this case, there are a total of $2^4 = 16$ genotypes. One genotype is 0000 which has the 0 allele for each of its four genes. An adjacent genotype would be 1000 that is similar since only the allele of the first gene has been changed. We can diagram the relationships between these possibilities using a four-dimensional Boolean hypercube shown in Figure 13.2.

Each vertex of this hypercube stands for one of the sixteen possible genotypes. The number of neighbors each genotype has in the hypercube is equal to the number of genes, which in this case is four. The direction which one travels in the cube corresponds to a dimension or change in one of the alleles. If we start with 0000 in the upper right corner then traveling left produces 1000, so left correspond to a change in the first allele. Moving down takes us to 0010 so downward motion means flipping the allele of the third gene. Moving forward changes the allele of the second gene resulting in 0100. Finally, moving inwards to the corner of the smaller cube embedded inside the larger (the fourth dimension) yields 0001 and changes the last allele.

Each genotype has a fitness value which determines how likely it is that an organism with that particular makeup will survive and reproduce. Since we don't know what these traits are, for the sake of the model we are going to start off by randomly assigning fitness values to each. These values will range from zero to one, with values closer to zero being less fit and values closer to one being more fit. For simplicity, we rank order the genotypes based on their fitness values, with the highest fitness being a 16, the next lowest fitness being a 15, etc. The second hypercube in Figure 13.2 shows the rank order fitness of each genotype.

What results is a *random fitness landscape*. The landscape is the space of possible genotypes that a population of organisms can transition through. To illustrate, imagine that we have one organism with a genotype of 0000. If this is a haploid organism, then there are four mutations it can undergo corresponding to a change in one of its four alleles. But a mutation will only be selected for if it has a higher associated fitness value. This means that the next generation of organism can "move" only to adjacent locations with higher fitness. In the hypercube example, 0000 can go down or inward but not left or forward. Moving down brings it from a fitness of 6 to 13. Moving inward raises the fitness to 16. If it moved left or forward, the fitness would drop to 4 or 2, respectively. Figure 13.2 has arrows indicating directions toward greater fitness.

This movement through the fitness landscape is called an *adaptive walk*. The rules for an adaptive walk are simple. Start at any point. Mutate an allele at random. If the mutation produces a higher fitness, accept the change and move to that new genotype. If the mutation does not produce a higher fitness, do not go to that location. Instead, randomly mutate again and go to this genotype if it is fitter. In reality, a "bad" mutation with a lower fitness would probably kill the resulting organism in which case the parent organism has a second chance by producing another offspring. This type of walk is essentially what happens in evolution by natural selection.

Notice that the walk stops if the starting point has no path to higher fitness. These points are the local optima. They represent the highest fitness that

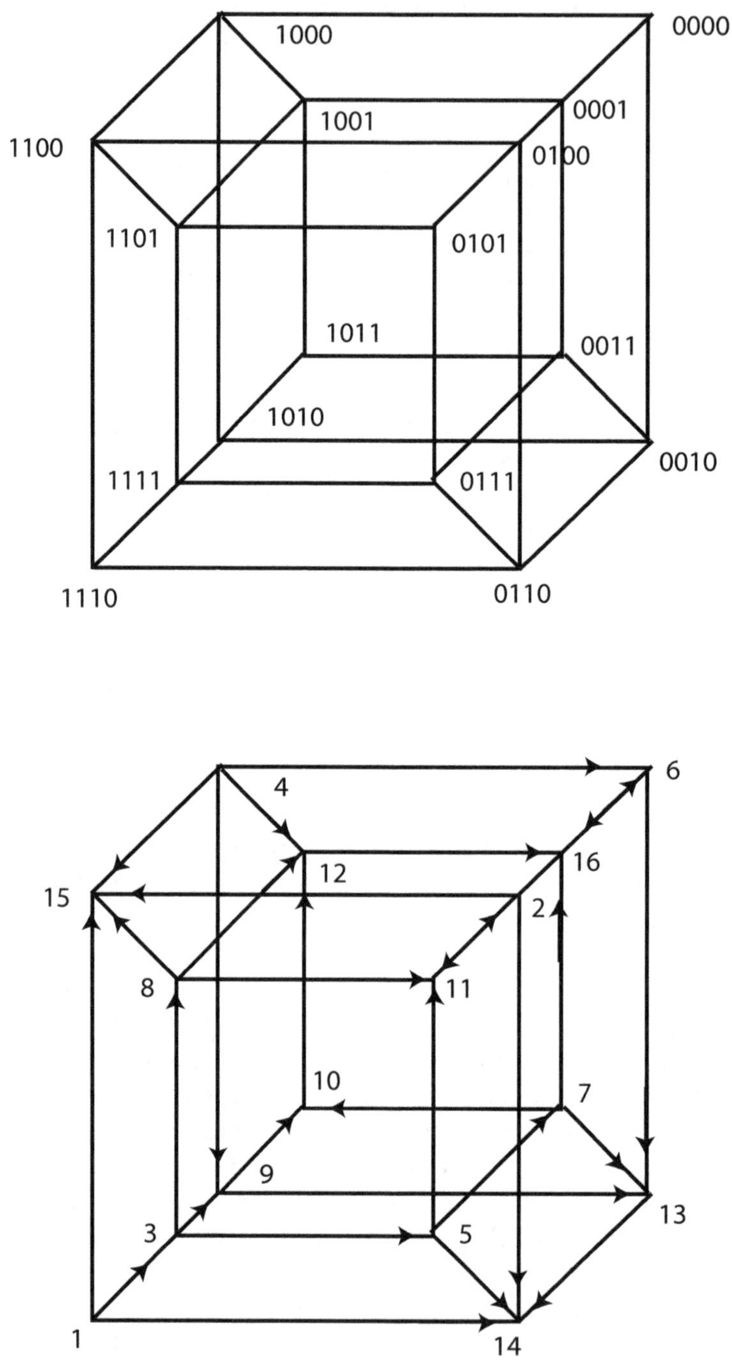

Figure 13.2 *Two four-dimensional Boolean hypercubes. In the top figure each corner or vertex is labeled with a different genotype. The bottom figure shows their associated fitness values. Evolution can be considered a walk along the edges from one vertex to another always traveling toward higher fitness (after Kauffman, 1993, 1995).*

can be achieved given a particular starting point. For any given space, there is also a global optima, which is the highest possible fitness that can be attained in the space. The 0001 has a rank order fitness of 16 and so is the global optima in this instance. 1100 and 0110 are both local optima with values of 15 and 14. For every starting point the walk will eventually settle on a local optima. The adaptive walk is thus always "uphill".

Correlated Fitness Landscapes

Adaptive walks will always march upwards until they reach a local peak. This peak represents the highest point in the immediate region but can be much lower than the highest peak in the entire landscape that represents the best solution. Organisms are "trapped" on these peaks because they cannot travel down decreasing fitness paths to find another summit. The top portion of Figure 13.1 that was pointed out earlier shows a fitness landscape with arrows indicating the direction of adaptive walks.

Locating the best solution in random fitness landscapes is difficult for another reason. These landscapes contain a huge number of local peaks, so many in fact, that it would take longer than the history of the universe to search them with even a moderate number of genotypes. So uphill searches are not an effective strategy in this type of terrain. Evolution cannot operate successfully on random landscapes.

Kauffman (1995) argues that evolution can only operate well on *correlated fitness landscapes*, terrains where changes in height are more gradual and predictable. In this geography an adaptive walk stands a decent chance of finding a good solution. Continuing to follow a path uphill is a reasonable strategy to adopt on this type of ground, since there is a good chance it will bring one to a high peak. Correlated landscapes can be smooth, with gentle rolling hills or more rugged with steeper mountains and valleys.

The *NK* Model

In reality, traits don't exist independently of one another. They work together to contribute to the organism's fitness. Let us take two traits, bone density and body weight and assume each is governed by two alleles. One allele for bone density codes for high density or thick bones, while the other codes for low density, lightweight bones. Likewise, the two alleles for body weight code for high and low body weight. Clearly, there are some combinations that will go well together, these being high bone density and high body weight and low bone density and low body weight. An animal stuck with low-density bones and a heavy body is at a disadvantage. Their skeleton may not be able to support their weight.

This linkage between genes is called *epistatic coupling*. Any given gene in this view must be considered not just in term of how it alone acts, but how it acts in concert with the other genes to which it is coupled. In effect, we have a genetic network with links between genes that influence each other. The fitness of a particular gene is a consequence of both itself and the others to which it is connected.

The *NK* model takes this into consideration. It allows any *N* genes to be coupled with *K* other genes. The fitness contribution of a gene is then a consequence of the fitness of its allele plus the alleles of its connected neighbors. It is difficult to know the joint fitness of such allele combinations, so an easier approach to the problem is to just assign each combination a random fitness value. The average fitness of the organism then becomes the average of these fitness values.

We can create a fitness space for the *NK* model the same way we did for the random fitness landscape. For example, we could start with three genes ($N = 3$) each with two alleles ($K = 2$). The eight (2^3) resulting genotypes would be randomly assigned fitness values. These genotypes can be represented as vertices, this time on a three-dimensional Boolean cube. Movement between these corners along directions of increasing fitness ends up on local optima, just as before.

So what is so important about the *NK* model? It is special because varying the amount of coupling varies the number of peaks and their ruggedness in the landscape. The more coupling there is, the higher the value of *K*, the greater the number of conflicting restraints. A conflicting constraint is a combination of alleles that is detrimental and lowers fitness, as in the light bones-heavy body combination mentioned earlier. The result is a landscape with many compromise solutions and few outstanding global solutions, one with lots of short local peaks and few very tall mountains.

When $K = 0$ and genes act independently, we have a single tall peak with a smooth gradual slope. Adaptive evolutionary walks work well here, although they take longer to move up this high mountainside. When *K* is maximal at $N - 1$, the result is the by now familiar random fitness landscape with rugged terrain, filled with lots of short local peaks. In the natural world, *K* is seldom near these extremes. Real world landscapes are typically at intermediate values in-between and have varying degrees of ruggedness.

There is another fascinating landscape property that emerges when we vary the amount of epistatic coupling. When *K* is at moderate values, high peaks can be accessed from a large number of starting positions. In other words the landscape at these values has good solutions that are easily reached. The mountains all have a large surface area, so that adaptive walks that begin on them are all capable of reaching the top. This appears to be an emergent characteristic of the *NK* models. The implications are enormous. It means that adaptive processes seem almost "designed" to work well with the sorts of ways that organisms change.

An Expanded View of Evolution

Is it mere coincidence that evolutionary adaptation works best in *NK* landscapes with the right *K* value? Probably not. Evolution itself may have selected for the right amount of epistatic coupling. After all, genes and their interactions are exactly what changes in natural selection when there is a change in the environment. However, Kauffman (1995) argues that selection is limited and may not be the only factor influencing the shape of fitness landscapes.

He believes that the second key ingredient is self-organization. This natural tendency for the world to form into ordered, stable and persistent structures generates the forms upon which selection acts. This type of stability is necessary. If things changed too quickly, there would be no substrate for evolution to select. An organism needs to stay the same long enough so that it can be adapted gradually over generations.

Complex systems on the edge of chaos can be in ordered or chaotic regimes. When they are ordered, they are stable and better suited to adaptation on smooth landscapes. But what about when they are chaotic? Chaotic systems are unstable. They seem better suited to optimizing on random uncorrelated landscapes. A chaotic system can undergo large changes in response to small environmental perturbations. This could enable it to jump from one local optimum where it is stuck to another location where it might stand a better chance of finding a higher peak. These "leaps" across a landscape in conjunction with some degree of order could allow a system and to find a better solution than if it was completely ordered.

The right balance between order and chaos probably depends on the nature of the system. In biological evolution chaotic change is not possible. An organism can't produce radical alterations in its genome to escape a local peak. But there are other domains where this can be done. One example is human-problem solving. Here, an individual can radically change an idea to see if it poses a better solution to a problem. We talk more about landscapes and their role in problem solving in the next section.

Mental Landscapes and Problem Solving

Evolutionary processes are a specific instance of problem solving. The problem is survival, or more specifically, adaptation at the species level to the environment. The variation found across individuals in a species represents possible solutions to this problem. Successful solutions are preserved through selection, which acts as a feedback process. All of these elements are present in human problem solving. There are candidate solutions that are generated. These are applied to the problem to evaluate their effectiveness. The solutions that work are used and the ones that don't are discarded.

So the conceptual framework of fitness landscapes, adaptation and optimization can all be used to study human problem solving. There are a number of interesting questions that arise. First, what are the spaces that underlie different problems? It is probably the case that the structure of these spaces differs depending on the kind of problem. The landscape must be correlated if an adaptive method is to work. A correlated landscape has candidate solutions that are similar to each other and that progressively lead toward a good or optimal solution.

One type of problem solving that has been studied by cognitive psychologists is *analogical reasoning* (Gick & Holyoak, 1980). Here, a schema or existing knowledge structure is mapped onto the problem and used to solve it. For example, if someone has never traveled by train, they could call on their bus travel knowledge to figure out what to do. They might then know to go to the ticket window in the station, buy a ticket, find out what platform and when the train departs, etc. Each of these elements were part of the their bus travel schema and have been applied to the current novel situation.

For any given situation, there are a family of candidate analogies that could be brought to bear, each varying in their similarity to the target situation and to each other. The greater the similarity of these analogies to one another, the closer they are in the landscape. The similarity metrics or dimensions of the space need to be discovered. In our example, travel by bus may be more similar to trains in that both are land-based forms of travel. But travel by air could also be considered similar to train travel in that they are used more often for long distance voyages.

Another consideration is the starting point. Individuals will vary in terms of where they start in the problem space. The closer to the solution they start, the more quickly they should converge on it, all other things being equal. A variety of factors including personal history will influence starting point location (see heuristics below). Another important issue is the type of feedback the problem solver receives. They need both negative and positive feedback. Negative feedback in the sense used here tells them that a solution won't work and that they should try something else very different. Positive feedback tells them that a solution is on the right track and that they should try something else but along similar lines.

One problem solving issue noted by Gestalt psychologists is fixation. *Functional fixedness* occurs when a problem solver focuses on one aspect of the problem and consequently persists in applying inadequate solutions (German & Defeyter, 2000). One classic example is the candlestick problem (Dunker, 1945). Participants are asked to attach a candle to a door using materials given to them in a box. Many fail to realize they can place the candles inside the box, seeing it only as a container to hold all the materials. Another example is the nine-dot problem. Here, participants are asked to connect all the dots by drawing several straight lines. Many fail to realize they can go outside the rectangular boundary imposed by the dot array.

Functional fixedness in common parlance means failing to think "outside the box". In the landscape model, it is equivalent to being stuck in one region of the problem space. The solution is to travel in a different direction. So we have another question: why do problem-solvers fail to try new solutions when the ones they are using fail? Rephrased another way, why are some local optima easier to "back out of" than others? An examination of the space in that region of the landscape might reveal some answers.

Clearly, human problem solving does not perfectly fit the adaptive walk model. In an adaptive walk, all steps are small and between adjacent solutions. Steps are also all uphill so one cannot back down from a local peak. These limitations are not present in human problem solving. It is possible to try a radically different approach, equivalent to a leap from one point in the landscape to another that may be far distant. It is also possible to head downhill, backtracking from an unsuccessful solution path.

These observations suggest that humans use different optimization procedures when problem solving. Experimental work might reveal what these are. It may be that people alternate back and forth between a variety of different techniques. For example, one could start off by randomly sampling the space of possibilities until finding a candidate solution that seems right and then performing an adaptive walk in that region. If the walk fails to yield a good solution, one could resort back to sampling over wider areas. This technique of switching back and forth between global and local search strategies might prove to be very effective.

There are many other forms of problem solving in addition to analogical reasoning. The Gestalt psychologists studied *insight learning*, where a solution comes suddenly after some time spent away from the problem (Kohler, 1976). Researchers in cognition and artificial intelligence have also approached problem solving using the idea of a path through a problem space (Anderson, 1985). Much of this work is based on the notion of *means-end analysis*, where the ultimate solution is treated as a goal that is approximated through the formation and attainment of sub-goals. Many of these approaches are amenable to study using the landscape model.

Computer models implementing the means-end approach rarely perform an exhaustive search through the entire space of possible solutions. That is because the number of such possibilities as we noted earlier is astronomically high. In chess, for instance, it would take years for a computer to search exhaustively through even a small portion of the game tree. Instead, researchers, taking a cue from human problem solving, have the program start searching in places that are likely to yield answer. This involves the application of a heuristic.

A *heuristic* is an informal "rule of thumb" that does not guarantee a solution but is faster and easier to use than systematic search. If you lost your keys, rather than looking in every possible location, you would probably first look in those places that you have most likely left them, such as your desk or the table near the door. Heuristics imply some knowledge of the landscape, some way of knowing where the best peaks are to be found. This is based on experience. New optimization techniques should further explore experience-based knowledge of this sort as well as other strategies used by human problem solvers.

CHAPTER 14
MENTAL ECOLOGY

Mental Ecology

An *ecosystem* is a collection of living organisms interacting with one another and with the nonliving aspects of a particular environment. Many different ecosystems exist on earth. Some, such as rainforests and coral reefs contain an extraordinary number of coexisting species. Others, like deserts and tundra contain fewer species. The distribution of species in an ecosystem can be quite varied, with some species found in great abundance while others are rare. Researchers are just beginning to discover the complex interactions between species in many ecosystems.

In this section we wish to draw an analogy between ecologies made up of living organisms and what we call "mental ecologies". There is substantial fertile theoretical ground here. Much of what has been learned in the study of traditional ecosystems in the external world outside of our bodies can be applied fruitfully to help us understand our rich internal mental world.

A *mental ecology* is a theoretical construct for understanding the dynamics of mind. Many of the notions derived from traditional ecology can be mapped onto neural and cognitive domains. For example, an individual organism in traditional ecology can be considered as activity in a neuron or group of neurons in the neural domain or as a single idea or memory in the cognitive domain. A species is then the equivalent of activity in a neural population looked at from a neural perspective or a collection of related ideas or memories in the cognitive framework. Table 14.1 shows a comparison between the three areas.

In what follows we summarize important findings in several different areas of ecology. We then take these findings and apply them to the brain and mind. The themes that become apparent are that ecosystems demonstrate self-regulation and emergent order. These are necessary if the ecosystem is to remain stable and to survive. We see evidence implying that ecosystems may be able to regulate their physical environments, keeping conditions like temperature and humidity constant. Ecosystems, regardless of the particular species that make them up, also show universal characteristics in their food webs, their defense against intrusion and the way they react to missing members or habitat destruction.

Homeostatic Environment Regulation

According to the *Gaia hypothesis*, the entire earth can be considered as a single large organism made up of the interconnected collection of its ecosystems. In the traditional biological view, the organisms on earth are believed to be passive. Changes in the environment serve as selection pressures that act on species, altering them over generations in the process of evolution. However, there

Characteristic	Ecology	Brain/Mind
Environment regulation	Carbon dioxide, temperature, humidity (possible)	Blood brain barrier, active transport, glial cells, active vision
Network organization	Food webs and species interdependency	Neural networks, propositional networks, module connectivity
Immunity and protective effects	Species defense and intruder immunity	Interference effects in artificial networks and memory studies
Competitive effects	Species competition for shared resources	Similarity-based interference in working memory and language tasks
Reorganization and coherence following item removal	Recovery following species deletion	Filling in processes, amodal completion, illusory contours, reconstructive memory
Reorganization following damage	Habitat fragmentation and extinction	Recovery following brain damage, collateral sprouting, release from inhibition

Table 14.1 *The many parallels between ecosystems and mental systems.*

is some evidence to suggest that this interaction is two-way and that species can also change their environments.

The British biologist James Lovelock proposed the Gaia hypothesis (Lovelock, 2000). Support for this notion comes from a computer model called Daisyworld that he constructed. In this model there are both dark and white daises. The dark daises grow best in dim cold conditions because they trap the heat and serve to warm the virtual "planet". White daises grow better in warmer conditions where there is more sunlight. They reflect the sun and promote a cooling effect. If the temperature in this model is gradually increased the dark daises flourish at first but then die off when temperatures get too high. At these higher temperatures the white daises flourish, but they themselves will diminish in number if temperatures rise even higher.

If the temperature in this model is increased in the absence of any daises, there is a steady linear increase. If it is increased within a given range with the two populations of daises it remains relatively constant. The result is that the dark daises boost the temperature in the lower range while the white daises lower the temperature in the upper range. The flowers thus act as a homeostatic mechanism. They operate in much the same way as a home thermostat that senses temperature and acts by turning a heater on or off. In this case the "heater" is the number of dark daises and their heat-trapping properties while the turning off of the heater corresponds to the number of white daises with their heat-reflecting properties.

The Daisyworld model shows that in principle it is possible for an ecosystem to maintain an environment that is optimal for the life in it. But could such a mechanism be at work in the real world here on earth? A number of scientists have criticized Lovelock. They argue that the model is simple, having too few species and only basic interactions between them. Real world ecologies can have many thousands of species interacting in very complex ways.

In response to these critiques, Lovelock has created a version of the program that incorporates three trophic or feeding levels. It brings in twenty species of daises, five species of rabbits to eat the daises and three species of foxes to eat the rabbits. Instead of showing chaotic or random behavior, the new model demonstrated stability. It regulated the temperature at a near constant level, just as before. The model was even stable under perturbation. Killing off some of the daises caused a temporary drop in the rabbit and fox populations and a brief fluctuation in temperature, but the populations recovered and the temperature returned to constancy.

The jury is still out on ecosystem regulation of the environment. However, researchers are now more accepting of two-way interactions between biological systems and their physical and chemical environments. Articles published years after Lovelock first suggested Gaia refer to biological mechanisms for lowering carbon dioxide levels and global temperature.

Self-Regulation and the Brain/Mind

Just as the species in an ecosystem need a proper physical environment to function properly, so does the brain. There are a number of support mechanisms in place that allow the brain to function correctly. The blood-brain barrier prevents harmful substances from entering the brain from the bloodstream. It also provides for the active transport of glucose, the brain's only fuel supply, across the barrier. The skull provides protection against impacts that might occur during a fall or fight. The spaces defined by the meninges and ventricles are filled with cerebrospinal fluid that provides structural support and service other functions.

Glial cells also play a huge role in brain regulation. They manufacture myelin to help speed the conduction of the action potential down the axon and are involved in response to injury or disease. Glial cells are no longer viewed as passive bystanders to neurons. Some release neurotransmitters on their own. Others play a role in clearing transmitter from the synaptic cleft. Yet others produce molecules that modify the growth of axons and dendrites. These functions suggest a two-way cyclical feedback flow between neurons and glia, where each communicate to the other to regulate and maintain proper nervous system operations.

If we consider a supportive environment more broadly, then we must take into account not only the body's internal physical conditions that support the brain but the external informational conditions conducive to the proper operation of a mind. Mind could not function without access to the world through the senses. Perception brings information about the world into the mind and serves as the foundation for concept formation. Without percepts, there would be few concepts. Without concepts there would be little linguistic and reasoning ability.

But does the mind self-regulate the kind of perceptual input it receives? The answer is yes. Saccades made during object viewing direct the eyes to areas of maximal information, such as contours. Fixations and saccades made while reading show that the eyes move back to previously read material in order to disambiguate and provide meaning. Much of this happens rapidly and subconsciously.

The ability to gather scene information dynamically and selectively is called *active vision*. This capacity shows that the mind is not a passive receiver of data, but guides and controls perception. Information flow is thus bidirectional. Stimuli come in from the senses producing perception and triggering motor commands that change future perception. There is a cyclic relation between the environment and the mind. Environment structures mind, but mind regulates the kind of environment it perceives.

Food Webs

Food webs, or food chains as they are sometimes called, describe the feeding relationships between species in an ecological community. Typically these are depicted graphically with boxes representing individual species and arrows between the boxes representing a feeding relationship such as prey, predator or parasite. Food webs have been studied for some time. One of the interesting findings to come out of this research is that the food webs for different ecologies share many characteristics in common. The length of the food chains, measured as the number of links between who eats whom from the bottom to the top, and the ratio of predator to prey species vary within a fairly narrow range across many different ecological communities.

Why should this be the case? The number of possible types of relationships that could exist is much greater than what is actually observed. One answer is that these relationships are the consequence of emergent order. They arise naturally from the dynamics of species interactions within the ecology because they convey some advantage to the system as a whole. This type of ordering in all likelihood promotes stability. For example, having too many competitors for the same food source would drive all species trying to access this food to extinction. On the other hand, having too few competitors might cause a population explosion and have devastating downstream effects on other species.

Webs and the Brain/Mind

The study of ecology food webs has some important ramifications for the study of mind. There is a long history in cognitive science of using web-based models to understand all sorts of cognitive phenomena (Friedenberg & Silverman, 2006). In a *process model*, a box typically stands for a particular computation while arrows stand for information that is input to and output from the computation. Process models can effectively convey the broad stages of information processing in a cognitive system. A classical example is the Atkinson and Shiffrin (1968) modal model of memory. Others include the capacity model of attention (Kahneman, 1973) and numerous models depicting the various stages of visual pattern recognition.

Researchers studying artificial neural networks of course adopt a web-based approach to modeling. In these models, circles typically stand for neurons and links as the synaptic connections between them (McClelland & Rumelhart, 1987). *Semantic networks* have been used to represent how items in long-term memory are organized. Here the nodes in the network are concepts such as those denoted by words and the links are the relationships between them. *Propositional networks* go one step further by representing a larger number of linguistic relationships such as agency, subject, and object. These can be used to stand for sentence-like information.

Food web models describe the energy flow through an ecosystem with energy being more diffuse at lower levels in such organisms as plankton and plants and becoming more concentrated into fewer species such as carnivorous animals at higher levels. There is a rough parity here with the brain. Energy in the form of glucose is fed diffusely to the brain from the circulatory system. From there it is concentrated in specific neural regions depending on the type of cognitive task being performed. Imaging techniques such as positron emission tomography (PET) are designed to track this energy flow through the brain.

In many cognitive models, spreading activation travels through the network, activating the nodes it encounters and tracing out the course of information processing. This spreading activation is energy. At the neural level it corresponds to post synaptic and action potentials. At the cognitive level it is equivalent to the flow of ideas or the stream of consciousness. The more fine-tuned our ability to track and record energy flow in the brain, the better our understanding of its dynamic processes will be.

Species Defense

Post and Pimm (1983) have built computer models of ecological systems. In one of these models they first specify the needs of different species, giving them behaviors, the amount of territory they need, what types of and how much food they need, and what other species they prey on or are predated upon. They then added each species into the virtual ecosystem one at a time and examined how well the system functioned as a whole.

They found that new species could be added to ecosystems with a small number of existing species, those with about twelve. Systems with larger numbers of species were difficult to invade. It was as if the ecosystem created an "invisible protective network" that resisted intrusion by newcomers. The effect was even stronger for older ecosystems. How could this be? One explanation has to do with individual species. It could be that over time through natural selection, some species became better adapted to the environment and so would be more fit than newer invading organisms.

However a finding by Drake (1990) shows the effect is more general. In one model, Drake took 125 species of plants, herbivores and carnivores. He then varied the order in which they were added to the system. What he found was that no matter what species were involved, the ecosystems all became resilient to invasion. This shows that the type of species involved is irrelevant. Any ecosystem that is sufficiently large, in this case about 15 species worth, will develop this type of resistance. These results suggest that intruder "immunity" is an emergent property of ecosystems. It makes sense that ecosystems would evolve toward species stability. If this weren't possible, life on Earth might be too chaotic or competitive to persist for very long.

Species Defense and Interference

There is an interesting analogy between adding species to an ecosystem and certain effects in learning and memory. In artificial neural networks, *interference* occurs when two sets of information that are similar in content interfere with one another (Green & Vervaeke, 1996). Small networks trained to learn large numbers of patterns demonstrate interference. They have difficulty distinguishing similar patterns from one another. In some cases, networks trained on one set of patterns and then on a second set later, will demonstrate *catastrophic interference.* The new information erases what was acquired previously (French, 1999).

Interference effects are also commonly found in memory experiments with human participants. Many classic early studies in memory have shown that items can interfere with one another (Klatzky, 1980). Memory for words in the middle of a list is typically worse than for those near the beginning and the end. The hypothesized explanation is that there is both proactive interference from the earlier words and retroactive interference from the latter words upon those in the middle.

Species defense suggests nature may have found an emergent solution to the problem of interference. Let us think of the intruding species as new information and the existing ecosystem as previously learned material. The new material becomes consolidated in such a way that it is difficult for new information to gain a foothold. How? One explanation posits that the memory items in the old set after learning have become highly associated with one another. The result is that these items cohere well but have few ways left to relate to novel material. It is as if each item had a fixed number of associative links. Once association with items in the first set has used up these links, they cannot easily be connected to new incoming items.

Species Competition

One way to visualize the dynamic interactions between species is to use a *coupled map lattice.* This consists of a discrete spatial grid of sites that can be occupied by one or more species populations (Kaneko, 1992). The rate at which the species grow and spread through the grid and the extent to which they compete with each other are varied as parameters.

Sole, Bascompte and Valls (1992) used coupled map lattices to show how two species in the same region interact. When the species don't compete with each other and can coexist peacefully, they spread out over time and come to occupy the entire grid. But when the competition parameter is set sufficiently high they come to dominate mutually exclusive regions of the grid. If the population of a dominant species is high in one area, the population of the second species will be correspondingly low, and vice versa. The result is a three-dimensional plot that resembles a photographic negative. Peaks for one species become the valleys of the other.

Species Competition and Interference
Species compete because they must access finite shared resources. The more similar two species are to one another, the more they will compete. That is because similar species have similar needs. They eat the same food, are preyed upon by the same predators, and range over the same type of terrain or geography. One can make the same argument for cognitive processes. Generally, the more similar two items are to one another, the more they will interfere with each other and the harder each will be to process.

Similarity-based interference effects have been found for working memory and language comprehension. Gordon *et al.* (2006) found that syntactical processing suffered when participants were asked to retain two similar noun phrases in working memory. A number of studies demonstrate retroactive interference in short-term memory across multiple codes such as conceptual/semantic, tactile, verbal/phonological, and visual (Logie, Zucco & Baddeley, 1990).

It would be interesting to apply coupled map lattice analysis to the study of similarity-based interference in memory. One could define the dimensional space for the particular task and see to what extent the items occupy which areas of the space. For example, it may be possible based on brain imaging or neurophysiological recording techniques, to identify the locations where the items are stored. If the items are similar and stored in the same region, then they may dynamically settle on mutually exclusive regions in much the same way species populations do in the computer modeling studies mentioned previously.

Species Deletion and Coherence

Another demonstration of self-regulation and emergent order in ecologies comes from cases where a single species is removed from a functioning ecosystem. In one study, a "keystone" species, one that regulates the abundance of other species, was removed from a rocky intertidal community (Paine, 1966). Instead of collapsing, the ecosystem reorganized and stabilized once more. The new community had a different group of species in it but the distribution of those species was nearly identical to the previous one. Although the actual members of the community had changed, its collective properties remained the same.

This is a good example of *coherence* in complex systems. Coherence occurs when a system reorganizes after some alteration such as a deletion of one of its parts. The reorganization preserves the overall function of the system. We see coherence in the business world. A company could have a completely different set of employees after several decades of operation. Yet, if these new people perform the same tasks in generally the same way as their predecessors, the company from a global perspective has not changed. Coherence emphasizes the relationships between parts rather than their identity. If the relations are preserved the functional operation of the whole is maintained.

Species Deletion and Filling In Processes

The reconfiguration of an ecosystem following removal of a species is reminiscent of filling-in and reconstructive processes in cognition. There are numerous instances in perception and memory where these occur. The most obvious is the *blind-spot phenomenon*. There are no photoreceptors located on the retina where the optic nerve leaves the eye. Despite this, we do not see a black spot. The visual system dynamically fills in the missing region (Ramachandran, 1992).

Another example from perception is *amodal completion*. When presented with an object that is occluded or covered by another, we perceive the covered figure as complete. This effect disappears when the occluding object is removed. In this case, we perceive a truncated figure, one that appears to have been chopped off. Another instance of filling-in during visual perception are *illusory contours*. Here, we perceive a complete object even thought the lines that

Figure 14.1 *The Kanizsa triangle is seen as part of a filling-in process where illusory contours are created.*

make up its border are absent. Figure 14.1 shows the famous Kanizsa triangle. The visual system creates the perception of a triangle based on the inducing figures, which are the straight lines and pac men shapes.

The reconstruction of previously learned events in memory has been well studied. In a classic experiment, Loftus (1975) showed participants videos of an automobile crash. Participants were then asked how fast they remembered the cars traveling. The questions were phrased to bias either slower or faster speeds. For example, one group was asked how fast the cars were going when they "bumped" into each other. Another was asked how fast they were going when they "smashed" into one another. The faster the speed implied by the verb in the question, the faster the participant's estimates. In *reconstructive memory* of this sort, information that comes subsequent to an initial memory is integrated into the previously learned items and changes their content.

The literature on filling in processes in perception and memory shows that cognition is dynamic and holistic. We don't reproduce a literal version of what the senses provide at a given instant in time. Instead, we actively create percepts and memories "after the fact". Previous history plays a role here. Our knowledge of what a person looks like probably helps us to fill in his missing parts. But context and grouping cues are also important. In the Kanizsa triangle, the missing contours are interpolated based on good continuation of the lines in the pac men figures.

A percept or concept then, is not just the sum of its parts, but a gestalt that takes into account context, part relations and history. There is information both in the stimulus itself, but also outside of it, in the form of the scene or pattern surrounding the stimulus and in the experience of the organism interacting with it. This and other information in a sense "over-specify" the stimulus and contain information on how to construct it if pieces are missing. Similarly, an ecosystem is more than the sum of its species. The relationships between the species themselves and with the surrounding physical environment contain information on how to reconstruct the ecosystem if a part is missing. Complex systems cohere. They are robust to degeneration within limits.

Habitat Fragmentation

Humankind presents an increasing problem for the survival of species on our planet. More now than at any other time in history, human populations are intruding on animal habitats. The destruction of these habitats by human action can in many cases lead to species extinction. A number of researchers have used computer modeling techniques to try and understand the effects of this damage.

These models typically start with a *metapopulation*, a group of local populations that can move from one site to another on a two-dimensional grid. Each site on the grid can be in one of three states. If the site is destroyed, no population can exist there. If it is intact, then it can be empty with no species there, or it can be occupied by a species. Bascompte and Sole (1996) ran a metapopulation model where species could only migrate locally, meaning they could only move with a certain probability to one of four adjacent positions from their current location. They allowed habitat loss to occur at random in progressively larger amounts and measured the resulting effects on the populations.

Surprisingly, an increase in the percentage of habitat destroyed did not result in a proportionate linear decrease in species population. There was a critical percentage of destroyed habitat. Below this threshold, a large percolating cluster connects many points on the grid and populations can migrate to occupy empty sites. In this domain, there is a gradual and continual decrease in occupied sites. Above the threshold, the percolating cluster is destroyed and the site fragments into disconnected islands. Migration is restricted and there is a rapid drop in occupied sites resulting in ultimate species extinction.

The above example had only one species. More complex models have been run that simulate habitat fragmentation on multiple different populations. These models assign a different rate of colonization to each species and allow for species interaction, i.e., for more than one type of animal to occupy the same site. They also allow for the possibility of immigration, where new species from outside the lattice can enter and occupy sites. As before, we progressively destroy more and more of the habitat in a random fashion and observe the effects.

Below the critical destruction threshold most of the species survive. Their numbers fluctuate up and down, characteristic of actual population numbers in the wild. But beyond the threshold most of the species die out. Only a few remain, including those that enter by immigration, but few actually survive.

Habitat Fragmentation and Brain Damage

The neural equivalent of habitat fragmentation is brain damage. Just as a species reorganize following habitat destruction, neurons and cognitive function also reorganize after brain damage. Damage to neural tissue can result from physical trauma, as when a person injures their head in a car accident, or from a stroke, when collapse of an artery denies a region access to the circulatory system. It can also be induced experimentally as is the case when researchers deliberately lesion or destroy neural areas.

In the habitat fragmentation model a species can survive by migrating to a new location, one that is still intact. In brain damaged patients an analogous process may occur. A copy of the neural circuit that was damaged or destroyed may "migrate" in the sense that it could be duplicated elsewhere in the brain or recreated from other parts of the memory that are undamaged.

There is evidence showing that multiple copies or codes of a given memory are distributed across many different brain locations. In a classic study, Lashley (1950) trained monkeys on a task and then systematically destroyed greater portions of their cortex. He found that even with most of their cortex removed, the monkeys still retained some residual memory for the task. He concluded that the *engram* or physical location of a memory in the brain could not reside in just one place. Also, the function of the hippocampus is believed to be that of an integrator, to combine or link together the different fragments of a memory stored in different regions into a single unified whole (Treves & Rolls, 1994). If one part of a memory is damaged, the hippocampus could reconstruct the whole based on the remaining portions.

The somatosensory cortex processes information from the skin surface of the body. It contains a topographical map of the body surface, with more neural tissue devoted to processing areas with a greater density of receptors, such as the lips and hands. Pons *et al.* (1991) cut sensory neurons that supplied information to the arm region of the primary somatosensory cortex in monkeys. Ten years later, they found that the cortical face representation had expanded into the original arm region, growing across its border by over a centimeter. The somatosensory neurons that had originally processed arm information, since they failed to receive inputs from the arm, now switched over and were being used for the face, which was still active. Other researchers have directly destroyed the region of somatosensory cortex in monkeys that responded to touch on the palm. Several weeks later they found that neurons next to the lesion site were now responding to the palm.

A number of mechanisms have been hypothesized to account for neural regeneration. The first of these is *collateral sprouting*. If one of two parallel pathways feeding a set of target neurons is severed, the remaining pathway will sprout lateral axonal branches to feed the adjacent target neurons. There is also the possibility that new neurons may grow in a damaged area such as the hypothalamus, restoring functionality.

Another plausible explanation is a phenomenon called *release from inhibition*. In this theory there is competitive innervation of neurons from more than one sensory area. If an area receives most of its activation from say region A and only a small amount from another region B, then A will effectively inhibit input from B. But if A's activation decreases the region will become activated more strongly from B. Evidence exists to support both of these processes. Release from inhibition probably accounts for short-term effects that have been demonstrated only minutes after the experimental manipulation. Collateral sprouting best accounts for long-term effects that happen days, weeks, or months later.

It would be revealing to apply the modeling techniques of habitat fragmentation to real case studies of patients with brain damage. One could modify the parameters of the program to account for the observed recovery of these patients following stroke or accident. If the brain or some portions of it have a percolating cluster connecting widespread regions, then we would obtain a critical threshold effect like that seen in the habitat studies. Patients with damage below this threshold would show graded loss of ability with an increase in damage and have a correspondingly greater chance of recovery. Those past threshold would show marked and dramatic loss of function. The result would be fewer intact abilities or death.

Conclusions

What lessons have we gleaned from our study of ecology and its application to the mind? It must be more than mere coincidence that so many major findings in ecology have an equivalent finding in neuroscience and cognition. Both ecologies and brains are complex systems. Both demonstrate emergent order and self-organization. What is remarkable is that these similarities persist despite the fact that the physical substrate and mechanisms within them differ so radically. After all, an ecosystem consists of individual organisms that move, eat and reproduce while a brain is a network of neurons that transmits electrical signals. It is tempting to conclude that emergent properties like environment regulation and reorganization following damage may be universal to all systems of a sufficient degree of complexity regardless of their physical makeup. However, further study is needed to see if such abstract properties are shared in common by a wider variety of complex systems.

CHAPTER 15
MODELS

Modeling Complexity

One aspect that sets the complexity sciences apart from other more traditional sciences is its emphasis on modeling. We have already read about several kinds of models, such as Boolean network models and the genetic optimization model from the previous chapter. Here, we present an extended discussion of models and outline the specific kinds used in the study of complexity. Investigators in the neurosciences rely on oscillatory models, since neurons exhibit oscillatory behavior. We then look at models of self-organization, including differential equation, cellular automata, and agent-based models. Where possible, we provide a critique of each model, summarizing its strengths and weaknesses.

General Features of Models

A *model*, in the most general sense, is a formal way of representing some aspect of the world. Most models typically capture elements and relations of a real physical system. So if we wanted to build a model airplane, we would reproduce certain aspects of the real thing. These would include for example parts such as the wings and fuselage. However, the model we build will vary drastically depending upon which aspect of the real system we wish to capture. If we wanted our model to fly under its own power, this would dictate how we would construct it, forcing us perhaps, to attach a working engine and propeller. A model that serves only to look like an actual airplane need not have such requirements. Similarly, a cognitive model that wishes to capture just one aspect of a cognitive system will end up being quite different from another model seeking to capture two or more aspects.

Holland (1998) mentions several features of models. First, a model does not have to look anything like what it is modeling. Mathematical equations describing the orbit of the planets about the sun don't look anything like real planets. Even so, they can do a better job describing this process than billiard balls or rocks that look more like planets. Second, models allow us to anticipate or predict what will happen. A model of stress on a steel girder tells us at what load it will crack or break. Third, a model is a simplified version of the thing it describes. A good model will capture much of the behavior of a system, but with much of the detail "sheared away".

Models can be both specific and general. Specific models attempt to reproduce some aspect of a particular system. For example, there are models that describe narrow visual capacities like pattern recognition, depth, motion or color perception. General models attempt to organize and explain more widespread cognitive phenomenon. Newell's SOAR system is a good illustration. It attempts to provide a framework within which all human cognitive problems can be solved (Newell, 1991). Universal models of this sort are sometimes called a *cognitive architecture*, because they specify the structure and function of many different cognitive systems. A real challenge is to try and develop general cognitive architectures that can account for all the specific models within their domain.

Broad Categories of Models

Holland (1998) mentions three broad categories of model used throughout human history. First, we explained the world according to the personalities of the gods. A storm for instance might be the manifestation of an angry god. Then we created mechanical mechanisms such as gates, pumps, and wheels to control the world. These helped us to do things such as move around and grow crops. In the final stage we have used mostly computer-based models. These have been applied to a wide variety of domains ranging from product testing to stock market prediction.

The model airplane described above is an example of a mechanical model. These are physical constructions intended to help understand or mimic some aspect of another physical system. In cognitive science and in science in general there are processes models (described below), computer models, and mathematical models. A computer model is a software program that can be fed data and run to simulate some aspect of a system's behavior. An advantage of computer models is that they can be tweaked or rewritten with relative ease to account for new findings. A mathematical model is simply an equation or set of equations that describe the system. Mathematical models can also be altered in various ways, for instance by changing a parameter or the value of a constant.

Different fields can construct models that are quite different from each other. A neuroscientist might develop a model of how information is encoded into memory by mimicking as closely as possible the anatomy and physiology of the hippocampus, the brain structure that underlies this capacity (Treves & Rolls, 1994). A computer scientist might develop a very different model of memory encoding that is designed to operate on a mobile robot. An interdisciplinary comparison of such models is very fruitful. It can reveal their strengths and weaknesses and provide guidelines for improvements.

A fourth general class of model are interpretive or explanatory models. Rather than simply attempt to reproduce the function of a given system, these models are teleological in nature. That is, they attempt to describe the goal-oriented nature of a system. These are especially useful in biology, where the functional and adaptational nature of a system needs to be explained. They are of great importance in neuroscience, where ideas like the efficient coding hypothesis have led to a number of insights, especially for the visual system (Field, 1994).

Models are only as good as their ability to represent, functionally reproduce, or explain their real target systems. They must be constantly tested against empirical data. If a new experiment shows the system is capable of doing something the model can't, then the model must be revised. Similarly, models sometimes yield results that inform hypothesis testing in science. So optimally, there should be a synergistic, two-way relation between modeling and experimentation. Each should inform and alter how the other is conducted.

Process Models

As noted elsewhere in this work, traditional approaches to mind have focused on its static nature. A dynamical systems approach would instead investigate how mental processes unfold over time. This requires a different set of tools from those used in the traditional cognitive science disciplines. Gone would be the emphasis on structural models and statistical methods. These would need to be replaced with new techniques that allow us to measure temporal change and to formulate dynamical models of brain and cognitive function.

Most existing cognitive models are known as *process models* or flow diagrams. They use boxes to represent computational units and arrows to represent the flow of information between units. Classical examples of these include the modal model of memory where each box stands for a different type of memory (Atkinson & Shiffrin, 1968). In the modal model one starts with a stimulus and sequentially works forward through the system from sensory to working and finally to long-term memory (Figure 15.1). These models are good in that they show the overall structure of a cognitive system, what it must do and the rough order in which it gets done.

But process models are lacking on a number of counts. Each of the units is to a certain degree a "black box". The computations these boxes perform are often not specified. If they were, they would require another process model inside of each box showing its internal workings. One encounters a homonuculus problem here, because each box would need another model inside of it until some underlying satisfactory level of detail, neural or smaller, is reached. Some of the dynamical techniques that show oscillatory behavior at different frequencies get around this issue, because they suggest that different mechanisms are operating at different speeds at different levels of spatial scale.

Another problem with process models is that many are inherently sequential, with a specific starting point, usually a stimulus, and a specific endpoint, typically a categorization, identification, or overt response. Most neural activity is massively parallel and recurrent. This simultaneous and cyclical action is not well captured in process models. Some of the new dynamical models however do take distributed processing and recurrence into account.

Lawrence Ward in his 2002 book titled "Dynamical Cognitive Science" outlines many different classes of cognitive model. He discusses static vs. dynamical models and how they might be combined. Ward then explores modeling of stochastic processes, noise, and oscillators. He also introduces several techniques from other disciplines such as Markov chains and Graph theory that could be put to good use by cognitive scientists. In the immediate sections that follow, we introduce these concepts, as they constitute a first step toward how we might represent and understand the more complex dynamics of mind.

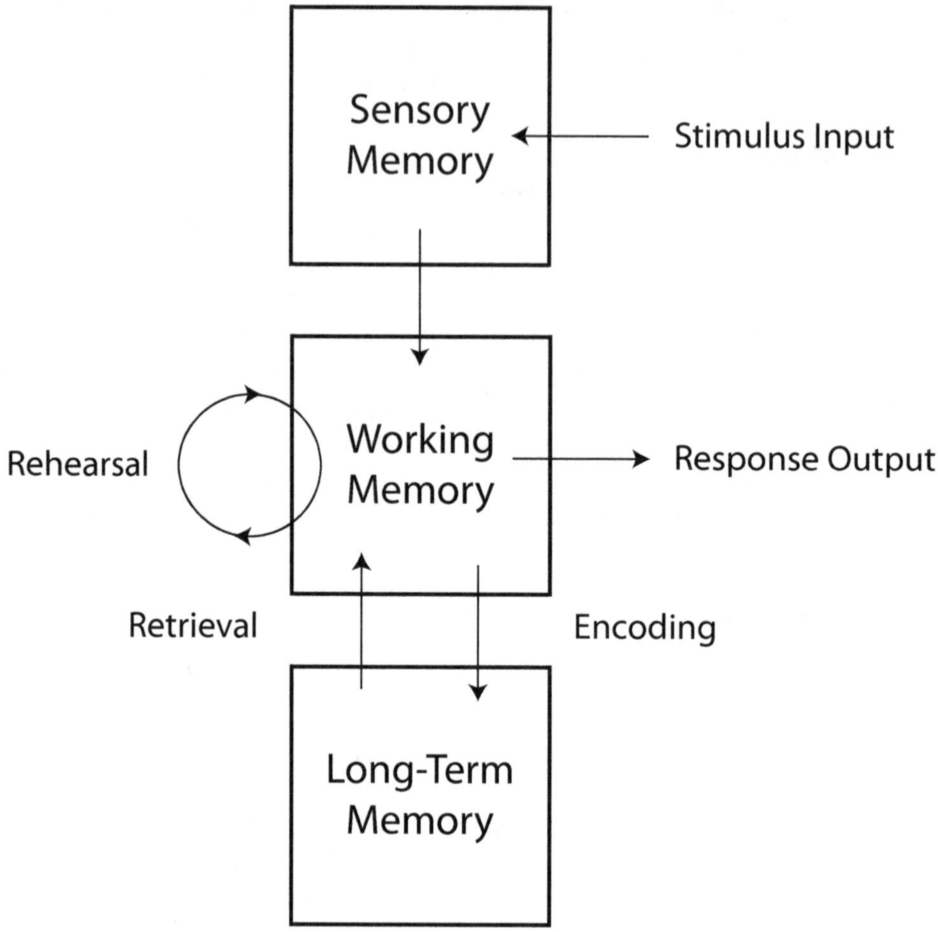

Figure 15.1 *The Atkinson & Shiffrin (1968) conception of memory is a classical example of a cognitive process model.*

Statical and Dynamical Models

A *statical model* assumes that the relevant state of the modeled system remains constant. It specifies the relations between the various state variables of the system. Formal statical models can be expressed with an equation. If so, they reduce to mathematical models. A good example of a formal statical model is Steven's power law from psychophysics: $P = Ks^n$. This relates the perceived magnitude of a subject's response to the magnitude of a stimulus and is described in greater detail in the section on fractals. The process models described above are also statical models because they specify only the type of relations in the system and not how they change with time.

In contrast, a *dynamical model* is concerned with change and describes the succession of states in the modeled system. A formal dynamical model can also be expressed with an equation. A differential equation of the form $dx/dt = at$ is a simple example. The dx/dt is the derivative of x with respect to t and represents how the variable x changes with time t. This is a linear growth model, because as time increases, there is a corresponding increase in the variable x. In fact, it will continue to increase forever with time. Most cognitive systems don't actually act this way. They may increase up to some limit and then asymptote or decrease. More complex changes with time may be oscillatory in which case they fluctuate in a regular manner.

Most of the models in biology, psychology and sociology are statical. They frame specific areas of study and define the important concepts and phenomena. Once this is done, a researcher can measure change in a specific variable knowing how it relates to the larger explanatory framework. Most models in physics and economics are dynamical. These fields are inherently quantitative and more focused on change. Part of the reason for this schism was outlined earlier where we noted that among other differences, psychological researchers are trained in statistical methods rather than for example, calculus.

More on Dynamical Models

According to Holland (1998), the major goal in constructing a dynamical model is to find the unchanging laws that give rise to the changing states of a system. He likens these laws to the rules of a game such as chess. The rules say how the states change as different moves are made. If we know the current configuration of a chessboard and we know what moves are allowable, then we can anticipate what the next state of this system will be. In this case, we can predict what the next board arrangement will be.

Holland (1998) produces a general description of dynamical model building. A first step is to select a level of detail of the system that we wish to describe. Then we have to capture the laws that describe change at that level. Of course, our selection of detail is crucial. It may be that there are no simple rules to describe change for a given level of detail. When it comes to the weather, we may not be able to predict the exact temperature for a region, but we can predict the range, defined by a high and low value.

Next, we define the states the system can be for this level of detail and figure out how the laws produce transitions between them. A *transition function* does just this. It assigns to each state the state that will occur next under the laws of change. A transition function is typically a partial differential equation or computer algorithm. Each system state can be thought of as taking an input from outside the system. This input, in combination with the current state of the system causes the next state.

Our dynamical model is always attempting to describe change in the real world. In Figure 15.2, we see a depiction of the world at time t. The laws of the natural world then cause the system to change to a new state at time $t+1$. Our model, based on an observation of the world at time t, represents that state. The transition function then specifies how the model will change and assume a new state at time $t+1$. The model's new state is meant to reflect what the real world state will be at that next step in time. In other words, it attempts to predict what will happen. Making a second observation comparing the second model state to the second world state can then test this prediction. There is an obvious parallel between transition functions and physical laws. The functions are the model's way of representing the real causal laws. The former produces a state change in the model. The latter produces a state change in the world.

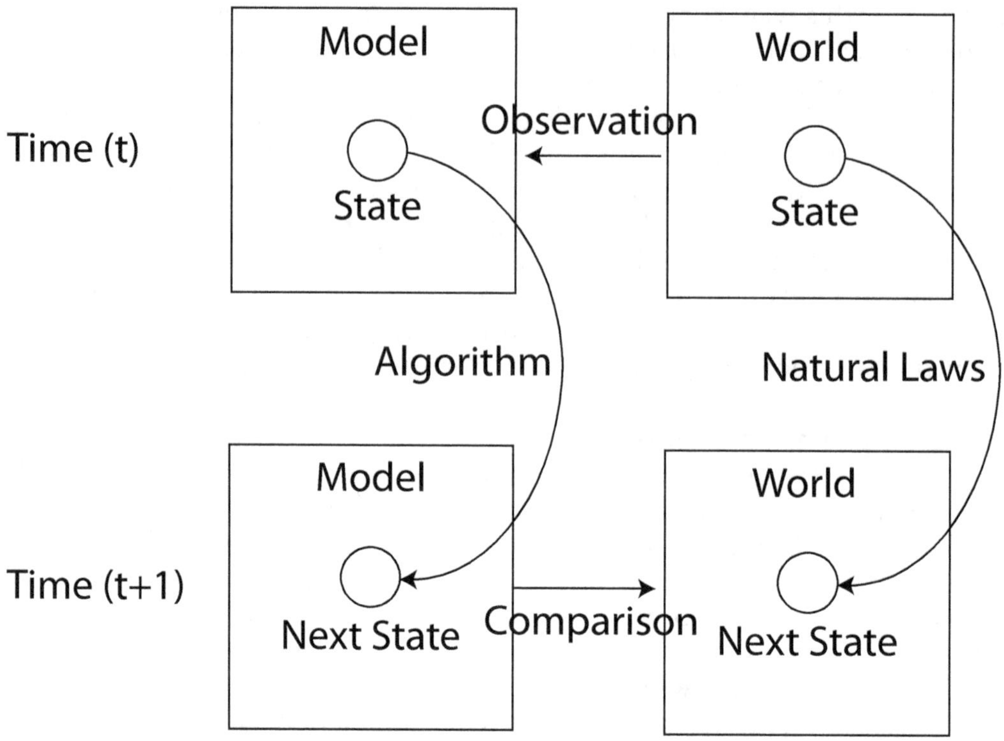

Figure 15.2 *A dynamical model is updated at each time step by an algorithm or transition function that specifies what the next state will be. This can then be compared against its corresponding state in the world (after Holland, 1998).*

This sort of model allows for prediction. If we know the current state and the next input, we can determine the succeeding state. Prediction in the real world however, seldom allows for long-term predictability, given the sensitivity to initial conditions found in complex dynamical systems. This so called sensitivity however, could be a result of several uncertainties: not knowing sufficient details of the system, having picked the wrong level of detail, or using a wrong or inaccurate transition function.

A perfect model is one that predicts successfully all the time. Holland (1998) states this differently by saying the model should satisfy a criterion called *commutativity of the diagram*. Commutativity is the idea that you can vary the order of operations and end up with the same result. One order of operation is as follows: an observation of the world at time *t* by the model generates a prediction about what the world will be like at time *t*+1. A second order of operations is if we watch the world change across this time step and then measure the result according to the model. If the prediction and the world result are the same based on either operation, the model can be said to be perfect.

Deterministic and Stochastic Models

A *deterministic model* is one that does not have any random elements. A random process, you will recall, is one where all events have an equal likelihood of occurring. However, in reality it is very difficult to distinguish a random event from a deterministic one. Random number generators in computers are a deterministic process governed by a generating algorithm. If the seed or starting value for these algorithms is known, one can recreate the random number or sequence they generate. We also mentioned earlier that it is extremely difficult to differentiate in reality a random process from a chaotic one. Many chaotic processes act in ways that make them seem random.

Stochastic models do contain random elements. A stochastic process is a random one, that is, one not governed by deterministic rules. Ward (2002) gives the following psychophysical equation as an example: $\log R_i = d \log S_i + \log c + E_i$ The last term, *E*, is a random variable. It typically consists of values drawn from a normal probability distribution. The purpose of the random variable is to represent variability not captured by the other terms in the equation.

Most social scientists analyze their data using statistical tests such as analysis of variance or multiple regression. Variance not accounted for in these tests is considered noise. It is usually ignored if it is sufficiently small. In psychology, noise is often attributed to uncontrolled subject variables like fluctuations in mood and personality differences or to uncontrolled environmental variables like time of day and temperature, etc. However, from a dynamical perspective, noise is something that should be studied in its own right because it may reflect underlying mechanisms that are part of the phenomenon under investigation.

Oscillatory Models

An oscillator is a process that changes in a regular way over time. The hour, minute and second hands of a clock oscillate because they move the same amount in a given interval. Many human behaviors are oscillatory, for example running and walking consist of component actions like lifting the legs, swinging forward etc. that are repeated. An *oscillatory model* then, is one that attempts to represent this kind of behavior. Equations that describe the characteristics of waves such as their amplitude, frequency and phase can be considered oscillatory models. A simple example is the basic sine wave or sinusoid function: $y = A\sin(Wt + \theta)$ where *A* is the peak deviation from center or amplitude, *W* is the angular frequency, *t* is time and θ is the phase shift.

If an oscillatory system is undamped and is being driven by an outside force it will continue indefinitely. This energy input from outside the spring system is known as forcing. A spring hanging from a wall with a mass attached at the opposite end under these conditions will repeatedly trace out a sinusoid. A damped spring has a damper attached that counteracts the force exerted by the spring. The motion of damped oscillators dies out over time, their amplitude gradually decreases. All real world oscillators encounter some amount of resistance and are thus damped. They are also forced to varying degrees, since they cannot be isolated from the rest of the universe.

Ward (2002) believes that a particular type of oscillator, the van der Pol relaxation oscillator, can serve as the basis for modeling a wide range of neural and cognitive phenomena. These types of oscillators when combined together can mimic the electrical functioning of the heart and even produce chaotic behavior (van der Pol & van der Mark, 1927). A relaxation oscillator, when driven by another forcing oscillator, will phase lock with it causing the two to fire at the same frequency but without a change in amplitude. This may be the basis for neural synchrony discussed elsewhere which seems to underlie a number of phenomena such as perceptual binding and consciousness.

Neural Oscillatory Models

One example of an oscillatory model involves a linear chain of locally coupled oscillators with different types of connections. Chain type architectures seem to exist in the brain as part of sensory motor cortex and in the hippocampus. Borisyuk *et al.* (2001) have modelled the dynamics of chained oscillators. In one study, they linked 101 neural elements together in a linear chain, all locally coupled. The center 51st element was driven by an external stimulus to oscillate continuously and the resulting propagation of activity was measured. For weak or moderate connection strengths and with continuous stimulation, the wave of oscillations that spread through the chain was localized. Elements on either side of the initiating element became activate, but the activity failed to spread through most of the chain.

But when the coupling strength was increased and the stimulus was time-locked, i.e., turned on and then off for a fixed duration, a different pattern of activity emerged. In this case, activity spread out to either side of the initiating element but instead of dying out, it triggered successive waves of activity. The new waves started at the endpoints of previous waves, in some cases, where these overlapped. The result was a fractal-like pattern in which subsets of similar activity appeared inside one another. So even in as simple an architecture as a linear chain, it is possible to observe complex, hierarchical patterns of activity.

A neural oscillator can be formed by two interacting populations of excitatory and inhibitory neurons. Each population can be reduced to a single neural element that is itself either excitatory or inhibitory (Figure 15.3). Borisyuk *et al.* (1995) examined the behavior of two coupled oscillators. When connections were weak, opposite connection types (excitatory to inhibitory or vice versa) lead to periodic in-phase oscillations, producing synchrony. Same connection types (excitatory to excitatory and inhibitory to inhibitory) lead to periodic anti-phase oscillations, producing asynchrony.

When connection strengths in this coupled model were of intermediate strength the network entered quasiperiodic or chaotic regimes. It also exhibited multi-stability, the existence of several different types of oscillatory behaviors. For excitatory to excitatory connections, oscillations occurred at the same frequency but with some amount of phase shift. The phase shift could occur across a broad range depending on various parameters as coupling strength. Note that although this model is quite simple, consisting only of two connected populations that can be reduced to just two neurons, the number of possible behaviors that can arise from it are great.

An Oscillatory Model of Hippocampal Function

Borisyuk and Hoppensteadt (1999) developed a detailed oscillatory model of hippocampal function. The hippocampus is a brain structure responsible for

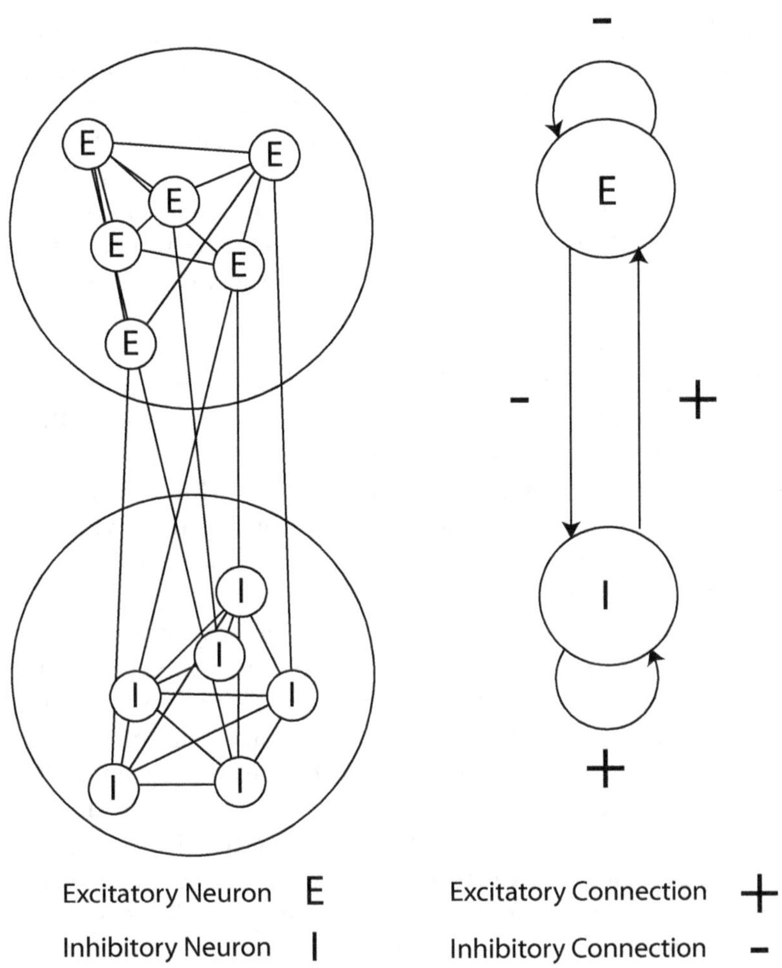

Figure 15.3 *A neural oscillator formed by an interacting excitatory and inhibitory neuron. An oscillating circuit can also exist between two populations of excitatory and inhibitory neurons (after Sole & Goodwin, 2001).*

the transfer of information from short-term to long-term memory. The neural architecture of the hippocampus can be modeled as a three-dimensional grid of neural elements, arranged in a cube-like fashion. Imagine that at each corner of these cubes are an excitatory and inhibitory neural oscillatory pair that constitute a node. The excitatory unit represents a pyramidal cell neuron and the inhibitory unit an interneuron. Each pyramidal cell receives six excitatory connections from its nearest pyramidal neighbors. Each pyramidal cell also receives seven inhibitory connections, one from the interneuron in its node and six from its nearest interneuron neighbors. The interneurons in each node receive one excitatory connection from the pyramidal cell in their node and six inhibitory connections from their nearest interneuron neighbors.

Two stimulatory inputs are then fed into this system. The first represents input from the entorhinal cortex and comes from one side of the grid. The second represents input from the septum and comes from the opposite side of the grid. The dynamics of the model were studied under different phase deviations between the two inputs. When the deviation was small, at 30 msec, the network showed regular periodic theta rhythm oscillations. *Theta rhythms* are 5-8 Hz waves that influence the establishment of synaptic strengthening underlying memory formation. When the deviation was larger, at 130 msec and the input was varied, elements in the middle of the grid failed to oscillate. Borisyuk and Hoppensteadt (1999) conclude that the phase deviation between these two inputs is a key parameter regulating cyclical hippocampal activity.

An Oscillatory Model of Attention.

Kryukov (1991) has created an oscillatory model for how attention might operate. It specifically explains how one might attend to one of two different objects and then switch the focus of attention between them. In this model, there is a central oscillator (CO) that is coupled with several other peripheral oscillators (PO) by recurrent or two-way connections. Although this might be a simplified model of the real brain architecture underlying attentional function, it captures much of the relevant dynamics.

Two groups of POs each represent the objects, with different oscillators in each group coding for different features of the object. Attentional focus happens with synchronous activity between the CO and some or all of the POs. There are at least three states of the network. In global synchronization, all oscillators are in synchrony. This occurs when attention if focused on both objects. Partial synchronization is the state where the CO and the group of POs corresponding to one of the objects are oscillating together. In this mode, attention is focused on one or the other of the two competing stimuli. In the no-synchronization mode, there is no synchronization whatsoever. This is when the attentional focus is unformed, i.e., when attention is not being focused at all.

Simulations using this model also suggest a new type of dynamic that occurs when there is an attentional shift from one object to the other and when focus is destroyed or removed before being shifted. The CO in this model may correspond to the central executive of an attentional system, such as the metacognitive component in some models (Brown, Bransford, Ferrara & Campione, 1983). Changes in the natural frequency of the CO could correspond to voluntary changes underlying focus.

Conclusions Concerning Oscillatory Neural Models

So far the models described are based on regular oscillations. However, synchronized oscillation can also be generated by chaotic activity. Borisyuk & Borisyuk (1997) created a network of excitatory and inhibitory neurons with global connectivity that demonstrated spatially coherent chaotic oscillations. The behavior of this network alternated between bursts of activity and periods of quiescence. It showed periods of global synchronization even though the output of its single neurons was irregular and chaotic.

Borisyuk, Borisyuk, & Kazanovich (2001) make some general conclusions regarding oscillatory neural models. They note that this approach, in contrast to traditional connectionist models, is in better agreement with neurophysiological evidence but that it requires more complicated analysis. Also, traditional artificial neural networks demonstrate convergent dynamics where activity slows until the network settles into a stable state. Oscillatory and chaotic dynamics show continuing activity over time, which is what we see in the brain (Braspening 1995).

Models of Self-Organization

A model is only as good as the real-world process it attempts to capture. Models must be subject to testing if they are to be valid. Camazine *et al.* (2001) describe how models need to be embedded as a step in the overall experimental method. In their view, one first starts with observation of and experimentation on a collective or self-organizing system. This results in the identification of the subunits and their interactions. The subunits are the basic functional agents of the system. They might be insects in a colony or neurons in a brain. One then hypothesizes a mechanism to explain how these units interact. A model is next constructed to test the hypothesis. The results of the model are compared against the real world system. If there is agreement, the model is supported. If not, the model and hypotheses can be altered. The process then repeats.

One problem with this approach is that it is difficult to know exactly how to alter a model to fit the data. There can be numerous reasons why a model's results fail to map onto a system's real-world behavior. The model may be using the wrong parameters. A more severe problem is that the mechanisms in the model may be wrong. If the model employs agents, it could be that the algorithm or decision-making process employed by the agents is incorrect. Or, it could be that the type of agent interactions that occur are wrong. In short, there is no way to know which aspect of the model needs changing. The best course of action in these circumstances is to make changes systematically, noting the effect of these changes and keeping the ones that produce the best fit with empirical data. Models that fail to conform to these sorts of changes ought to be abandoned and new ones with different organizations then tried.

The Differential Equation Model

Camazine *et al.* (2001) then suggest a set of specific models that can be used to understand self-organizing systems. A *differential equation model* provides a description of the activity or information flow between different compartments. In their illustration, they use this model to describe the collective foraging behavior of ants. The system is one where ants leave a nest, search for food, locate food, and bring it back to the nest. Certain initial conditions are assumed, such as the total number of ants in the system, where they are located, etc. The number of ants at the different locations and the rates at which they perform the various actions are variables in the differential equations. Solving these produces solutions showing the number of ants in each of the compartments over time.

The ant model can very easily be converted into a cognitive or neural model. We replace ants with neurons. Instead of ants moving from location to another we have communication from one neuron to another via an action potential sent down an axon. We plug in the number of neurons in each location and the rate at which impulses are transmitted between locations. The resulting functions now tell us the number of impulses or amounts of information transmitted between compartments over time, providing us with a global measure of the system's behavior.

Figure 15.4 shows a cognitive model of visual search that can be implemented using differential equations. In this example we want to describe the

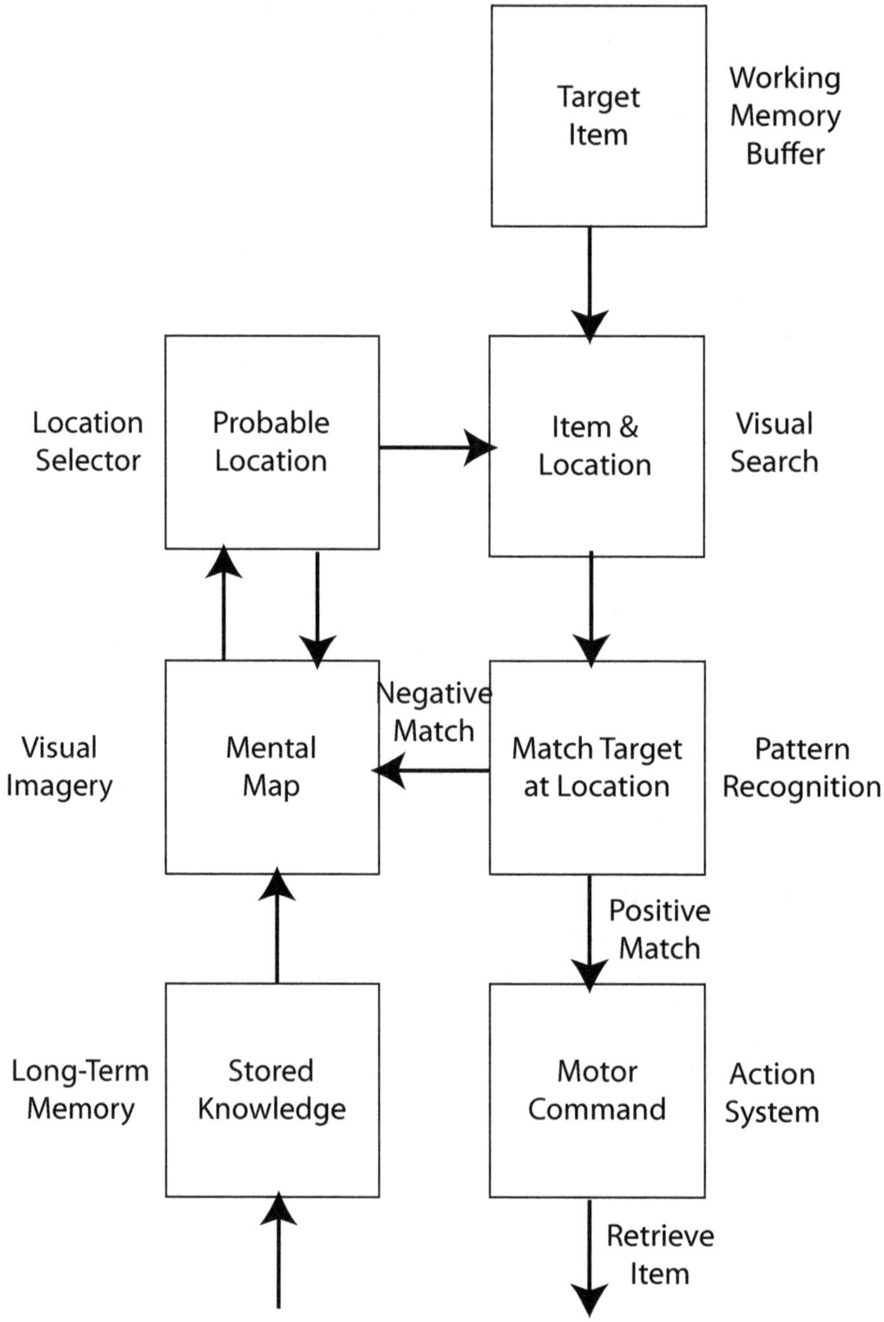

Figure 15.4 *A differential equation model of visual search. Neuron populations can be substituted for the modules shown here.*

cognitive or neural systems that govern searching for a lost pair of keys somewhere in an apartment. Information in long-term memory is used to construct a mental map of the environment. The map is fed to a location selector that chooses probable locations (desk, coffee table...) and rejects others (toilet, kitchen cupboard...). A single location is then selected and fed to a search process. The search mechanism after performing the search outputs its results to a comparison or matching process that determines if the desired target is at the current location. If so, the target has been located and an action, such as walking over to the location, reaching and grasping for the keys, is performed. If the target is not at the location, the entire process iterates.

In this example, there are many ways the search can be performed. For instance, a search for a blue pair of shorts could be directed on the basis of color rather than location (Chiao, Cronin & Osorio, 2000). We only posit this as one possible way such a search could be conducted. The differential equation model could be adapted to any number of different search models.

This example uses representations and computations and so may be seen as supporting the constructivist or classical cognitive science view. This however is not the point. One could just as easily substitute the ecological notion of the optic array for maps of the room or images of objects at the searched location. The model only needs to have some degree of compartmentalization, neural or cognitive activity that occurs at those locations and the transmittal of signals between them. The size of the compartments can be reduced and their number increased to more closely approximate a distributed network model.

The point is that the global and collective action of the system can now be described. The picture we get is dynamic rather than static. Rate and amount of activity are captured and the time course of this activity can be monitored when different parameters or contextual conditions are changed. Emergent properties of the system may then be observed. For example, it could be that oscillations at different frequencies develop between certain compartments or that new circuits arise. This sort of activity may never have been predicted on the basis of a static model alone. Once dynamical models become complicated enough they will provide us with examples of emergent behavior that we can then analyze and implement in even more complex and powerful models.

Camazine *et al.* (2001) summarize the pros and cons of differential equation models. They state that these models have been commonly used in the biological sciences. They allow researchers to obtain a precise and quantitative description of how a system changes over time. One can use them to analyze the different states of a system including equilibrium conditions and bifurcations. A problem is that it may difficult to find equations that accurately describe the behavior of the self-organizing system under study. They also lend themselves better to continuous rather than discrete processes. Also, they may be difficult to use for investigators who are not mathematically inclined.

Cellular Automaton Models

A *cellular automaton* (CA) is an abstract array of cells that is programmed to carry out a set of rules. Each cell has a finite number of states, for example, it can be either "on" or "off". Time is discrete in this system and the rules are executed repeatedly with every tick of the clock creating a new generation or pattern. The state of a cell at any given time is a function of the states of a finite number of its neighborhood cells.

One of the most famous of all CA examples is John Conway's "Game of Life". He employed a two-dimensional grid of cells that followed three very simple rules. First, if an empty square has three occupied neighbors (diagonal and adjacent) the cell turns on. Second, if a square has two occupied neighbors, it does nothing and remains unchanged. Third, if an occupied square has less than two neighbors or greater than three neighbors, it turns off, essentially "dying" of either loneliness or overcrowding.

Based on these simple rules, an amazing amount of complex patterned activity occurs. Figure 15.5 shows a frame from the Game of Life. Distinct shapes emerged with names such as Beehive, Snake, Long Ship, Honey Farm, Pulsar and the Glider. The system could even continue to grow without limits, as a Glider Gun shape was found that could produce a Glider every thirty iterations (Gardner, 1971). This resulted in the addition of five new "live" pixels to the configuration. The parallels between the Game of Life and natural living organisms are strikingly clear. The shapes correspond to different types of or-

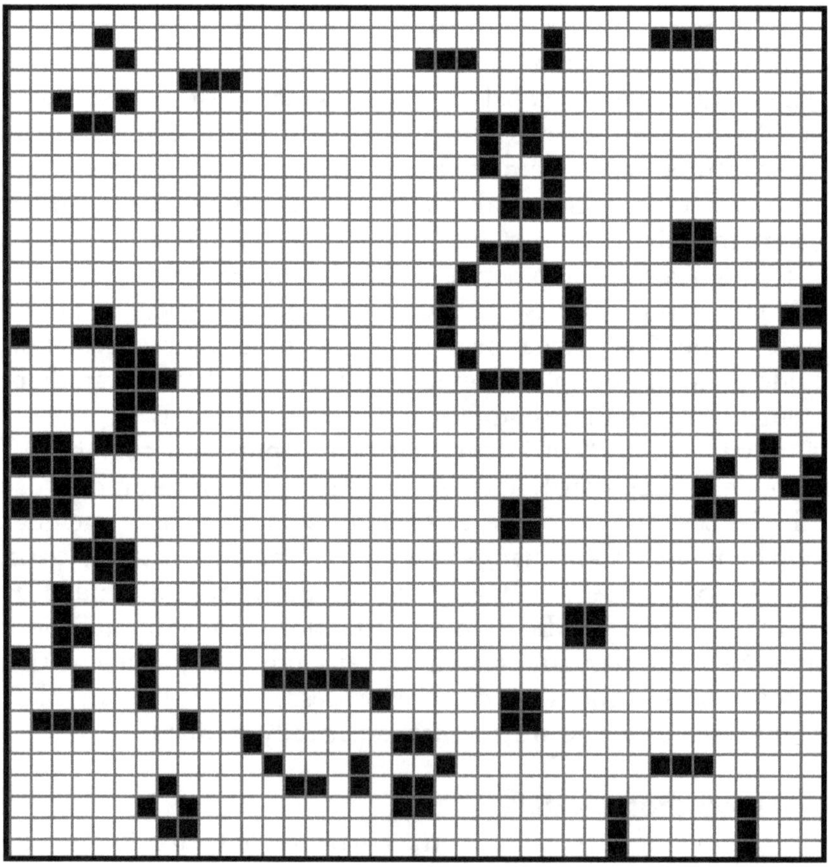

Figure 15.5 *A frame from the cellular automaton Game of Life.*

ganisms, some of which move around, interact with each other, reproduce and evolve.

The Game of Life has proven to be a very useful tool in the study of complex natural processes. Computer simulations employing variations of this CA have been used to model biological development, chemical reactions, crystal growth, and the structure of snowflakes (Coveney & Highfield, 1995). It has provided insights into how these processes arise from the execution of simple local rules.

Wolfram (2002) has investigated one-dimensional automata where each of the squares or sites are arranged along a single starting line. He would randomly determine the initial state of the system. The rules were then applied in successive iterations that produced new patterns on the lines below. These were displayed graphically to show how the system evolved.

He found he was able to categorize the resulting CA into four types, no matter what local rules were used. In Class I, the pattern disappears or stays the same over time. These are the equivalent of fixed-point attractors in dynamical systems. Class II CA form structures that repeated indefinitely. These evolve to limit cycle attractors. Class III CA are chaotic, they contain little order or regularity and are associated with chaotic attractors. Finally, Class IV showed complex patterns that expand and contract irregularly. These CA patterns are equivalent to systems on the "edge of chaos" and fit in-between Classes II and III in terms of complexity. Figure 15.6 shows examples of each class.

CA are very useful for modeling self-organizing processes. The components in the model can stand for the functional units of a self-organizing system. They can stand for fish, ants, or termites in biological systems and for neurons, memories or ideas in psychological systems. CA models have been employed successfully in biology (Ermentrout & Edelstein-Keshet, 1993) but much less so in the cognitive sciences. A cognitive CA model could for instance be used to model how spreading activation travels through a neural or semantic network.

CA models have both advantages and disadvantages (Camazine *et al.*, 2001). They are typically easier to implement than those involving differential equations. A real plus for them is that they provide a visual representation of a process. One can actually see how activity spreads in a network or how clusters or other emergent structures develop. However, there are some downfalls to this method. The simulation slows down dramatically with increases in grid size because computations are made over every cell in the array. They also require some degree of expertise in computer programming.

Cellular Automata and Computation

A *Turing machine* is an abstract symbol-manipulating device that can simulate the logical operations of any known computer. A *universal Turing machine* is one that can simulate any other Turing machine. It has been shown that a CA is equivalent to a universal Turing machine. Suitably modified, it can perform the logical and mathematical computations that underlie much of what our everyday computers do, things like manipulating text, images, and other forms of data. Although the original Turing machine was a serial device, CA are inherent-

Figure 15.6 *Wolfram's (1986) four types of cellular automata.*

ly parallel. The arrays in which they operate can be divided into separate regions where computations occur simultaneously. This is the same style of computation we see in real biological brains.

In the classical cognitive science view, human thought is essentially computation carried out in a biological substrate. The brain can therefore be viewed as a CA that generates thought and consciousness. The units of the arrays are not two-dimensional squares as we see in the Game of Life, but neurons arranged in a three-dimensional spatial configuration. Just as cells in the computer CA models turn on or off, the neurons also have discrete states. They can be firing at different rates. Neurons also have neighborhoods. Their activity is determined mostly by the state of cells in their immediate vicinity.

The rules neurons follow are at heart quite simple. They receive excitatory and inhibitory inputs from their neighbors, summate this activity and compare it against a threshold value. If the inputs exceed this threshold, an action potential is generated and transmitted down its axon. CA rules are simple but give rise to complex organization and activity. Just as the Game of Life gives rise to emergent structures, neurons do as well. Their combined activity leads to patterns of neural activation that replicate, reproduce, and change over time.

Researchers at MIT have been hard at work creating a CA computer (Toffoli & Margolus, 1987). Instead of a single processor as is found in most current computers, their work is trending toward a processor for every unit or cell in the simulation. This is limited by practical constraints. One version, the Cellular Automata Machine 8 (CAM-8) still has multiple cells per processor. It has however, proven to be a success and has been used to model liquid flow, chemical reactions and crystallization.

Agent-Based Modeling

Elsewhere, we defined complex adaptive systems (CAS) as collections of interacting agents that adapt to a changing environment. Let's summarize again some of the essential ingredients of a CAS. A CAS is complex. Its aggregate behavior cannot be understood as a simple linear summation of its constituent agents. A CAS is adaptive. It is resilient to perturbation and can change along with changes in its environment. In addition to adaptation, a CAS exhibits other emergent properties such as communication, specialization, spatial and temporal organization and reproduction. A CAS can also be self-similar, the agents as well as the system as a whole can share features in common.

A *multi-agent system* (MAS) is simply a system composed of multiple interacting agents. A MAS does not have to be complex, nonlinear, adaptive, self-similar, or demonstrate emergent properties. It simply refers to the class of all systems that have interactive agents. We can thus consider a CAS as a special kind of MAS. A MAS is the more general case, the CAS is a specific example or subset of this larger category.

One way to investigate systems such as these is to implement them in a computer model. Researchers have created virtual agents, imbuing them with different rules and then setting them loose in simulated environments. One can then track the behavior of this system over time as the agents interact, looking for examples of complex or emergent phenomena. This technique is called *agent-based modeling* (ABM) and has been applied to solve problems in a wide range of different fields including supply chain optimization and logistics, consumer behavior, workforce, traffic, and portfolio management.

Epstein (1999) describes five features of ABM. The first is heterogeneity, which means that the agents operate according to different sets of rules.

This mimics human societies, where people have different values and preferences and so act differently, even in the same situations. Secondly, the agents are autonomous, they can be thought of as acting of their own accord. There is no top-down control imposed in these models. Order is present in the rules these agents follow, but macroscopic order appears spontaneously as the result of agent interaction.

Another feature of ABM is explicit space, meaning that the evolution of the model takes place on some defined landscape or n-dimensional lattice. Local interactions are another characteristic. All agents interact only with their neighbors and not with agents that are located far away on the virtual landscape. This also mimics natural human societies where people mostly interact with those who live near them. Finally, ABM agents exhibit bounded rationality. Their decisions and actions are based on imperfect knowledge. This also reflects decision making in the real world where people don't act in a perfectly rational way.

Sugarscape

Epstein and Axtell (1996) were among the first to implement and study agent-based models. They created a model called "Sugarscape". This was a two-dimensional landscape where each cell held a different concentration of sugar. Agents in this world were programmed to find, eat, and store sugar. Agents that are unable to locate sufficient sugar will die. Each agent in this model is provided with three capabilities. They can see, move, and metabolize. The agents are capable of seeing sugar concentrations at varying distances across the landscape. They are allowed to move in four directions, either up, down, left or right. Finally, they can consume the sugar they encounter and use it for energy.

A Sugarscape simulation can begin with a random distribution of agents about the landscape. The agents are each given different abilities. Some can see farther than others, some can move faster than others, etc. This helps reproduce the kind of variability we see among animals in nature. It is then run for a given number of iterations.

Sugarscape has demonstrated several emergent features. One finding was that the distribution of "wealth", as measured by the sugar store of the agents, was concentrated among a small number of individuals. This mirrors the skewed distribution of income found in many economies. The agents were also found to cooperate to produce global behavior incapable of being performed by any of the agents acting in isolation. A group of agents, through cooperative interaction, could move together in a diagonal direction, whereas each agent on their own could only move perpendicularly. Various versions of Sugarscape with different variables added or subtracted or different parameters altered have replicated a wide range of social phenomena including migration, pollution, combat, and disease transmission.

Evaluating Agent-based Models

Epstein and Axell (1996) point out that agent-based models have several advantages over other kinds of computer models. First and foremost, they are realistic. They incorporate elements and actions that closely mimic their real world counterparts. As mentioned above, the agents in their models also act only on the basis of local information, much the way real agents do. Rarely does an animal or even a person have a godlike or "birds eye view" of their environment. Most of the time we make decisions based only on what information we have available, which is usually limited, i.e., local.

The main goal in ABM simulations is to examine how agent interactions give rise to more complex and global patterns. One might, for instance, wish to examine how corporations form in a virtual economy. Implicit in these models is the idea of "bottom-up" processing. In bottom-up systems, the actions of the small- scale individual agents give rise to larger structures.

However, it is probably the case that "top-down" organization also develops and influences ABM behavior. Previously, we discussed the notion of upward and downward causation in the context of the mind. We said that local interactions could give rise to global structure, but that global structure could also feed back downward and influence the actions of smaller elements in the system. In sufficiently complex ABM top-down forces may evolve, producing a synergy between global and local levels. In addition, there may be several levels of organization that emerge in these models, each with unique forms of interaction. Some of these interactions may be lateral, or within a given level. More attention needs to be devoted to the possible two-way interaction between multiple emergent levels in ABM research.

Although interactions are local in these simulations, this isn't always the case, even in the real world. For example, people can have long distance telephone conversations or be aware of events that happen on the other side of the world. Information transfer, although mostly local is also global, and can affect agent action. In small world networks, this communication takes place through explicit long distance connections. Different classes of ABM could be run investigating how system behavior changes with increased amounts of or different types of long distance interaction.

It must also be acknowledged that ABM cannot hope to capture all of the relevant variables affecting complex systems in the real world. The addition of just one additional variable in a model could cause it to act in a completely different way. One way to overcome this is to introduce new variables in a step-wise faction and study how they interact with and shape system behavior.

Constrained Generating Procedures

Holland (1998) proposes an alternate conception of an agent-based model. He calls it a *constrained generating procedure* (cgp). A cgp gets its name from three sources. Its actions are constrained by limitations of the "game" or environment it exists within, much the same way the rules of a game limit possible board configurations. It generates an action or behavior and it does so according to a rule or procedure.

There are several points to keep in mind when discussing cgps. To begin, the mechanisms respond to actions or information that they take as input. These produce resultant actions as output. Second, in Holland's scheme, individual mechanisms are linked together into networks, the networks being the cgps. It is the interaction of the mechanisms that produces complex organized behavior. The complexity of a cgp model increases with the number and type of mechanisms it contains. Just as in multi-agent models, mechanisms interact both with their environment and with other agents. The output of one agent can serve as the input for another. Note the similarity here of cgps with agents and of organized collections of cgps as multi-agent systems.

The state of the entire cgp is defined in terms of the states of its individual mechanisms. We can call this its global state. The change from one state to the next is then defined by the transition function we introduced earlier. An understanding of the mechanisms that make up a cgp and of the transition function is enough to allow us to predict what its next state will be.

A cgp is organized hierarchically. Each mechanism can itself consist of other mechanisms and so on. However, low-level mechanisms are simple. More complex ones get their complexity because they contain hierarchically ordered layers of simpler mechanisms within them. There are many examples of these sorts of nested systems. The molecule-cell-organ-organism systems from biology can be considered one. The neuron-network organization found in the brain is a specific instance of this kind of natural cgp.

Holland (1998) describes how we can build a cgp in order to form a model. We start with the selection of a set of mechanisms or primitives. Two mechanisms are connected to each other when the state sequence of the first mechanism determines a sequence of values for one of the second mechanism's inputs. In other words, the behavior of one mechanism is dependent upon another. In this way we can build up a complex cgp by wiring primitives together into particular configurations. Figure 15.7 shows several basic cpgs formed from primitives. Notice that each basic mechanism can take inputs from others or from outside environmental factors, what are called free inputs. A cellular automata can be considered an example of a model cgp.

Modeling Chaotic Brain Activity

Control and Anti-Control of Chaos

Chaos may be essential to proper physiological function. In the brain, it could allow for the flexible transition between ordered and disordered states. This type of dynamic may underlie cognitive processes such as perception and memory. The loss of chaos in neural tissue could lead to epileptic brain seizures and in cardiac tissue to heart failure (Schiff *et al.*, 1994). If we could steer a dynamical system that has deviated toward more orderly and periodic behavior back to the chaotic realm, we may be able to prevent or reduce such problems. Toward this end, a number of researchers have created mathematical and/or network models to enhance and maintain chaos (Gupte & Amritkar, 1996). These procedures are known as *anti-control of chaos.*

Ramaswamy, Sinha and Gupte (1999) for example, have utilized a method of adaptive control in which a feedback loop drives a system parameter or parameters to a desired value. Their procedure allows a dynamical system to be directed to a desired target state, such as a value of the local *Lyapunov exponent.* This exponent reflects the tendency for two starting points in a dynamical

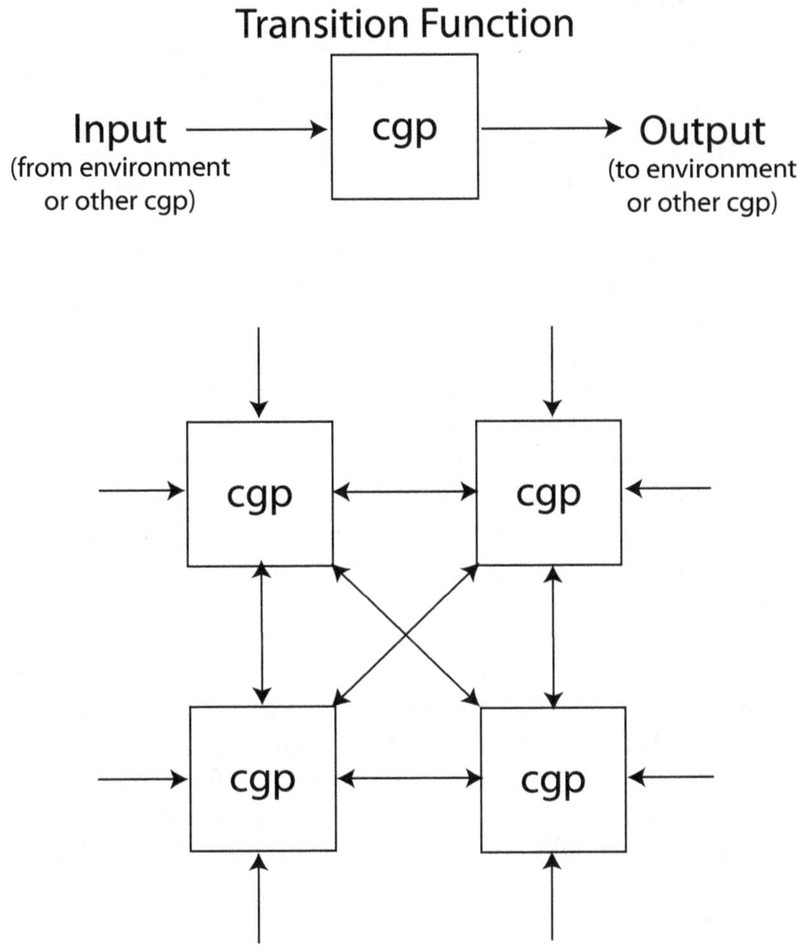

Figure 15.7 *A system model formed from constrained generating procedures (after Holland, 1998).*

system to diverge. If the exponent is negative the space contracts points within it and the system stays localized to a particular region of its state space. If the exponent is positive, the system is expansive and points may travel through a large region of the state space.

On the other hand, there are times when a system is too chaotic and we wish to reign in its behavior. Chaos can lead systems to catastrophic failures. The presence of chaos in neural networks can also interfere with learning. A number of *control of chaos* techniques have therefore been developed. Ott, Grebogi and York (1990) were among the first to do so. In their procedure, the behavior of a system is monitored until its trajectory comes near a desired orbit. The value of a parameter is then changed a small amount. This changes the location of the orbit and forces the next iteration back to the local stable direction of the original fixed point. Other techniques involve altering the threshold function and the use of constant pinnings in coupled map lattices (Kurian & Joseph, 1999; Parekh & Sinha, 1999).

Chaotic Computing

One of the more interesting applications of chaotic modeling has been the utilization of chaotic procedures for computing. Sinha and Ditto (1998) used a lattice of threshold coupled chaotic elements to emulate logic gates, encode numbers and execute arithmetic operations like addition and multiplication. This suggests that one could create a fully programmable chaos computer, one that could perform all the operations that a modern computer can, i.e., one that is a universal Turing machine. In their simulation, the programming was accomplished by feeding in as input different threshold values for the chaotic elements.

Sinha (1999) points out that this model differs from that of traditional artificial neural networks, which do not have any intrinsic dynamics. The nodes or neurons in these traditional models are functionally simple, whereas the elements in their model are chaotic and more behaviorally complex. He also indicates that their adjustment of the threshold values is not analogous to the adjustment of weights to match some desired response. They instead are able to precisely specify the coupling required for the system to perform a particular computation. Their chaotic computing model thus does not have to go through a set of learning trials.

Traditional neural networks are built to perform specific tasks such as pattern recognition and classification. In contrast, chaotic computers of this sort are general purpose and may be able to handle a wide variety of computing scenarios (Sinha, 1999). The implications of this work are significant, as they may allow us to construct chaos computers capable of performing many computations faster than traditional artificial neural networks. Because these systems are parallel, they may additionally serve as a model for how the brain performs dynamical computation.

A Review of Dynamical Modeling

Abraham (1995) makes several general comments on dynamical modeling procedures. He notes that there are many advantages for psychology if these techniques are adopted. They allow us to understand more complex multivariate phenomena and not have to isolate variables from what he refers to as "extraneous" influences. In other words, we can include outside variables in our analyses rather than try to control for them by holding them constant or randomizing them in an experimental design.

In addition, the concept of bifurcation allows us to link what we may have thought were disparate phenomena. We can now realize that a system undergoes dramatic change when it bifurcates and that the conditions before and after this change are really part of the same system. This way we don't have to formulate separate theories and models for the conditions on "either side" of these bifurcation points. Instead of creating disconnected linear models for these stable attractor regions, we can formulate nonlinear differential or difference equations that may explain all of them together.

Dynamical modeling therefore provides us with a parsimonious account of the natural world. Instead of using many different variables to explain what we thought were distinct phenomena, we can now use a single control parameter and/or smaller number of order parameters. It is ironic, but dynamical models, although more complex, actually simplify. However, Abraham mentions that we need to avoid the tendency to oversimplify. Many phenomena that we think are low dimensional may actually be more complex and contain a greater number of dimensions. Einstein's dictum applies well here: "Make everything as simple as possible but not simpler" (Calaprice, 2005).

The use of computer simulations in dynamics allows us to explore what a system is like under various conditions, reflected in different simulations with new parameters added or subtracted and with variations in parameter values. In addition, graphical computer techniques enable us to visualize complex relationships and to see in a fast and intuitive manner how they change over time. Computer modeling also allows us to study how a much larger set of variables. This makes the model more like the real world, where many interacting variables are usually at play. We are no longer limited to the introduction of just two or three independent and dependent variables as we find in so many psychology experiments.

CHAPTER 16
CONCLUSIONS

A Summary

We have covered a lot of ground in this book. Let's pause for a moment and summarize the salient points. The central thesis of this work is that the scientific perspective has changed radically within the last few decades. This has happened on a number of fronts. These fronts include the paradigms of complexity science, self-organization, dynamical systems and network science. We briefly recap here the theoretical orientations of these perspectives and their import for psychology.

The most interesting phenomena in the world, and the ones that cannot easily be explained are complex systems, where there is unfortunately no commonly agreed upon definition of complexity. Complexity science is the study of these systems and seeks to understand complex behavior in terms of simple rules. Behavior is nonlinear and arises as a result of complicated part interactions where the parts can be rule-governed agents. The emphasis here is not on prediction, but instead on explanation, being able to understand how complex causal links in a system give rise to global action. A hallmark of complexity science is the use of mathematical and computer models. These models are best used synergistically with empirical research methods.

The human brain is the most complex system known. It is complex by almost any criteria we wish to adopt. It has a large number of parts, performs a large number of different functions, has great structural variability as well as informational and computational complexity. In fact, the brain is a complex adaptive system, since it adjusts its own behavior in response to changes in the environment. Brains are obviously capable of learning and of storing memories for later use. Brains and other adaptive systems such as ecologies share a number of features in common. We can also draw an analogy between brains and the process of biological evolution. Both can be viewed as systems attempting to optimize fitness on a changing landscape.

Another break with traditional notions in science has to do with how order gets created. Classical science cannot account for how order gets into the universe. According to the principle of self-organization though, the universe naturally creates order. This tendency comes from accumulated energy that seeks to flow. More ordered systems are able to dissipate their energy faster. This has led to the evolution of systems with greater and greater order. Transitions to more efficient energy dissipation are accompanied by structural and behavioral reorganizations. These sudden shifts correspond to bifurcations and the emergence of new levels of organization.

Self-organized systems are dynamic, have emergent properties, demonstrate stability in the face of change, and show complicated behavior that arises from parts that interact according to simple rules. They are also able to coordinate their activity despite the absence of a central leader. The brain fulfills all these criteria. Although the brain does not contain agents that move, the neurons and neural groups of which it is composed can be modeled as agents. A number of new models have emerged that can be successfully used to simulate brain behavior. The simplest of these are Boolean nets and cellular automata, which despite the simplicity of the rules they follow can still produce complex yet ordered behavior. Oscillatory models are also a very useful method to simulate and explore mental processes, since these more closely map onto the way real neurons function.

The dynamical systems approach is poised to replace many classical ideas in science. Specifically, natural phenomena are seen as nonlinear, interdependent, and open systems governed by recurrent processes. Although many systems may be determined and can even be described using mathematical formulas they can only be predicted over short time spans.

The behavior of a complex system is best depicted using a state space, where the axes of this space define relevant dimensions. A trajectory or path through this space shows behavioral change over time. The course of a trajectory is best explained with the concept of an attractor that "pulls" on or captures the trajectory. Attractors can range from being simple, such as points that represent constant or unchanging behavior, through periodic or cyclical patterns, to very complex and chaotic paths. It is helpful to visualize a state space as a landscape with hills and valleys and the trajectory as a ball traveling along its surface. In this analogy the valleys correspond to attractors and the hills to repellors that "push" the system state away.

State space concepts have been used extensively in all fields of psychology. They have been used to describe the development of walking and other abilities early in life, romantic relationships, the ways in which organizations change, job performance, and the emergence of leaders. They have served as a useful way to conceptualize personality traits, psychological defenses, and the development of anxiety, mood, and dissociative identity disorders, as well as schizophrenia and learning disabilities. In the field of cognition we see even greater use of dynamical models and techniques to describe perception, attention, memory, language, decision-making and problem solving.

In the dynamical systems approach a control parameter is varied that drives the system through qualitatively different states. The system's behavior in response to this manipulation is observed using an order parameter. The resulting changes are not always smooth and gradual but can be characterized by sudden and dramatic shifts. The system is said to bifurcate to a new state, the bifurcation marking the dramatic change. The psychological literature is replete with examples of fast dramatic change. They have been found in motor behavior to name just one case.

The complex system we have concerned ourselves with in this book is the human brain. The brain is a network and so the study of networks is extremely relevant for neuroscience. Findings from network science like synchrony, coupling, and message propagation have been found to be general features of many other networks in addition to brains. Neuroscience has shown great advances recently in its use of network concepts to analyze and model brain processes. It has been especially adept at utilizing fractal network concepts where neural activity at different scales is linked to explain coordinated action.

Although dynamical systems, complexity science, the study of self-organization and network science might seem like separate disciplines, they really share a lot in common. All seek to understand the operation of complex systems. All also utilize models as well as experimental techniques. That is why we have presented them together in this book. Each discipline though, has a different emphasis. Dynamics focuses on change, complexity science focuses on part interactions, self-organization focuses on the creation of order, and network science focuses on a particular type of system, namely networks. A truly unified science could perhaps merge these together into a single meta-theoretic paradigm with methods abstracted from each.

The Problem with Defining Emergent Levels

Most complex systems, as we have seen, are characterized by having organization at several different levels. The brain can be studied at the level of a neuron, neural circuits or brain areas. A society can be studied at the level of the individual, family, school or state. Each of these levels may be thought of as emerging from the elements below it but as also possessing new properties and behaviors not shared by the levels beneath it. A scientific study of complexity requires that we be able to define formally what we mean by an emergent level. If this were possible, we could analyze how levels form, examine what levels share in common with each other, and compare levels across disciplines. In this section we describe several ways to describe emergent levels, but show they all suffer from problems.

Carving Systems at their Joints

Goldstein (2002) shows how difficult it is to define what we mean by an emergent level. Perhaps the simplest way to determine the boundaries between levels is draw them where we actually see differences in organization, characteristics, and behavior. In cellular automata, the individual cells form one level of analysis while the larger forms that emerge from them constitute another. These forms, such as the glider, are shapes that persist over time, move across the screen, and display features not seen in single cells.

Similarly, we can turn to nature to see where boundaries between levels form. Atoms form into molecules. These molecules act quite differently from their atomic constituents. These molecules in turn can be assembled into cell organelles and cells, which form tissues, organs, and whole organisms. These organisms form into ecosystems. Notably, at each level, we have an entirely different discipline devoted its study. Starting at the lowest level and progressing upward, we have physics, chemistry, biochemistry, biology and ecology. This approach has been described as "carving nature at her joints".

Although this approach seems the most natural and intuitive, it cannot yield a formal analysis of emergence or levels. Take the cellular automata example. The larger forms actually do share some features in common with the cells that make them up. Both they and the larger forms translate across the screen. The same is true in nature. Many substances break down into their parts when heated. However, in nature, the shared similarities between levels varies considerably depending on the rules by which they are assembled. Molecules that form a membrane in the shape of a flat sheet will possess very different properties from the very same molecules that form a closed sphere.

Alternate Conceptions of Emergent Levels

Entropy, a measure of the disorder in a system, has been suggested as a way to discern emergent levels (Bar-Yam, 1997). But entropy tells us little about the dissimilarities of levels in the structure of a complex system (Wolpert & Macready, 2000). It also fails to tell us much about how complex a system is. Bennett (1988) measured the entropy of the human body and found it to be in between that of a crystal and a gas. Although this may be accurate in a gross sense, it doesn't say much about how the human body is ordered and how that order might differ from other bodies. Goldstein (2002) concludes that although entropy measures say something about the structure at a specific level, they are not useful in detecting emergent levels or in examining the richer structure in systems with multiple levels.

The nonlinear dynamical systems approach considers an emergent level to be essentially the same as the formation of an attractor (Goldstein, 2002). During an attractor state the system is ordered and displays new properties. A bifurcation can thus tell us when an emergent level has formed. However, dynamical systems theory still falls short. It has no real way of measuring or describing novelty (Goldstein, 2001).

Clark (1996) suggests that changes in parameters can indicate the presence of emergent levels. As we have seen, in many systems, there are sudden and dramatic changes that take place when the value of an order parameter reaches some value. Since new organizations occur at these break-points, they could be indicative of emergent levels. However, according to Anderson (cited in Hartman, 2000) there is no well-defined function that acts as an order parameter under the non-equilibrium conditions where emergence often takes place. Order parameters tell us that order has begun, but not what the nature of that order is like.

Another feature that seems to correlate with order is the rate at which a process occurs. Generally speaking, smaller events take place more rapidly. Atomic processes happen very rapidly in comparison to most of those taking place on a biological or ecological scale. In the brain, neural integration and synaptic action are fast in relation to messages passed between neurons in a network or cortical regions. However, Goldstein (2002) points out that this is not always the case. A drop in the speed of processing cannot always be taken to indicate the formation of an emergent level. It could simply be the result of performing a more difficult computation.

Aggregates Again

One assumption often made when looking at levels in systems is that the elements at the lower level remain distinct. This distinctiveness allows us to see the differences between them and the larger wholes that they might form. Certainly this is true in some cases. A human society, for instance, can be thought of as a collection of individual people. A brain can be thought of as a collection of neurons. But it is rare for these lower-level elements to act the way they would if they were in isolation. Social psychology tells us that people act very differently when they are in groups. Neuroscience tells us that neurons act another way when they are connected to others.

In fact, it is difficult to find a system made up of parts where this is not the case. Perhaps a pile of rocks or a beach made of sand particles come close. As previously discussed, these wholes can be considered as aggregates, collections of elements where the grouped elements act the same or similar to the way they would if they were in isolation. In most systems, especially in complex ones, the elements lose their distinctiveness and take on new properties. Their organization and behavior alters after they join a group. This means we may have to abandon our traditional view of systems as consisting of nested hierarchies in which groups at one level are classified in terms of elements at a lower level. It is better to think of the two levels as being confounded and inseparable from each other (Goldstein, 2002).

Informational Complexity and Emergence

So we return to our former question. Is it possible to formalize the concept of an emergent level so that we can study them using logical, mathematical, and scientific methods? One issue that must be addressed is that of novelty. Emergent levels are not just repetitions of their elements, but some unique combination of them. The shuffling of genes and their selective recombination during sexual reproduction creates novelty in biological evolution. The result is offspring that are similar but not identical to their parents.

Goldstein (2001) uses the term *self-transcending constructions* (STC) to describe novelty generation in *artificial life* (A-life) scenarios. A-life simulations are computer models that follow specific rules to generate life-like forms. Cellular automata and the Sugarscape models discussed earlier are examples of A-life simulations although there are others that are specifically designed to allow the resulting creatures to reproduce under selection pressures (Ray, 1991).

One way to reproduce in A-life simulations is through the use of a *genetic algorithm*. These algorithms mimic biological reproductive processes. Two creatures in an A-life simulation, for instance, could alter their binary genetic "codes" by swapping them in a crossover operation. The result would be two new creatures with different features from the previous generation. A genetic algorithm can be considered an example of a STC, since it produces novel structures.

Logical depth is one way of determining the complexity of an object in an A-life scenario. It is defined as the time required by a standard universal Turing machine (general computing device) to generate the object from an input that is algorithmically random (Bennett, 1988). Logical depth is similar to the notion of algorithmic information content (AIC) discussed previously. The notion is that the more complex an object or system, the longer its informational description or "build" instructions.

But, as it turns out, logical depth is also an incomplete measure of emergent structure. It doesn't, for instance, detect multiple levels within a system. According to Goldstein (2002), a description generated by an STC could be logically deep but not naturally emergent. However, he admits that if a natural system were logically deep, it was probably formed through an emergent process.

An Empirical Approach

So we see that none of the proposed methods provide us with a good way to formalize an emergent level. Perhaps the only method left is an empirical one. We could study the circumstances under which emergent levels form and attempt to describe them in as much detail as possible. This description would include the context or environmental conditions under which the transformation takes place as well as the particular causal factors that give rise to it.

This has already been done for the formation of Bénard cells of the Belousov-Zhabotinski reaction. If similar attempts were made to explain emergence in other domains, there may be general physical rules at work in each. In such a way we could perhaps uncover the general properties of emergent level formation, those that are invariant across all or many domains.

Evaluating the Reduction-Emergence Debate

In the introductory chapter we gave a description of both the reduction and emergence approaches. Both have certain advantages but also suffer from problems. In particular, we showed that reduction has failed more often than not to link disciplines or to provide a satisfactory explanation of a system's behavior. Emergence puts the focus on relations and not parts, but it too has failed to provide valid explanations for many phenomena. And, as we have just elaborated, there are significant problems in defining emergent levels. So what is a scientist to do? We advocate that both methods are necessary. Reduction has and will continue to be a powerful technique that can tell us much about the natural world. But it must be used together with emergent procedures in order to be truly effective. In what follows, we propose three approaches that may serve to unify these two methods.

The Interdisciplinary Approach

One possible solution is to adopt a multidisciplinary systems approach where the content knowledge, perspectives, and methodologies of different disciplines are integrated and brought to bear in understanding a given phenomenon. At a minimum, we could have researchers from two disciplines work together to try and understand the interface between their specialized domains. For instance, we could have a cell biologist and a chemist collaborate to determine the role protein synthesis plays in neurotransmitter regulation. This type of collaboration might link processes and explanations at the molecular level with those at the cellular level. Of course, in many cases, it is better to introduce more than two perspectives. In this example, it would be useful to have the contributions of a mathematician and a physicist as well.

A great example of multidisciplinary cooperation is the new hybrid field of cognitive neuroscience. Here, cognitive psychologists and neuroscientists collaborate. Cognitive psychologists focus on information-processing models of mind while neuroscientists specialize in neuron function and brain activity. A particular function, such as working memory, can be described at a more abstract cognitive level that specifies what a working memory must do, what representations and computations must be performed, etc. This can then be wed to a neuron or brain-based level of explanation that involves particular brain regions and types of neural action. The cognitive level specifies what must be explained, the neuroscience level how it is actually instantiated in physical brains. The results from both fields mutually constrain the types of explanations that can be formulated.

Multiple-Level Consideration

If we assume that there are legitimate boundaries between descriptive levels, then it is clear that reductionists and emergentists define their context too narrowly. In most cases, only two levels are considered, one that is either "above" or "below" the level under consideration. Reduction considers whether the upper can be reduced to the lower. Emergence considers whether the upper is something more than what is below it. A more integrated and holistic view would consider the multiple interactions between three or more levels. Take chemistry as an example. The reductionist would study whether it could be reduced to physics or another lower level description. The emergentist would examine whether chemical phenomena could arise from physics. A complete analysis necessitates the addition of a level above chemistry, such as biology. We can then study the downward causation of biology on chemistry as well as the upper or emergent influence of chemistry on biology. In this scenario we have four linkages between three levels instead of two linkages between two levels. Admitting the third level introduces additional explanatory power.

In fact, it may be possible to systematically isolate the effects of different levels and study their influence on a particular system. Take the example of a neuron. The upper levels to which a neuron belongs are those structures and organizations in which it is embedded and plays a functional role. Examples of these global levels may be the nuclei or brain regions of which the neuron is a part. The local levels that the neuron itself encompasses include the cellular nucleus inside of it, the cytoskeletal structure, etc. We could examine the neuron's function by holding constant the influence of one level while varying the others, i.e., we could see how the neuron operates when it is removed from the brain or connected to it, when it is connected in one way but not another and so on. This is nothing more than the application of the scientific method to manipulate nested or hierarchical influences instead of influences at just one level of spatial scale.

Redefining Parts and Wholes

We have just seen in our discussion of emergent levels that there are many different ways to define parts and wholes. Is a part a physical structure that occupies a given region of space? Does it have to perform a specific function? Do parts have to be made of other parts? In the human body we can consider internal organs to be parts. But is it legitimate to consider the stomach and small intestine in isolation from each other when they are connected and form part of the digestive system? A redefinition of parts and wholes may help the reductionists. It could be that the behavior of many large-scale systems can be effectively reduced when parts and wholes are defined appropriately. Similarly, emergent explanations may become possible with a valid conception of parts and wholes.

One approach to this problem is to define parts and wholes in a given system in different ways and see which procedure yields a more effective explanation. One could, for example, define them structurally, in terms of how parts fit inside one another. Alternatively, one could define them functionally, in terms of the particular functions they serve. It may even be possible to combine part-whole definitions in different ways to develop a meta-theoretic conception. For example, instead of defining the neuron as the primary "part" of the brain, we could use neural populations that are coupled or synchronize. We could then look at how these larger groups interact.

The Unification of Psychology

Psychology, perhaps more than most disciplines, is fragmented into different areas and subareas. There is biological, cognitive, clinical, developmental, social psychology and more. Within each of these are multiple more specialized areas. For instance, cognitive psychology can be further subdivided into perception, attention, memory, language, decision-making, and problem solving, as the sections of this book attest. The differences between these areas are demarcated not only by the subject matter. In many cases they adopt different theoretical perspectives and methodologies.

This fractured landscape has a positive side, as it allows for a plurality of viewpoints that stimulate argumentation and debate. However, it also isolates researchers who more often than not fail to understand each other's work, preventing them from exchanging information and collaborating. There have been several historical paradigms that have provided some coherence to this interdisciplinary mayhem. More recent paradigms of this sort we have already mentioned and include the ecological approach (Gibson, 1986) and connectionism (McClelland & Rumelhart, 1986). The cognitive science approach is inherently cross-disciplinary, but lacks an overarching theory of mind (Friedenberg & Silverman, 2006). Other earlier approaches mentioned by Abraham (2004) include the Gestalt approach (Kohler, 1965), field theory (Lewin, 1951), behaviorism (Skinner, 1953), drive reduction theory (Hull, 1943), and statistical learning theory (Estes, 1959). None of these have succeeded in uniting psychology.

There is some hope that the dynamical approach will succeed where these others have failed. Its methodology is universal, like statistics and research methods, it can be applied to any area of study. Indeed, the dynamical approach is much more general than psychology and can serve as a new unifying approach for all the sciences. Abraham (1995) mentions several other reasons. The mathematical language of dynamics allows investigators to deal with the complexity of real life situations, but is also fairly easy to communicate, especially in the visual and geometric terms provided by state spaces, trajectories and attractors.

The dynamic approach also frames the free will determinism debate in a new light. It shows us that complex phenomena can be completely determined yet difficult to predict. Freedom of choice can be incorporated into dynamic models by movement in the control space or by restarting initial conditions such that a trajectory can be selected that leads to a desired attractor (Abraham, 1995). So in a sense, dynamics allows us to have our cake and eat it to. It accommodates both determinism and freedom.

Cognitive Science, Dynamical Systems, and Evolutionary Psychology

Kenrick (2001) argues that we should build an integrative paradigm for psychology based on cognitive science, evolutionary psychology and dynamical systems theory. In particular, evolutionary and dynamical theorists seem to know little about each other, yet both provide unifying frameworks for psychology. The evolutionary approach applies not just to biology, but to many different natural processes as well. We see random variation and selection at the level of neurons, creative thinking, learning over the life span, and species evolution. Kenrick (2001) says we ought to taxonomize these processes and explain how emergent phenomena at one level combine into patterns at other levels.

We also need to study how dynamic processes affect the operation of adaptive mechanisms. In his own research Kenrick (2000), using cellular automata models, has found that mating decision rules used by men and women have a strong influence on the global distribution of mating behavior. If women shy away from unrestricted sexual behavior but men do not, the majority of mating habits in the population are monogamous, surrounding small pockets of individuals who "sleep around". This is similar to what we see in existing large heterosexual populations. If both sexes are instead willing to adopt unrestricted sexual behavior, there is widespread sleeping around at the population level, similar to what is seen in real world homosexual populations.

One way to blend these paradigms is to implement a layered rule system. At the individual level, a modeler could instantiated cognitive decision-making rules that guide behavior. These behaviors at the next level up would be one of the factors determining global behavior, such as proximity in space, that could be simulated using cellular automata or multi-agent systems rules. Evolutionary forces could also be applied at different time scales. For example, we could apply the rules of natural and sexual selection to determine the characteristics of offspring in a second generation.

In the above example, cognitive science determines the mating decisions individuals make at the smallest time scale, dynamical systems describes their spatial distribution over an intermediate time span, and evolutionary psychology details what successive generations will be like over longer periods of history. This type of model would allow us to look at interesting interactions between levels in a controlled and simplified setting not possible in the real world. For example, pockets of unrestrictive maters, if unmarried, will reproduce at a lower rate than monogamous groups, causing the dissolution and reformation of pockets at each generation.

References

Abraham, F. D. (1995). "A postscript on language, modeling, and metaphor," in F. D. Abraham and A. R. Gilgen, (eds.), *Chaos Theory in Psychology*, ISBN 0-275-95140-5, pp. 311-336.

Abraham, F. D. (2002). "Chaos, bifurcations, and self-organization: Dynamical extensions of neurological positivism and ecological psychology," in A. Combs, B. Goertzel, and M. Germine (eds.), *Mind in Time: The Dynamics of Thought, Reality, and Consciousness*, ISBN 1572732563, pp. 111-144.

Abraham, F. D., Abraham, R. H. and Shaw, C. D. (1990). *A Visual Introduction to Dynamical Systems Theory for Psychology*, ISBN 0942344097.

Aftanas, L. I. and Golocheikine, S. A. (2002). "Non-linear dynamic complexity of the human EEG during meditation," *Neuroscience Letters,* ISSN 0304-3940, 330(2) : 143-146.

Aftanas, L. I., Lotova, N. V., and Kosharov, V. I. (1998). "Non-linear dynamical coupling between different brain areas during evoked emotions: An EEG investigation," *Biological Psychology*, ISSN 0301-0511, 48(2): 121-138.

Aftanas, L. I., Lotova, N. V., Kosharov, V. I., Popov, S. A., and Makhen, V. P. (1997). "Nonlinear forecasting measurements of human EEG during evoked emotions," *Brain Topography*, ISSN 0896-0267, 10(2): 155-162.

Aks, D., Spott, J. C. (1996). "Quantifying aesthetic preference for chaotic patterns." *Empirical Studies of the Arts*, ISSN 0276-2374, 14: 1-16.

Albert, R., Jeonq, H. Barabasi, A. (1999). "Diameter of the world wide web," *Nature* ISSN: 0028-0836, 10(September): 40, 130-131.

Alexander, D. M. (1993). "A neural architecture with multiple scales of organization," in P. Leong, and M. Jabri (eds.), *Proceedings of the Fourth Australian Conference on Neural Networks*, ISBN 0867586508, pp. 157-160.

Alexander, D. M., Globus, G. G. (1996). "Edge-of-chaos dynamics in recursively organized neural systems," in E. M. Cormac and M. I. Stamenov, (eds.), *Fractals of Brain, Fractals of Mind: In Search of a Symmetry Bond*, ISBN 1556191871, pp. 31-74.

Anderson, J. R. (1985). *Cognitive Psychology and Its Implications (2nd ed.)*, ISBN 978-0716701101.

Anderson, R. S. (1990). "Eolian ripples as examples of self-organization in geomorphological systems," *Earth-Science Reviews*, ISSN 29: 77-96.

Anderson, C. M. and Mandell, A. J. (2002). "Fractal time and the foundations of consciousness: Vertical convergence of 1/f phenomena from ion channels to behavioral states," in E. M. Cormac and M. I. Stamenov (eds.), *Fractals of Brain, Fractals of Mind: In Search of a Symmetry Bond*, ISBN 1556191871, pp. 75-126.

Anokhin, A. P., Birbaumer, N., Lutzenburger, W., Nikolaev, A., Elger C. E., and Lehnertz, K. (1996). "Age increases brain complexity," *Electroencephalography and Clinical Neurophysiology,* ISSN 1388-2457, 99(1): 135-144.

Atkinson R. C., and Shiffin R. M., (1971). "The control of short term memory," *Scientific American*, ISSN 0036-8733, 225: 82-90.

Atlan, H. (1983). "Information theory: Basic elements and recent developments" in R. Trappl (ed.) *Cybernetics: Theory and Applications,* ISBN 0891161287, pp. 9-41.

Baddeley, A. D. (1992). "Working Memory," *Science,* ISSN 0036-8075, 255: 556-559.

Baggot, J.E. (1992). *The Meaning of Quantum Theory,* ISBN 019855575X.

Bak, P. (1996). *How Nature Works: The Science of Self-Organized Criticality,* ISBN 0387987385.

Bak, P., Tang, C. and Wiesenfeld, K. (1987). "Self-organized critically: An explanation for 1/f noise," *Physical Review Letters,* ISSN 0031-9007, 59: 381-384.

Bar-Yam, Y. (1997). *Dynamics of Complex Systems,* ISBN 978-0201557487.

Barabási, A. L. (2002). *Linked: The New Science of Networks,* ISBN 0738206679.

Baran, P. (1964). "Introduction to distributed communications networks," in report Memorandum RM-3240-PR (RAND Corporation, Santa Monica, CA), available for download at: http://www.rand.org/publications/RM/baran.list.html

Barkely, R. A. (1997). "Behavioral inhibition, sustained attention and executive functions: Constructiong a unified theory of ADHD," *Psychological Bulletin* ISSN 0079-2993, 121: 65-94.

Barton, S. (1994) . "Chaos, self-organization and psychology," *American Psychologist,* ISSN 0003-066X, (January): 5-13.

Bascompte, J. and Solé, R. V. (1996). "Habitat fragmentation and extinction thresholds in spatially explicit models," *Journal of Animal Ecology,* ISSN 0021-8790, 65: 465-473.

Bateson, G. and Bateson, M. C. (1990). *Angels Fear,* ISBN 1572735945.

Bennett, C. (1988). "Dissipation, information, computational complexity, and the definition of organization," in D. Pines (ed.), *Emerging Syntheses in Science,* ISBN 978-0201156867, pp. 215-231.

Bernander, O., Douglas, R. J., Martin, K. A. C., Koch, C. (1991). "Synaptic background activity determines spatial-temporal integration in pyramidal cells," *Proceedings of the National Academy of Sciences,* ISSN 0027-8424, 88: 11569.

Bickle, J. (2001). "Understanding neural complexity: A role for reduction," *Minds and Machines,* ISSN 0924-6495, 11: 467-481.

Biederman, I. (1987). "Recognition-by-components: A theory of human image understanding," *Psychology Review,* ISSN 0033-295X, 94: 115-147.

Bollt, E., Stanford, T., Lai, Y., Zyczkowski, K. (2000). "Validity of threshold crossing analysis of symbolic dynamics from chaotic time series," *Physical Review Letters,* ISSN 0031-9007, 85: 3524-3527.

Bonin, V., Mante V. and Carandini, M. (2006). "The statistical computation underlying contrast gain control," *Journal of Neuroscience,* ISSN 0270-6474, 26(23): 6346-6353.

Borisyuk, R., Borisyuk, G. and Kazanovich, Y. (2001). "Temporal structure of neural activity and modeling of information processing in the brain," in S. Wermter, J. Austin and D. Willshaw (eds.), *Emergent Neural Computational Architectures Based on Neuroscience: Towards Neuroscience-Inspired Computing*, ISBN 978-3540423638 , pp. 237-254.

Borisyuk, G., Borisyuk, R., Khibnik, A. and Roose, D. (1995). "Dynamics and bifurcation of two coupled neural oscillators with different connection types," *Bulletin of Mathematical Biology,* ISSN 1522-9602, 57(6): 809-840.

Borisyuk, R. M., Borisyuk, G. N. Kazanovich, Y. B., Strong, G. (2000). "Oscillatory neural networks: Modeling binding and attention by synchronization of neural activity" in D. S. Levine, V. R. Brown and V. T. Shiry (eds.), *Oscillation in Neural Systems*, ISBN 0805820663, pp. 261-284.

Borisyuk, R. and Borisyuk, G. (1997). "Information coding on the basis of synchronization of neuronal activity," *BioSystems*, ISSN 0303-2647, 40(1-2): 3-10.

Borisyuk, R. and Hoppensteadt, F. (1998). "Memorizing and recalling spatial-temporal patterns in an oscillator model of the hippocampus," *BioSystems* ISSN 0303-2647, 48(1): 3-10.

Borisyuk, R. and Hoppensteadt, F. (1999). "Oscillatory model of the hippocampus: A study of spatiotemporal patterns of neural activity," *Biological Cybernetics*, ISSN 0340-1200, 81(4): 359-71.

Bosssomair, T. and Green, D. (1998). *Patterns in the Sand: Computers, Complexity, and Everyday Life*, ISBN 0738201723.

Brabender, V. (1997). "Chaos and order in the psychotherapy group," in F. Masterpasqua and P. Perna (eds.), *The Psychological Meaning of Chaos: Translating Theory into Practice*, ISBN 1557984298, pp. 225-252.

Braspenning, P. J. (1995). "Neural cognidynamics," in P. J. Braspenning, F. Thnijsnman, and A. Weitjters (eds.), *Artificial Neural Networks: An Intro to ANN Theory and Practice*, ISBN 3540594884, pp 247-272.

Breakspear, M. (2002). "Nonlinear phase desynchronization in human electroencephalographic data," *Human Brain Mapping*, ISSN 1065-9471, 15(3): 175-198.

Breakspear, M. and Terry, J. R. (2002). "Detection and description of non-linear interdependence in normal multi-channel human EEG data," *Clinical Neurophysiology*, ISSN 1388-2457, 113(5): 735-753.

Brooks, R. A. (1991). "Intelligence without representation," *Artificial Intelligence*, ISSN 0004-3702, 47: 139-159.

Brooks, R. J. (2002). *Flesh and Machines: How Robots Will Change Us*, ISBN 037572527X.

Brown, A. L., Bransford, J. D., Ferrara, R. A., and Campione, J. C. (1983). Learning, remembering and understating. In J. H. Flavell and E. M. Markman (eds.), *Handbook of Child Psychology*, ISBN 0471272878, pp. 77-166.

Buchman M. (2002). *Small Worlds and the Groundbreaking Theory of Networks*, ISBN 0393324427.

Busemeyer, J. R. and Townsend, J. T. (1993). "Decision field theory: A dynamic-cognitive approach to decision making in an uncertain environment," *Psychological Review*, ISSN 0033-295X, 100(3): 432-459.

Bütz, M. R. (1990). "Chaos: An omen of transcendence in the psychotherapy process," presented at the *Symposium on Chaos and Psychology*, Springfield, MA.

Bütz, M. R. (1997). *Chaos and Complexity: Implications for Psychological Theory and Practice*, ISBN 1560324198.

Calaprice, A. (2005). *The New Quotable Einstein*, ISBN 0691120757.

Camazine, S. and Deneuburg, J. L. (2001) *Self-Organization in Biological Systems*, ISBN 0691012113.

Carver, C. S. and Scheier, M. F. (1994). "Situational coping and coping dispositions in a stressful transaction," *Journal of Personality and Social Psychology*, ISSN 0022-3514, 66: 148-195.

Chaitin, G. (1987). *Information, Randomness & Incompleteness: Papers on Algorithmic Information Theory*, ISBN 9971504790.

Chappell, M. and Humphreys, M. (1994). "An auto-associative neural network for sparse representations: Analysis and application to models of recognition and cued recall," *Psychological Review*, ISSN 0033-295X, 101(1): 103-128.

Chino, C., Cronin, T. W., and Osorio, D. (2000). "Color signals in natural scenes: Characteristics of reflectance spectra and effects of natural illuminants," *Journal of the Optical Society of America A*, ISSN 1084-7529, 17: 218-224.

Chung S. H., Raymond S. A. and Lettin, J. Y. (1970). "Multiple meaning in single visual units." *Brain, Behavior, and Evolution*, ISSN 0006-8977, 3: 12-101.

Clark, A. (1996) "Happy couplings: Emergence and explanatory interlock," in M. Boden (ed.), *The Philosophy of Artificial Life*, ISBN 0198751559, pp. 262-281.

Clark, A. and Chalmers, D. (1998). "The extended mind," *Analysis*, ISSN 0003-2638, 58: 7-19.

Clayton, K. and Frey, B. (1997). "Studies of mental noise," *Nonlinear Dynamics, Psychology, and Life Sciences*, ISSN 1090-0578, 1(3): 173-180.

Cohen, A. H. (1992). "The role of heterarchical control in the evolution of central pattern generators," *Brain, Behavior, and Evolution,* ISSN 0006-8977, 40(2-3): 112-124.

Colarusso, C. A. and Nemiroff, R. A. (1987). "Clinical implications of adult development theory," *American Journal of Psychiatry*, ISSN 1535-7228, 144: 1263-1270.

Colwell, K. A. (2003). "Emergence and evolution of silence networks for commercializing radical innovations: The case of nanotechnology." Unpublished doctoral dissertation, University of Oregon.

Combs, A. and Winkler, M. (1995). "The nostril cycle: a study in the methodology of chaos science," in R. Robertson and A. Combs (eds.), *Chaos Theory in Psychology and the Life sciences*, ISBN 0805817360, pp. 51-60.

Condon, W. S. and Ogston, W. D. (1967). "A segmentation of behavior." *Journal Of Psychiatric Research*, ISSN 0022-3956, 5: 221-235.

Coveny, P. and Highfield, R. (1995). *Frontiers of Complexity. The Search for Order in a Chaotic World*, ISSN 0571179223.

Cutting J. E. and Ganin, J. J. (1987). "Fractal curves and complexity," *Perception & Psychophysics*, ISSN 0031-5117, (Oct): 42(4), 365-370.

DeMaris, D. (2000). "Attention, depth gestalts, and spatially extended chaos in the perception of ambiguous figures," in D. Levine and V. Brown (eds.), *Oscillations in Neural Systems: The International Neural Networks Society Series*, ISBN 0805820663, pp. 239-260.

Dennett, D. (1991). *Consciousness Explained*, ISBN 0316180653.

Derrickson-Kossman, D., Drinkard, L. (1997). "Dissociative disorders in chaos and complexity," in F. Masterpasqua and P. A. Perna (eds.), *The Psychological Meaning of Chaos: Translating Theory Into Practice*, ISBN 1557984298, pp. 117-146.

Diederch A. (2003) ."MDFT account of decision making under time pressure," *Psychonomic Bulletin and Review*, ISSN 1069-9384, 10(1): 157-166.

Dittman, A., T. and Llewellyn, L. G. (1969). "Body movement and speech rhythm in social conversation," *Journal of Personality and Social Psychology*, ISSN 0022-3514, 11(2): 98-106.

Dobrunz, L. E. and Stevens, C. F. (1997). "Heterogeneity of release probability facilitation, and depletion at central synapses," *Neuron*, ISSN 0896-6273, 18: 995.

Drake, J. A., (1990). "The mechanics of community assembly and succession," *Journal of Theoretical Biology*, ISSN 0022-5193, 147: 213-233.

Duncker, K. (1945). "On Problem Solving," *Psychological Monographs*, ISSN 0096-9753, 58: 1-112.

Edwards, F. A. (1995). "LTP – A structural model to explain the inconsistencies," *Trends in Neuroscience*, ISSN 0166-2236, 18: 250-255.

Elman, (1998). Language as a Dynamical System," in R. F. Port and T. van Gelder (eds.), *Mind as Motion: Exploration in the Dynamics of Cognition*, ISBN 0262661101.

Epstein, J. M. (1999). "Agent-based computational models and generative social science," *Complexity*, ISSN 1076-2787, 4(5): 41-60.

Epstein, J. M. and Axtell, R. L. (1996). *Growing Societies: Social Science From the Bottom Up*. ISBN 9780262550253.

Erdös, P. and Rényi, A. (1959). "On random graphs," *Publicationes Mathematicae*, ISSN 0033 - 3883, 6: 290-297.

Ermentrout, G. B. and Edelstein-Keshet, L. (1993). "Cellular automata approaches to biological modeling," *Journal of Theoretical Biology*, ISSN 0022-5193, 160: 97-133.

Estes, W. K. (1959). "The statistical approach to learning theory," in S. Koch (ed.), *Psychology: A study of a Science. Vol. 2. General Systematic Formulations, Learning and Special Processes*, ISBN 9657328799, pp. 380-491.

Fell, J., Klaver, P., Lehnertx, K., Grunwald, T., Schaller, C., Elger, C. E. and Fernandez, G. (2001). "Human memory formation is accompanied by rhinal-hippocampal coupling and decoupling," *Nature Neuroscience*, ISSN 1097-6256, 4(12): 1259-1264.

Ferri, R., Chiaramonti, R., Elia, M., Musumeci, S. A., Rgazzoni, A. and Stam, C. J. (2003). "Nonlinear EEG analysis during sleep in premature and full-term infants," *Clinical Neurophysiology*, ISSN 1388-2457, 114(7): 1176-1180.

Ferri, R., Elia, M., Musumeci, S. A. and Stam, C. J. (2001) "Non-linear EEG analysis in children with epilepsy and electrical status epilepticus during slow-wave sleep (ESES)," *Clinical Neurophysiology*, ISSN 1388-2457, 112(12): 2274-2280.

Ferri, R., Parrino, L., Semerieri, A., Terzano, M. G., Elia, M., Musumeci, S. A., Pettinato, S. and Stam, C. J. (2002). "Non-linear EEG measures during sleep: Effects of the different sleep stages and cyclic alternating patterns," *International Journal of Psychophysiology*, ISSN 0167-8760, 43(3): 273-286.

Festinger, L, Schachter, S. and Back, K. (1950). *Social Pressures in Informal Groups, Volume 5*, ISBN 0804701733.

Field, D. J. (1994). "What is the goal of sensory coding?," *Neural Computation*, ISSN 0899-7667, 6: 559-601.

Fodor, J. (1983). *The Modularity of Mind.* ISBN 0262060841.

Folkman, S. and Lazarus, R. S. (1985). "It it changes it must be a process: Study of emotion and coping during three stages of a college examination," *Journal of Personality and Social Psychology*, ISSN 0022-3514, 48: 150-170.

Forrest, S. B. and Haff, P. K. (1992). "Mechanics of wind ripple stratigraphy," *Science* ISSN 1095-9203, 255: 1240-1243.

Freeman, W. (1990). "Searching for the signal and noise in the chaos of brain waves," in S. Krasner (ed.), *The Ubiquity of Chaos*, ISBN 0871683504, pp. 47-55.

Freeman, W. J. (1987). "Simulation of chaotic EEG patterns with a dynamic model of the olfactory system," *Biological Cybernetics*, ISSN 0340-1200, 56(2-3): 139-150.

French, R. M. (1999). "Catastrophic interference in connectionist networks," *Trends in Cognitive Sciences*, ISSN 1364-6613, 3(4): 128-135.

Friedenberg, J. and Silverman, G. (2006). *Cognitive Science: An introduction to the Study of Mind*, ISBN 1412925681.

Friston, K. J. (1997). "Transients, Metastability, and Neuronal dynamics," *Neuroimage*, ISSN 1053-8119, 5: 164-171.

Friston, K. J. (2000). "The labile brain. I. Neuronal transients and nonlinear coupling," *Philosophical Transactions of the Royal Society of London*, ISSN 0261-0523, 355: 215-236.

Friston, K. J. (2001). "Brain function, nonlinear coupling, and neuronal transients," *The Neuroscientist.* ISSN 1073-8584, 7(5): 406-418.

Gaik, W. (1993). "Combined evaluation of inter-aural time and intensity differences: Psychoacoustic results and computer modeling," *Journal of the Acoustical Society of America*, ISSN 0001-4966, 94(1): 98-110.

Gaka V. and Meyer A. D. (2005). "The diffusion of corporate venture investing programs: What do perspective adopters observe?" Working Paper, *INSEAD*.

Gallagher, R. and Appenzeller, T. (1999). "Complex systems viewpoints article," in R. Gallagher and T. Appenzeller, T. (eds.), *Science*, ISSN 0036-8075,

284: 79-109.

Gammaitoni, L., Hanggi, P. and Marchesoni, F. (1998). "Stochastic resonance," *Review of Modern Physics*, ISBN 0034-6861, 70: 223-287.

Gardner, M. (1971). "On cellular automata, self- reproduction, The garden of Eden and the game of life," *Scientific American,* ISSN 0036-8733 (February): 224(2): 112-117.

Gell-Mann, M. (1994). *The Quark and the Jaguar: Adventures in the Simple and the Complex*, ISBN 0716727250.

Gentry, T. A. (1995). "Fractal geometry and human understanding," in F. D. Abraham and A. Gilgen (eds.), *Chaos Theory in Psychology*, ISBN 0275951405, pp. 145-155.

German, T. P. and Defeyeter, M. A., (2000). "Immunity to Functional Fixedness in Children," *Psychonomic Bulletin and Review,* ISSN 1069-9384, 7: 707-712.

Gibson, J. J. (1986). *The ecological Approach to visual perception*, ISBN 0898599598.

Gick, M. C. and Holyoak, K. J. (1980). "Analogical Problem Solving," *Cognitive Psychology*, ISSN 0010-0285, 12: 306-355.

Gilden, D. L. (1997). "Fluctuations in the time required for elementary decisions." *Psychological Science,* ISSN 0956-7976, 8: 296-301.

Gilden, D., L., Thornton, T., and Mallon, M. W. (1995). "$1/f$ noise in human cognition," *Science*, ISSN 0036-8075, 267: 1837-1839.

Gilden, D. L., Schmucker, M. A. and Clayton, K. (1993). "The perception of natural contours," *Psychology Review*, ISSN 0033-295X, 100: 460-478.

Glass, L. and Perez, R. (1973). "Perception of random dot interference patterns," *Nature*, ISSN 0028-0836, 246: 3603-362.

Globus, G. G. (1992). "Towards a non-computational cognitive neuroscience," *Journal of Cognitive Neuroscience,* ISSN 0898-929X, 4(December): 299-310.

Goerner, S. (1995). "Chaos, evolution, and deep ecology," in R. Robertson and A. Combs, (eds.), *Chaos Theory in Psychology and the Life Sciences*, ISBN 0805817360, pp. 17-38.

Goertzel, B. (1994). *Chaotic Logic*, ISBN 0306446901.

Goertzel, B. (2002). "Chance and Consciousness," in A. Combs, B. Goertzel and M. Germine (eds.), *Mind in Time: The Dynamics of Thought, Reality, and Consciousness*, ISBN 1572732563, pp. 46-84.

Goertzel, B. (2002). "On the algebraic structure of consciousness," in A. Combs, B. Goertzel and M. Germine (eds.), *Mind in Time: The Dynamics of Thought, Reality, and Consciousness*, ISBN 1572732563, pp. 13-46.

Goldstein, E. B. (2002). *Sensation and Perception*, ISBN 0534539645.

Goldstein, J. (2001). "Emergence, radical novelty, and the philosophy of mathemtics," in W. Sulis and I. Trofimova (eds.), *Nonlinear Dynamics in the Life and Social Sciences,* ISBN 1586030209, pp 133-152.

Goldstein, J. (2002). "The singular nature of emergent levels: Suggestion for a theory of emergence," *Nonlinear Dynamics, Psychology, and Life Sciences,* ISSN 1090-0578, 6(4): 293-307.

Gonzalez, M.C., Hidelgo C.A. and Barabasi, A. (2008). "Understanding individual human mobility patterns," *Nature*, ISSN: 0028-0836, 453: 779-782.

Gordon, P. C., Hendrick, R., Johnson, M., and Lee, Y. (2006). "Similarity-based interference during language comprehension: Evidence from eye-tracking during reading," *Journal of Experimental Psychology: Learning, Memory, and Cognition*, ISSN 0278-7393, 32: 1304-1321.

Govindan, R. B., Narayanan, K., Gopinathan, M. S., Pradhan, N., Sreenivasan, R. and Dwivedi, P. (1999). "The spectrum of unstable periodic orbits of the human brain," in N. Pradhan, P. E. Rapp and R. Sreenivasan, (eds.), *Nonlinear Dynamics and Brain Functioning*, ISBN 1560726482, pp. 345-377.

Graham, C. H. (1965). "Perception of movement" in C. Graham, (ed.), *Vision and Visual perception*, ISBN 0471321702.

Graham, D. J., Chendler, D. M. and Field, D. J. (2006). "Can the theory of 'whitening' explain the center-surround properties of retinal ganglion cell receptive fields?" *Vision Research*, ISSN 0042-6989, 46(18): 2901-2913.

Grassberger, P. (1986). "Do climatic attractors exist?," *Nature*, ISSN 0028-0836, 323: 609-612.

Gray, C. M., Korny P. and Singer, W. (1989). "Oscillatory responses in cat visual cortex," *Nature*, ISSN 0028-0836, 338: 334-337.

Green, C. D. and Vervake, J. (1996). "What kind of explanation if any is a connectionist net?" in C. W. Tolman, F. Cherry, R. van Hezewijk, and I. Lubek (eds.), *Problems of Theoretical Psychology*, ISBN 189669117X, pp. 201-208.

Grossburg, S. (1980). "How does a brain build a cognitive code?" *Psychology Review*, ISSN 0033-295X, 87(Jan): 1-51.

Guastello, S. J. (1981). "Catastrophe model of equity in organizations," *Behavioral Science*, ISSN 0005-7940, 26(January): 63-74.

Guastello, S. J. (1987). "A butterfly catastrophe model of motivation in organizations: Academic performance [Monograph]," *Journal of Applied Psychology*, ISSN 0021-9010, 72(May): 165-182.

Guastello, S. J. (1992). "Population dynamics and workforce productivity," in M. Michaels, (ed.), *Proceedings of the Annual Chaos Network Conference: The Second iteration*, Urbana, IL: People Technologies, pp. 120-127.

Guastello, S. J. (2001). "Nonlinear Dynamics in Psychology," *Discrete Dynamics in Nature and Society*, 6(1): 11-29, available for download at: http://www.hindawi.com/GetArticle.aspx?doi=10.1155/S1026022601000024&e=cta.

Guastello, S. J. (2007). "Nonlinear dynamics and leadership emergencies," *The Leadership Quarterly*, ISSN 1048-9843, 18(4): 357-369.

Guastello, S. J. and Bond, R. W. (2007). "A swallowtail catastrophe model for the emergence of leadership in coordination-intensive groups," *Nonlinear Dynamics, Psychology, and Life Science*, ISSN 1090-0578, 11(2): 235-251.

Guastello, S. J., Dooley, K. J. and Goldstein, J. A. (1995). "Chaos, organizational theory, and organizational development," in F. D. Abraham and A. R. Gilgen (eds.), *Chaos Theory in Psychology*, ISBN 0275951405, pp. 268-276.

Gupte, N. and Amritkar, R. E. (1996). "Enhancing chaos in chaotic maps and flows," *Physical Review A*, ISSN 1094-1622, 54, 4580-4585.

Gurney, K. (1997). *An Introduction To Neural Networks*, ISBN 1857285034.

Gutman, S. E. Gilroy, L. A. and Blake, R. (2007). "Spatial grouping in human vision: Temporal structure trumps temporal synchrony," *Vision Research*, ISSN 0042-6989, 47(2): 219-230.

Hannah, T. (1990). "Does chaos theory have application to psychology? The example of daily mood fluctuations," *Network*, 8(3): 13-14.

Hartman, H. (2000). "Symmetry breaking and the origin of life" in Y. Bar-Yam (ed.), *Unifying Themes in Complex Systems: Proceedings of the International Conference on Complex Systems*, ISBN 0813341221, pp. 248-257.

Hayashi, H. and Ishizuka, S. (1999). "Synchronization of neurons during complex and irregular activity in both neuronal network and real tissue," in N. Pradhan, P. E. Rapp and R. Sreenivasan, *Nonlinear Dynamics and Brain Functioning,* ISBN 1560726482, pp. 379-400.

Heath, R. A. (1993). "A nonlinear model for human associative memory based on error accumulation," in P. Leong and M. Jabri (ed.), *Proceeding of the Fourth Australian Conference on Neural Networks*, ISBN 0805832009, pp. 130-133.

Heath, R. A. (2000). *Nonlinear Dynamics. Techniques and Applications in Psychology*, ISBN 0805831991.

Hebb, D. O. (2002). *The Organization of Behavior*, ISBN 978-0805843002.

Hegde, J. and Felleman, D. J. (2007). "Reappraising the functional implications of the primate visual anatomical hierarchy," *The Neuroscientist*, ISSN 1073-8584, 13: 416-421.

Heiby, E. M. (1991). "Implications of chaos theory for the study of depression," presented at the *American Psychological Association National Convention*, San Francisco, CA.

Heiby, E. M. (1995a). "Assessment of behavioral chaos with a focus on transitions in depression," *Psychological Assessment*, ISSN 1040-3590, 7: 10-16.

Heiby, E. M. (1995b). "Chaos theory, nonlinear dynamical models, and psychological assessment," *Psychological Assessment*, ISSN 1040-3590, 7: 5-9.

Hobson, J. A. and Pace-Schott, E. F. (2002). "The cognitive neuroscience of sleep: Neuronal systems, consciousness and learning," *National Review of Neuroscience*, ISSN 1471-003X, 3(9): 679-693.

Hock, H. S., Kelso, J. A. S., Schöner, G. (1993). "Bistability, hysteresis, and los of temporal stability in the perceptual organization of apparent motion," *Journal of Experimental Psychology: Human Perception and Performance*, ISSN 0096-1523, 19(1): 63-80.

Holland, J. H. (1995). *Hidden Order. How Adaptation Builds Complexity*, ISBN 0201407930.

Holland, J.H. (1998). *Emergence: From Chaos to Order*, ISBN 0192862111.

Hopfield, J. J. (1982). Neural networks and physical systems with emergent collective computational properties." *Proceedings of the National Academy of Sciences*, ISSN 0027-8424, pp.2254-2258.

Horgan, J. (1995). From complexity to perplexity: Can science achieve a unified theory of complex systems?" *Scientific American*, ISBN 0036-8733, 284: 104-109.

Hubel, D. H. (1982). "Exploration of the primary visual cortex, 1955-1978," *Nature*, ISSN 0028-0836, 7(October): 299, 515-524.

Hubel, D., Wiesel, T. (1979). "Brain Mechanisms of Vision," *Scientific American*, ISSN 0036-8733, 241: 150-162.

Hull, C. L. (1943). *Principles of Behavior*, New York, NY: D. Appleton and Company.

Ishai, A., Underleider, L. G., Martin, A., Schouten, J. L. and Haxby, J. V. (1999). "Distributed Representation of objects in human ventral visual pathway," *Proceedings of the National Academy of Sciences*, ISSN 0027-8424, 96(16): 9379-9384.

Jaffe, J., and Feldstein, S. (1970). *Rhythms of dialogue*. ISBN 0123798507.

Jeong, J., Chae, J. H., Kim, S. Y. and Han, S. H. (2001). "Nonlinear dynamic analysis of the EEG in patients with Alzheimer's disease and vascular dementia," *Journal of Clinical Neurophysiology*, ISSN 07360258, 18: 58-67.

Jones, M., Boltz, M. (1989). "Dynamic attending and responses to time," *Psychological Review*, ISSN 0033-295X, 96(3): 459-491.

Kaas, J. H. (1982). "Why do sensory systems have so many subdivisions?" in W. Neff (ed.), *Contributions to Sensory Psychology, Volume 7*, ISBN 0121518019, pp. 201-240.

Kahneman, D. (1973). *Attention and Effort*, ISBN 0130505188.

Kahneman, D. and Tversky, A. (1979). "Prospect Theory: An analysis of decision under risk," *Econometrica*, ISSN 0012-9682, 47(2): 263-291.

Kampen , N. G. (1987). "Some theoretical aspects of noise," in C.M. Van Vilet (Ed.), *Ninth International Conference on Noise in Physical Systems*, ISBN 9971503972, pp. 3-10.

Kaneko, K. (1992). "Focus issue on coupled map lattices," *Chaos*, ISSN 1054-1500, 2: 279-408.

Kauffman, S. A. (1993). *The Origins of Order: Self-Organization and Selection in Evolution*, ISBN 0195058119.

Kauffman, S. A. (1995). *At Home in the Universe: The Search for the Laws of Self-Organization and Complexity*, ISBN 0670847356.

Kelso, J. A. (1995). *Dynamic Patterns: The Self-Organization of Brain and Behavior*, ISBN 0262112000.

Kelso, J. A. S., Case, P., Holroyd, T., Horvath, E., Rączaszek, J., Tuller, B. and Ding, M. A. (1994). "Multistabilty and metastability in perceptual and brain dynamics," in M. Stadler and P. Kruse (eds.), *Multistability in Cognition*, Berlin, Germany: Springer-Verlag.

Kenrick, D. T. (2001). "Evolutionary psychology, cognitive science and dynamical systems: Building an integrative paradigm," *Current Directions in Psychological Science*, ISSN 0963-7214, 10(1): 13-17.

Kenrick, D. T., Li, N. and Butner, J. (2000). "Dynamical systems and the mating game," *Behavioral and Brain Sciences*, ISSN 0140-525X, 23(4): 607-608.

King, R., Barchas, J. D. and Huberman, B. A. (1983). "Theoretical psychopathology: An application of dynamical systems theory to human behavior," in E. Basar, H. Flohr, H., H. Haken, and J. Mandell (eds.), *Synergetics of the Brain*, Heidelberg: Springer-Verlag.

King, R., Barchas, J. D. and Huberman, B. A. (1984). "Chaotic behavior in dopamine neurodynamics," *Proceedings of the National Academy of Sciences of the United States of America (Neurobiology)*, ISSN 1091-6490, 81: 1244-1247.

Kingsbury M. and Finlay, B. (2001). "The cortex in multidimensional space: Where do cortical areas come from?," *Developmental Science*, ISSN 1363-755X, 48: 135-152.

Klatzky, R. L. (1980). *Human Memory: Sttructures and Processes*, ISBN 0716711133.

Knill, D. C., Field, D. and Kersten, D. (1990) "Human discrimination of fractal images," *Journal of the Optical Society of America A*, ISSN 1084-7529, 7(6):1113-1123.

Kobayashi, T., Misaki, K., Nakagawa, H., Madokoro, S., Ihara, H., Tusda, K., Umezawa, Y., Murayama, J. and Isak, K. (1999). "Nonlinear analysis of the sleep EEG," *Psychiatric Clinical Neuroscience,* ISSN 0940-1334 53(2): 159-161.

Kobayashi, T., Modokoro, S., Wada, Y., Misaki, K., Nakagawa, H. (2001). "Human sleep EEG analysis using the correlation dimension," *Clinical Electroencephalography*, ISSN 1388-2457, 32(3): 112-118.

Koch, C. and Crick, F. (1990). "Towards a neurobiological theory of consciousness," *Sem Neuroscience*, ISSN 2: 263-275.

Koch, C. and Segev, I. (1998). *Methods in Neuronal Modeling: From Ions to Networks.* ISBN 0262112314.

Koch, C. and Laurent G. (1999). "Complexity and the Nervous System,"*Science*, ISSN 0036-8075, 284: 96-98.

Köhler, W. (1976). *The Mentality of Apes*, ISBN 0871401088.

Köhler, W. (1965). *Dynamics in Psychology*, ISBN 0871400871.

Kolb, B., and Whishaw, I. Q., (1996). *Fundamentals of Human Neuropsychology*, ISBN 0716723875.

Kondakor, I., Brandeis, D., Wackmann, J., Kochi, K., Koenig, T., Frei, E., Pascual-Margui, Yagyu, T., and Lehmann, D. (1997). "Multichannel EEG fields during and without visual input: Frequency domain model source locations and dimensional complexities," *Neuroscience Letters*, ISSN 0304-3940, 226(1): 49-52.

Kosslyn S. M. (1980). *Image and Mind*, ISBN 067443659.

Kruse, P. and Stadler, M. (1990). "Stability and instability in cognitive systems: Multistability, suggestion and psychosomatic interaction," in Haken, H. and Stadler, M. (eds.), *Synergetics of Cognition*, ISBN 3540519297, pp. 201-215.

Kryukov, V. I. (1991). "An attention model based on the principle of dominanta" in A. V. Holden and Kryukov, V. I. (eds.), *Neurocomputers and Attention I. Neurobiology, Synchronization and Chaos.* ISBN 0471935050, pp. 319-352.

Kukkarni, D. R., Parikh, J. C. and Pratap, R. (1999). "Simulation and modeling of evoked response electroencephalograph signal," *International Journal of*

Modern Physics, ISSN 0218-2718, 10(4): 759-776.

Kumar, T., Zhau, P., and Glaser. D. A. (1993). "Comparison of Human performance with algorithms for estimating fractal dimension of fractional Brownian statistics," *Journal of the Optical Society of American A, Optic Image and Science*, ISSN 0740-3232, 10(6): 1136-1146.

Kunda, Z., and Thagard, P. (1996). "Forming impressions from stereotypes, traits, and behaviors: A parallel- constraint- satisfaction theory," *Psychological Review,* ISSN 0033-295X , 103: 284-308.

Kuramoto, Y. (1984). *Chemical Oscillations, Waves, and Turbulence.* ISBN 978-0486428819.

Kurian, G. and Joseph, K. B. (1999). "Control of Chaos in neural networks by altering the thresholding function," in N. Pradhan, P. E. Rapp and R. Sreenivasan (eds.), *Nonlinear Dynamics and Brain Functioning*, ISBN 1560726482, pp. 49-51.

Lago-Fernandez, L. F., Huerta, R. Corbacho, F. and Siguenza, J. A. (2000). "Fast response and temoral coherent oscillations in small-world networks," *Physical Review Letter*, ISSN 1079-7114, 84(12): 2758-2761.

Lakoff, G. (1987b). "Cognitive models and prototype theory," in U. Neisser (ed.), *Concepts and Conceptual Development: Ecological and Intellectual Factors in Categorization*, ISBN 0521322197, pp. 63-100.

Lakoff, G. (1987b). *Women, Fire, and Dangerous Things: What Categories Reveal About the Mind,* ISBN 0226468044.

Lamberts, J., van den Broek, P., Bener, J., van Egmond, J., Dirksen, R., Coenen, A. (2000) "Correlation dimension of the human electroencephalogram corresponds with cognitive load," *Neuropsychobiology,* ISSN 1423-0224, 41: 149-153.

Langton, C. G. (1990). "Computation at the edge of chaos: Phase transitions and emergent computation," *Physica D: Nonlinear Phenomena*, ISSN 0167-2789, 42(1-3): 12-37.

Langton, C. G. (1992). "Life at the edge of chaos," in C.G. Langton, C. Taylor, J. D. Farmer, and S. Rasmussen (eds.), *Artificial life II:Santa Fe Institute Studies in the Sciences of Complexity*, ISBN 0201525712, pp. 41-91.

Lashley, K. S. (1950). "In search of the engram,"*Society of Experimental Biology Symposium, No. 4: Psychological Mechanisms in Animal Behavior*, London: Cambridge University Press, 4: 454-482.

Laskoski, G. M. (1990). "Addressing scaling and plasticity problems with a biologically motivated self-organizing network," *Proceedings of the IEEE International Conference on Neural Networks*, ISSN 1098-7576, 2: 355-360.

Latora, V. and Marchiori, M. (2001). "Efficient behavior of small-world networks," *Physical Review Letters*, ISSN 0031-9007, 87: 198701.

Le Van Quyen, M. (2003). "Disentangling the dynamic core: A research program for neurodynamics at the large scale," *Biological Research,* ISSN 0100-879X, 36(1): 67-86.

Le Van Quyen, M., Martinerie, J., Navarro, V., Baulac, M. and Varela, F. J. (2001). "Characterizing neurodynamic changes before seizures," *Journal of Clinical Neurophysiology,* ISSN 1388-2457, 18(3): 191-208.

Lovelock, J. (2000). *Gaia: A New Look at Life on Earth*, ISBN 019217665X.

Levy, W. and Wu, X. B. (2000). "Some randomness benefits a model of hippocampal function," in P. Arhem, C. Blomberg and H. Liljenstrom (eds.), *Disorder Versus Order in Brain Function: Essays in Theoretical Neurobiology*, ISBN 9810240082, pp. 221-238.

Lewin R. (1999). *Complexity. Life at the Edge of Chaos*, ISBN 0226476553.

Lewin, K. (1947). "Frontiers in group dynamics," *Human Relations*, ISSN 0018-7267, 1: 5-41.

Lewin, K. (1951). *Field Theory in Social Science*, New York: Harper.

Lewis, M. D. (1995). "Cognition-emotion feedback and the self-organization of developmental paths," *Human Development*, ISSN 0018-716X, 38: 71-102.

Lewis, M. D. and Junyk, N. (1997). "The self-organization of psychological defenses," in F. Masterpasqua and P. A. Perna, (eds.), *The Psychological Meaning of Chaos: Translating Theory Into Practice*, ISBN 1557984298, pp. 41-74.

Liebovitch, L. and Todorov, A. (2000). "What causes ion channel proteins to open and close?," in P. Arhem, C. Blomberg and H. Liljenstrom (eds.), *Disorder Versus Order in Brain Function: Essays in Theoretical Neurobiology*, ISBN 9810240082, pp. 83-106.

Liljenström, H. and Wu, X. B. (1995). "Noise-enhanced performance in a cortical associative memory model," *International Journal of Neural Systems*, ISSN 0129-0657, 6: 19-29.

Lloyd, J. E. (1983). "Bioluminescence and communication in insects. *Annual Review of Entomology*, ISSN 0066-4170, 28: 131-160.

Loftus, E. F. (1975). "Leading questions and the eyewitness report," *Cognitive Psychology*, ISSN 0010-02857, 7(4): 560-572.

Logie, R. H., Zucco, G., and Baddeley, A. D. (1990). "Interference with visual short term memory," *Acta Psychologia*, ISSN 0001-6918, 75(1): 55-74.

Loring, D. W., Struss, E., Hermann, B. P., Perrine, K., Trenerry, M. R., Barr, W. B., Westfield, M., Chelune, G. J., Lee, G. P., and Mendor, K. J. (1999). "Effects of anomalous language representation on neuropsychological performance in temporal lobe epilepsy," *American Academy of Neurology* ISSN 1080-2371, 53(2): 260-264.

Losada, M., and Markovitch, S. (1990). "Group analyzer: A system for dynamic analysis of group interaction," in B. Shriver (ed.), *Proceedings of the 23rd Annual Hawaii International Conference on System Sciences*, ISBN 0818620099, pp. 101 – 110.

Lutzenberger, W., Elbert, T., Birbaumer, N., Ray, W. J. and Schupp, H. (1992). "The scalp distribution of the fractal dimension of the EEG and its variation with mental tasks," *Brain Topography*, ISSN 0896-0267, 5: 27-34.

Markram, H. (1997). "A network of tufted layer 5 pyramidal neurons," *Cerebral Cortex*, ISSN 1047-3211, 7(6): 523-533.

Marks-Tarlow, T. (1995). "The fractal geometry of human nature," in R. Robertson and A. Combs (eds.), *Chaos Theory in Psychology and the Life Sciences*, ISBN 0805817360, pp. 275-284.

McClelland, J. L. Rumelhart, D. E., and the PDP Research Group. (1987). *Parallel Distributed Processing, Vol 2: Psychological and Biological Models*, ISBN 978-0262631105.

McDonald, J. and Ward, L., M. (1998). "Nonlinear dynamics of event-related human brain activity," Unpublished manuscript, University of British Columbia.

Messick, D. M., and Liebrand, V. B. G. (1997). "Individual heuristic and the dynamics of cooperation in large groups," *Psychological Review*, ISSN 0033-295X, 102(1): 131-145.

Meyer A. D. (1982). "Adapting to environmental jolts," *Administrative Science Quarterly*, ISSN 0001-8392, 27(4): 515-537.

Meyer A. D., Gaba, V. and Colwell, K. A. (2005). "Organization far from equilibrium: Non-linear change in organizational fields," *Organization Science*, ISSN 1526-5455, 16(5): 456-473.

Meyer-Lindenberg, A. (1996). "The evolution of complexity in human brain development: An EEG study," *Electroencephalography and Clinical Neurophysiology* ISSN 1388-2457, 99(5): 405-411.

Micheloyannis, S., Flitzanis, N., Papanikolaou, E., Bourkas, M. Terzakis, D., Arvanitis, S., Stam, C. J. and Simos, P. G. (2002). "Ongoing electroencephalographic signal study of simple arithmetic using linear and non-linear measures," *International Journal of Psychophysiology*, ISSN 0167-8760, 44(3): 231-238.

Milgram, S. (1967). "The small world problem," *Psychology Today*, ISSN 0033-3107, 2: 60-67.

Miller, G. A. (1956). "The magical number seven, plus or minus two: Some limits on our capacity for processing information," *Psychological Review*, ISSN 0033-295X, 101(2): 343-352.

Molle, M., Marchall, L., Pietrowsky, R., Lutzenburg, W., Fehm, L., Born, J. (1995). "Dimensional complexity of the EEG indicates a right fronto-cortical locus of attention control," *Journal of Psychophysiology*, ISSN 0167-8760, 9: 45-55.

Mormann, F., Kreuz T., Andrezjak, R. G., David, P., Lenertz, K. and Elger, C. E. (2003) "Epileptic seizures are preceded by a decrease in synchronization," *Epilepsy Research*, ISSN 0920-1211, 53(3): 173-185.

Morowitz, H. J. (2002). *The Emergence of Everything: How the World Became Complex*, ISBN 019513513X.

Muller, V., Lutzenberger, W., Preissel, H., Pulvermuller, F. and Birbaumer, N. (2003). "Complexity of visual stimuli and non-linear EEG dynamics in humans," *Cognitive Brain Research*, ISSN 0926-6410, 16(1): 104-110.

Murray, J. D. (1981). "A prepattern formation mechanism for animal coat markings," *Journal of Theoretical Biology*, ISSN 0022-5193, 88: 161-199.

Muzalevskaya, N. I., Uritsky, V. M., Korolyov, E. V., Reschikov, A. M. and Timoshinov, G. P. (1993). "Stochastic control of living systems. Normalization of physiological function by magnetic field with 1/f power spectrum," in P. H. Handel and A. L. Chung (eds.), *Noise in Physical Systems and 1/f Fluctuations*, ISBN 1563962705, pp. 724-727.

Newell, A. (1991). *Unified Theories of Cognition*, ISBN 0674920996.

Newtson, D. (1994). "The perception and coupling of behavior waves," in R. R. Vallacher and A. Nowak (eds.), *Dynamical systems in Social Psychology*, ISBN 1572303530, pp. 139-167.

Nezlek, J. B. (1993). "The stability of social interaction," *Journal of Personality and Social Psychology*, ISSN 0022-3514, 65(5): 930-941.

Nezlek, J. B. and Wheeler, L. (1984). "Rochester interaction record analysis package," *Psychological Documents,* ISSN 0021-843X, 14(6): 2610.

Nicolis, J. S. and Tsuda, I. (1985). "Chaotic dynamics of information processing," *Bulletin of Mathematical Biology*," ISSN 1522-9602, 47(3): 343-365.

Noam, G. G., Powers, S. I., Kilkenny, R. and Beedy, J. (1990). "The interpersonal self in life-span developmental perspective: Theory, measurement, and longitudinal case analyses," in P. B. Baltes, D. L. Featherman and R. M. Lerner (eds.), *Life-span development and behavior*, ISBN 0805815078, pp. 59-104.

Noelle-Neumann, E. (1984). *The Spiral of Silence: Public Opinion - Our Social Skin*, ISBN 0226589366.

Novikov, E., Novikov, A., Shannahoff-Khalsa, D., Schwartz, B. and Wright , J. (1997). "Scale-similar activity in the brain," *Physical Review E*, ISSN 1063-651X, 56(3): R2387-R2389.

Nowak, A., and Lewenstein, M. (1996). "Modeling social change with cellular automata," in R. Hegselmann, K. Troitzch, and U. Muller (eds.), *Modeling and Simulation in the Social Sciences from the Philosophy of Science Point of View*, ISBN 0792341252, pp. 249-285.

Nowak, A., and Vallacher, R. R. (1998). *Dynamical Social Psychology*, ISBN 1572303530.

Nowak, A., Lewenstein, M. and Szamrej, J. (1993). "Social transitions occur through bubbles," *Scientific American*, ISSN 0036-8733, 12: 16-25.

Nowak, A., Szamrej, J., and Latane, B. (1990). "From private attitude to public opinion: A dynamic theory of social impact," *Psychological Review*, ISSN 0033-295X, 97: 362-376.

Odum, H. T. (1988). "Self-organization, transformity, and information," *Science*, ISSN 0036- 8075, 242: 1132-1139.

Osborne, A. R. and Provenzala, A. (1989). "Finite correlation dimension for stochastic systems with power-law spectra," *Physica*, ISSN 0031-8949, 35D: 357-381.

Otmakhov, N., Shirke, A. M. and Malinow, R. (1993). "Measuring the impact of probabilistic transmission on neuronal output," *Neuron*, ISSN 0896-6273, 10: 1101-1111.

Ott, E., Grebogi, C., and Yorke, J. A. (1990). "Controlling chaos," *Physical Review Letters,* ISSN 0031-9007, 64: 1196-1199 .

Paine, R. T. (1966). "Food web complexity and species diversity," *American Naturalist*, ISSN 0003-0147, 100: 65-75.

Palus, M. (1996). "Nonlinearity in normal human EEG: Cycles, temporal asymmetry, nonstationarity and randomness, not chaos," *Biological Cybernetics*, ISSN 0340-1200, 75(5): 386-396.

Parekh, N., and Sinha, S. (1999). "Suppressing spatiotemporal chaos in coupled map lattices using constant pinnings," in N. Pradhan, P. E. Rapp and R. Sreenivasan, (eds.), *Nonlinear Dynamics and Brain Functioning*, ISBN 1560726482, pp. 77-.

Pelli, D. G. (1981). Effects of visual noise. Ph.D. Thesis. Cambridge University.

Pereda, E., Manas, S., De Vera, L., Garrido, J. M., Lopez, S. and Gonzalez, J. J. "Non-linear asymmetric interdependencies in the electroencephalogram of healthy term neonates during sleep," *Neuroscience Letters*, ISSN 0304-3940, 337(2): 101-106.

Peschl, M. F. (1997)."The representational relation between environmental structures and neural systems: Autonomy and environmental dependency in neural knowledge representation," *Nonlinear Dynamics, Psychology, and Life Sciences*, ISSN 1090-0578, 2(November): 99-121.

Pettigrew, A. G., Bleasel, A. F. (1994). "Developmental and properties of spontaneous oscillations of the membrane potential in inferior olivary neurons in the rat," *Developmental brain research*, ISSN 0165-3806, 65(1): 43-50.

Phelan, S. E. (2001). "What is complexity science, really?" *Emergence: Complexity and Organization*, ISSN 1521-3250, 3(1): 120-136.

Pons, T. P., Garragty, P. E. Ommaya. A. K., Kass, J. H., Taub, E. and Mishlan, M. (1991). "Major cortical reorganization after sensory differentiation in adult macaques," *Science*, ISSN 0036-8075, 252: 1857-1860.

Port, R. F. and van Gelder, T. (1998). *Mind as Motion: Exploration in the Dynamics of Cognition*, ISBN 0262661101.

Post, W. M., and Pimm, S. L. (1983). "Community assembly and food web stability," *Mathematical Biosciences*, ISSN 1547-1063, 64: 169-192.

Pradhan, N. and Dutt, D. N. (1993). "Use of running fractal dimension for the analysis of changing patterns in electroencephalograms," *Computers in Biology and Medicine*, ISSN 0010-4825, 23(5): 381-388.

Raeymaekers, L. (2002). "Dynamics of Boolean networks controlled by biologically meaningful functions," *Journal of Theoretical Biology*, ISSN 0022-5193, 218(3): 331-341.

Ramachandran, V. S. (1992). "Blind spots." *Scientific American*, ISSN 0036-8733, 266: 86-91.

Ramaswamy, R., Sinha, S., and Gupte, N. (1999). "Enhancement and maintenance of chaos using adaptive anti-control," in N. Pradhan, P. E. Rapp and R. Sreenivasan (eds.), *Nonlinear Dynamics and Brain Functioning*, ISBN 1560726482, pp. 113-120.

Rapoport, A. (1957). "A contribution to the theory of random and biased nets," *Bulletin of Mathematical Biophysics*, ISSN 0007-4985, 19: 257-271.

Rapp, P. E., Zimmerman, I, D., Albano, A. M., Deguzman, G. C. and Gressbaum, N. N. (1989). "Dynamics of spontaneous neural activity in the simian motor cortex: The dimension of chaotic neurons," *Physics Letters*, ISSN 0015-9018, 110: 99-118.

Ray, T. S. (1991) "Approach to the synthesis of artificial life," in C. Langton, C. Yaylor, J. D. Farmer and S. Rasmussen (eds.), *Artificial Life II*. ISBN 0201525704, pp. 341- 408.

Read, S., J., and Miller, L.C. (eds.), (1998). *Connectionist Models of Social Reasoning and Social Behavior*, ISBN 0805822168.

Redies, C. (2007). "A universal model of esthetic perception based on the sensory coding of natural stimuli," *Spatial Vision*, ISSN 1568-5683, 21: 97-117.

Reeke, G. N. Jr., and Edelman, G. M. (1984). "Selective networks and recognition automata," *Annals of the New York Academy of Sciences*, ISSN 0077-8923, 426: 181-201.

Reicher, G. M. (1969). "Perceptual recognition as a function of meaningfulness of stimulus material," *Journal of Experimental Psychology*, ISSN 0096-3445, 81: 274-280.

Rieke, F., Warlund, D., de Ruyter van Steveninck, R. and Bialek, W. (1996). *Spikes: Exploring the Neural Code*, ISBN 0262681087.

Rinaldi, S. and Gragnani, A. (1998). "Minimal models for dyadic processes: A review," in F. Orsucci (ed.), *The Complex Matters of the Mind*, ISBN 9810233396, pp. 87-104.

Robinson, C. (1998). *Dynamical Systems: Stability, Symbolic Dynamics, and Chaos*, ISBN 0849384958.

Rourke, B. P., and Conway, J. A. (1997) "Disabilities of arithmetic and mathematical reasoning: Perspectives from neurology and neuropsychology," *Journal of Learning Disabilities*, ISSN 1354-4187, 30(1): 34-46.

Rusbult, C. E. (1980). "Commitment and satisfaction in romantic associations: A test of the investment model," *Journal of Experimental Social Psychology*, ISSN 0022-1031, 16: 172-186.

Rusbult, C. E. (1980). "A longitudinal test of the investment model: The development (and deterioration) of satisfaction and commitment in heterosexual involvement," *Journal of Experimental Social Psychology*, ISSN 0022-1031, 45: 101-117.

Rusbult, C. E. and Martz, J. M. (1995). Remaining in an abusive relationship: An investment model analysis of nonvoluntary dependence," *Personality and Social Psychology Bulletin*, ISSN 1552-7433, 21: 558-571.

Sabelli, H. C. and Carlson-Sabelli, L. (1990). "Psychogeometry: The dynamics of behavior," presented at the *34th Annual Meeting of the International Society for the Systems Sciences*, Portland, Oregon: ISSS, pp. 769-775.

Sabelli, H. C., Carlson-Sabelli, L. and Javaid, J. I. (1990). "The thermodynamics of bipolarity: A bifurcation model of bipolar illness and bipolar character and its psychotherapeutic applications," *Psychiatry*, ISSN 0033-2747, 53: 346-368.

Sabelli, H. C., Carlson-Sabelli, L., Patel, M., Levy, A. and Diez-Martin, J. (1995). "Anger, fear, depression, and crime: Physiological and psychological studies using the process method," in R. Robertson, and A. Combs (eds.), *Chaos Theory in Psychology and the Life Sciences*, ISBN 0805817360, pp. 65-88.

Sathian, K., Zangaladze, A., Hoffman, J., and Grafton, S. (1997). "Feeling with the mind's eye," *Neuroreport: An International Journal for the Rapid Communication of Research in Neuroscience*," ISSN 0959-4965, 8: 3877-3881.

Saunders, P. T. (1980). *An Introduction to Catastrophe Theory*, ISSN 052123042X.

Sammer, G. (1996). "Working-memory load and dimensional complexity of the EEG," *International Journal of Psychophysiology*, ISSN 0167-8760, 24: 173-182.

Shadlen, M.N., Newsome W.T. (2001). "Neural basis of a perceptual decision in the parietal cortex (area LIP) of the rhesus monkey," *Journal of Neurophysiology*, ISSN 0022-3077, 86(4): 1916-1936.

Scannell, J. W. (1997). "Determining cortical landscapes," *Nature*, ISSN 0028-0836, 386(April): 452-452.

Scannell, J. W. and Young, M. P. (1999). "Neuronal population activity and functional imaging," *Proceedings in Biological Science*, ISSN 1939-0084, 266(1422): 875-881.

Schall, J. D. (2003). "Neural correlates of decision processes: Neural and mental chronometry," *Current Opinion in Neurobiology*, ISSN 0959-4388, 13(2): 182-186.

Schank, R. and Abelson R. (1977). *Scripts, Plans, Goals, and Understanding: An Inquiry Into Human Knowledge Structures*, ISBN 0898591384.

Schiff, S. J., Jerger, K., Duond, D. H., Chang, T., Spano, M. L. and Ditto, M. L. (1994). "Controlling chaos in the brain," *Nature*, ISSN 0028-0836, 370: 615–620.

Schmid, G. B. (1998). "The six fundamental characteristics of chaos and their clinical relevance to psychiatry: A new hypothesis for the origin of psychosis," in F. Orsucci (ed.), *The Complex Matters of the Mind*, ISBN 9810233396, pp. 139-181.

Schrodinger, E. (1944). *What is life? The Physical Aspect of the Living Cell*, ISBN 0521427088.

Seeley, T. D. (1989). "Social foraging in honey bees: How nectar foragers assess their colony's nutritional status," *Behavioral Ecology and Sociobiology*, ISSN 1432-2013, 24: 181-199.

Segal, G. (2000). *A Slim Book About Narrow Content*, ISBN 978-0262194310.

Sheehan, S. (1982). *Is There no Place on Earth for Me?* ISBN 0395318718.

Silberstein, M. (2002). "Reduction, emergence, and explanation," in P. Machamer and M. Silberstein (eds.), *The Blackwell Guide to the Philosophy of Science*. ISBN 0631221081, pp. 80-107.

Singer W. (1996). Neural Synchronization: A solution to the binding problem? in R. Llinas and P. S. Churchland (eds.), *The Mind-Brain Continuum: Sensory Processes*, ISBN 0262121989, pp. 100-130.

Singer, J. L. (1995). "Mental processes and brain architecture: Confronting the complex adaptive systems of human thought," in H. J. Morowitz and J. L. Singer (eds.), *The Mind, the Brain and Complex Adaptive Systems*, ISBN 0201409860, pp. 1-10.

Sinha, S. (1999). "A 'Computing' Chaotic System," in N. Pradhan, P. E. Rapp and R. Sreenivasan (eds.), *Nonlinear Dynamics and Brain Functioning*, ISBN 1560726482, pp. 138-143.

Sinha, S. and Ditto, W. L. (1998). "Dynamics-based computation," *Physical Review Letters*, ISSN 0031-9007, 81(10): 2156-5159.

Skinner, B. F. (1953). *Science and Human Behavior*, ISBN 0029290408.

Smetters, D. (2000). "Noise in neural computation," in P. Arhem, C. Blomberg and H. Liljenstrom (eds.), *Disorder Versus Order in Brain Function: Essays in Theoretical Neurobiology*, ISBN 9810240082, pp. 107-116.

Smith, E. R. (1996). "What do connectionism and social psychology offer each other?" *Journal of Personality and Social Psychology*, ISSN 0022-3514, 70: 893-912.

Smith, J. C. and Feldman, J. L. (1990). "Organization of oscillatory motor pattern generation networks in the mammalian central nervous system," *Proceedings of the IEEE International Conference on Neural Networks*, ISSN 1098-7576, 2: 651-656.

Sole, R. and Goodwin, B. (2000). *Signs of Life: How Complexity Pervades Biology*, ISBN 0465019277.

Solé, R. V., Bascompte, J. and Valls, J. (1992). "Stability and complexity in spatially-extended two-species competition," *Journal of Theoretical Biology*, ISSN 0022-5193, 159: 469-476.

Spehar, B., Clifford, C. W. G., Newell, B. R. and Taylor, R. P. (2003). "Universal aesthetic of fractals" *Computers and Graphics*, ISSN 0097-8493, 27: 813-820.

Spivey, M. (2007). *The Continuity of Mind*, ISBN 0195170784.

Stam, C. J. (2003). "Chaos, continuous EEG, and cognitive mechanisms: A future for clinical neurophysiology," *American Journal of Electroneurodiagnostic Technology*, ISSN 1086-508X, 43(4): 211-227.

Stam, C. J. (2005). "Nonlinear dynamical analysis of EEG and MEG: Review of an emerging field," *Clinical Neurophysiology*, ISSN 1388-2457, 116(10): 2266-2291.

Stam, C. J., Jelles B., Achtereekte, H. A. M., Rombouts, S. A. R. B., Slaets, J. P. J. and Keunen, R. W. M. (1995). "Investigation of EEG non-linearity in dementia and Parkinson's disease" *Electroencephalography and Clinical Neurophysiology*, ISSN 1388-2457, 95(5): 214-224.

Stam, C. J., Tavey D. L. J., Jelles B., Achtereekte, H. A. M., Slaets, J. P. J. and Keunen, R. W. M. (1994)."Non-linear dynamical analysis of multi channel EEG data: Clinical applications in dementia and Parkinson's disease," *Brain Topography*, ISSN 0896-0267, 7(2): 141-150.

Stepien, R. A. (2002). "Testing for non-linearity in EEG signal of healthy subjects," *Acta Neurobiologiae Experimentalis*, ISSN 0065-1400, 62: 90-96.

Sternberg, S. (1966). "High-speed scanning in human memory," *Science*, ISSN 0036-8075, 153(3736): 652-654.

Stevens, C. F., Wang, Y. (1994). "Changes in reliability of synaptic function as a mechanism for plasticity," *Nature*, ISSN 0028-0836, 371: 704-707.

Strogatz,.S. (2003). *Sync: How Order Emerges from Chaos in the Universe, Nature, and Daily Life*, ISBN 0786887214.

Sutton, R. S. and Barto, A. (1998) *Reinforcement Learning: An Introduction*, ISBN 0262193981.

Swindale, N. V. (1980). "A model for the formation of ocular dominance stripes," *Proceedings of the Royal Society of London, Series B*, ISSN 0962-8452, 208: 243-264.

Swenson, R. (1989). "Emergent evolution and the global attractor: The evolutionary epistemology of entropy production," *Proceedings of the 33rd Anniversary Meeting of the International Society for the Systems Sciences*, ISSN 0016-6588, 3: 46-53.

Takakura, K., Kosugi, Y., Ikebe, J. and Musha, T. (1987). "1/f controlled transcutaneous electrical stimulation for pain relief," in C. M. Van Vliet (ed.), *Ninth International Conference on Noise and Physical Systems*, ISBN 9971503972, pp. 279-282.

Tech, M. C. (1989). "Fractal character of the auditory neural spike train," *IEEE Transactions on Biomedical Engineering*, ISSN 0018-9294, 36(1): 150-160.

Tesser, A. (1980). "When individual dispositions and social pressure conflict: A catastrophe," *Human Relations*, ISSN 0018-7267, 33: 393-407.

Tesser, A., and Achee, J. (1994). "Aggression, love, conformity, and other social psychological catastrophes," in R.R. Vallacher and A. Nowak (Eds.), *Dynamical Systems in Social Psychology*. ISBN 1572303530, pp. 96-109.

Theiler, J. (1991). "Some comments on the correlation dimension of 1/f noise," *Physics Letters*, ISSN 0015-9018, 115A: 480-493.

Thelen, E. and Smith, L. B. (1994). *A Dynamic Systems Approach to the Development of Cognition and Action,* ISBN 026270059X.

Toffoli, T. and Margolus, N. (1987). *Cellular Automata Machines.* ISBN 0262200608.

Tononi, G. and Edelman, G. M. (1998). "Consciousness and complexity," *Science*, ISSN 0036-8075, 282: 1846-1851.

Tononi, G., Sporns, O. and Edelman, G. M. (1996). "A complexity measure for selective matching of signals by the brain," *Proceedings of the National Academy of the Sciences*, ISSN 0027-8424, 93(8): 3422-3427.

Torre, C. A. (1984). "Problem solving and decision-making: An integration of cognitive, affective, and pragmatic operations," presented at the *Second Biennial International Conference on Thinking*, Harvard University, (August).

Torre, C. A. (1987). "Thinking, culture and education," presented at the *Third Biennial International Conference on Thinking*, University of Hawaii at Manoa, (January).

Torre, C. A. (1995). "Chaos in the Triadic Theory of Psychological Competence in the Academic Setting" in F. D. Abraham and A. R. Gilgen (eds.), *Chaos Theory in Psychology*, ISBN 0275951405, pp. 279-294.

Treisman, A. (1987). "Features and objects in visual processing," *Scientific American*, ISSN 0036-8733, 255(5): 114-125.

Treves, A. and Rolls, E. T. (1994) "Computational analysis of the role of the hippocampus in memory," *Hippocampus,* ISSN 1050-9631, 4(3): 374-391.

Tryon, W. W. (1995). "Synthesizing psychological schisms through connectionism," in F. D. Abraham, and A. R. Gilgen (eds.), *Chaos Theory in Psychology*, ISBN 0275951405, pp. 247-263.

Tsuda I. (2001). "Toward an interpretation of dynamic neural activity in terms of chaotic dynamical systems," *Behavioral and Brain Science*, ISSN 0140-525X, 24: 793-841.

Uttal, W. R. (1973). *The Psychobiology of Sensory Coding*, ISBN 1125942991.

Vaadia, E. Haalman, I. Abeles, M., Bergman, H., Prut, Y., Slovin H. and Aertsen, A. (1995). "Dynamics of neuronal interactions in monkey cortex in relation to behavioral events," *Nature,* ISSN 0028-0836, 373: 515-518.

Valdes, P., Bosch, J., Jimenez, J. C., Trujillo, N., Biscay, R. and Morales, F. (2001). "The statistical identification of nonlinear brain dynamics: A progress report," in N. Pradhan, P. E. Rapp and R. Sreenivasan (eds.), *Nonlinear Dynamics and Brain Functioning*, ISBN 1560726482, pp. 243-264.

Van der Pol, B., van der Mark, J. (1927). "Frequency demultiplication," *Nature*, ISSN 0028-0836, 120: 363-364.

Van der Malsburg, C. (1999). "The what and why of binding: The modeler's perspective," *Neuron*, ISSN 0896-6273, 24: 95-104.

Van Orden, G. C., Holden, J. G. and Turvey, M. T. (2005). "Human cognition and 1/f scaling," *Journal of Experimental Psychology: General*, ISSN 0096-3445, 134(1): 117-123.

Varela, F. J. (1995). "Resonant cell assembles: A new approach to cognitive functions and neuronal synchrony," *Biology Research*, ISSN 0716-9760, 28: 81-95.

Varela, F. J. (1999). "The specious present," in J. Petetot, F. J. Varela, B. Pachoud and J. M. Roy, (eds.), *Naturalizing Phenomenology*, ISBN 0804733229, pp. 266-316.

Velazquez, J. L., Cortez, M. A., Carter, Snead III, O. and Wennberg R. (2003). "Dynamical regimes underlying epileptiform events: Role of instabilities and bifurcations in brain activity," *Physica D,* ISSN 0167-2789, 186: 205-220.

Von Neuman, J. and Morgenstern, O. (1947). *Theory of Games and Economic Behavior*, ISBN 0691003629.

Voss, R. F. and Clarke, J. (1975). "1/*f* noise in music and speech," *Nature*, ISSN 0028-0836, 258 (5533): 317-318.

Ward, L. M. (2002). *Dynamical Cognitive Science*, ISBN 0262232170.

Ward, L. M. and Richard, C. M. (2001). "1/*f* noise and decision complexity," Unpublished manuscript, Vancouver, Canada: University of British Columbia.

Watts D.J. (2003). *Six Degrees: The Science of a Connected Age*, ISBN 0393041425.

Watts, D. J. and Strogatz, S. H. (1998). "Collective dynamics of 'small-world' networks," *Nature*, ISSN 0028-0836, 393: 440-442.

Weinberg, Gerald M. (1975). *An Introduction to General Systems Thinking*, ISBN 0471925632.

Westheimer, G. (1991). "Visual Discrimination of fractal borders," *Proceedings of the Royal Society of London,* ISSN 0962-8452, 243: 215-219.

Williams G. P. (1997). *Chaos Theory Tamed*, ISBN 0309063515.

Winkler, M., Combs, A. L. and Daley, C. (1994). "A chaotic systems analysis of rhythm in feeling states," *The Psychological Record*, ISSN 0033-2933, 44: 359-368.

Winkler, M., Combs, A. L., Dezern, D., Alstott, T., Burnham, J., Ran, B. and

Walker, S. (1991). "Cyclicity in moods: A dynamical systems analysis," presented at the *Society for Chaos Theory in Psychology*, San Francisco, CA.

Wolff, P. H. (1987). *The Development of Behavioral States and the Expression of Emotions in Early Infancy*, ISBN 978-0226905204.

Wolfram S. (2002). *A New Kind of Science*, ISSN 1579550088.

Wolpert, D. and Macready, W. (2000). "Self-dissimilarity: An empirically observable complexity measure" in Y. Bar-Yam (Ed.), *Unifying Themes in Complex Systems: Proceedings of the International Conference on Complex Systems*, ISBN 0813341221, pp. 625-643.

Wong, R. (1999). "Retinal waves and visual system development,"*Annual Review of Neuroscience*, ISSN 1471-003X, 22: 29-47.

Wynn, T. (1999). "The Evolution of Tools and Symbolic Behaviour," in A. Lock, C. Peters (eds.), *The Handbook of Human Symbolic Evolution*, ISBN 0631216902, pp. 263-287.

Xu, J., Tong, Q. and Liu, R. (1999). "Information transmission of human cerebral cortex and schizophrenia in N. Pradhan, P. E. Rapp and R. Sreenivasan (eds.), *Nonlinear Dynamics and Brain Functioning*, ISBN 1560726482, pp.

Xu, Y., Shen, M., Tony, S., Thakor, N., Lee, Y., Zhu, Y. (1990). "The nonlinear dynamical analysis of the EEG in schizophrenia with temporal and spatial embedding dimension," *Journal of medical engineering & technology*, ISSN 0309-1902, 25(2): 79-83.

Yagyu, T., Wackermann, J., Shigeta, M., Jelic, V., Kinoshita, T., Kocki, K., Julin, P., Almkvist, O., Wahlund, L. O., Kondarkor, I. and Lehmann, D. (1997). "Global dimensional complexity of multi-channel EEG in mild Alzheimer's Disease and age-matched controls," *Dementia and Geriatric Cognitive Disorders*, ISSN 1420-8008, 8: 343-347.

Zangaladze, A., Epstein, C., Grafton, S. and Sathian, K. (1999). "Involvement of visual cortex in tactile discrimination of orientation," *Nature*, ISSN 0028-0836, 401: 587-590.

Zera, D. A. and Lucian, D. G. (2001). "Self-organization and learning disabilities: A theoretical perspective for the interpretation and understanding of dysfunction," *Learning Disability Quarterly*, ISSN 0731-9487, 24(2): 107-118.

INDEX

A

B

C

D

E

F

G

H

I

J

K

L

M

N

O

P

R

S

T

U

V

W

www.ingramcontent.com/pod-product-compliance
Lightning Source LLC
LaVergne TN
LVHW062117200926
848484LV00007B/1703